The Uncertainty of Measurements

Physical and Chemical Metrology: Impact and Analysis

Also available from Quality Press

Reliability Methods for Engineers
K. S. Krishnamoorthi

HALT, HASS & HASA Explained: Accelerated Reliability Techniques
Harry W. McLean

The Handbook of Applied Acceptance Sampling: Plans, Procedures, and Principles
Kenneth S. Stephens

Statistical Quality Control Using Excel (book plus software)
Steven M. Zimmerman, Ph.D., and Marjorie L. Icenogle, Ph.D.

The Desk Reference of Statistical Quality Methods
Mark L. Crossley

Statistical Process Control Methods for Long and Short Runs, Second Edition
Gary K. Griffith

Glossary and Tables for Statistical Quality Control, Third Edition
ASQ Statistics Division

The Uncertainty of Measurements

Physical and Chemical Metrology: Impact and Analysis

Shri Krishna Kimothi

ASQ Quality Press
Milwaukee, Wisconsin

Library of Congress Cataloging-in-Publication Data

Kimothi, Shri Krishna, 1944-
The uncertainty of measurements : physical and chemical metrology : impact and analysis / Shri Krishna Kimothi.
p. cm.
Includes bibliographical references and index.
ISBN 0-87389-535-5 (hardcover : acid-free paper)
1. Physical measurements. I. Title.
QC39 .K52 2002
530.8--dc21

2002006517

10 9 8 7 6 5 4 3 2 1

ISBN 0-87389-535-5

Acquisitions Editor: Annemieke Koudstaal
Project Editor: Craig S. Powell
Production Administrator: Jennifer Czajka
Special Marketing Representative: Denise M. Cawley

Printed in the United States of America

Printed on acid-free paper

American Society for Quality
ASQ®

Quality Press
600 N. Plankinton Avenue
Milwaukee, Wisconsin 53203
Call toll free 800-248-1946
Fax 414-272-1734
www.asq.org
http://qualitypress.asq.org
http://standardsgroup.asq.org
E-mail: authors@asq.org

To

my wife,

my children,

and my grandchildren

Table of Contents

Foreword

The measurements made by the testing and calibration laboratories within a country as well as the quality system and product certification issued in that country are not always accepted in other countries. This has been considered a significant technical barrier to trade (TBT) by the World Trade Organization (WTO). The TBT agreement was therefore evolved, and it is now binding on all WTO member countries. It was recognized that mutual acceptability of test results and product certification could be achieved only if a mechanism could be evolved whereby the user has confidence in the technical competence of the laboratories and the soundness of their measurement and quality systems and of the product certification bodies operating in various countries. To sustain the confidence of all concerned, a global conformity assessment structure was evolved, and all members of the WTO are expected to establish such conformity assessment schemes within their countries.

The conformity assessment schemes envisage the following:

1. The test and calibration laboratories are to operate in accordance with the requirements stipulated in international standard ISO/IEC 17025. Evidence of their technical competence is to be provided through their accreditation by nationally recognized laboratory accreditation bodies. These accreditation bodies are to enter into multilateral recognition agreements (MRAs) with similar accreditation bodies in other countries or with an association of these bodies under the aegis of the International Laboratory Accreditation Cooperation (ILAC).
2. Manufacturing and service organizations should be conscious of customer requirements and should operate in accordance with the ISO 9000 series of international standards on quality management systems. These should also be certified by quality system certification bodies. Similarly, compliance of product quality with its specifications is to be certified by product certification bodies based on product testing in recognized laboratories. The quality system and product certification bodies should be accredited by government-recognized accreditation bodies. These accreditation bodies are to enter into MRAs with similar accreditation bodies in other countries or with an association of these bodies under the aegis of the International Accreditation Forum (IAF).

With the efforts of the WTO, economies are opening up for global competition and consumers are becoming more and more demanding with reference to the quality of products and services. Testing and calibration laboratories as well as the metrology infrastructure of manufacturing and service organizations are important resources that are helping to supply quality products to customers. To ensure quality, products need to be evaluated at various stages of their production, and measurement plays a key role in the evaluation process. Thus, the quality of the measurement results obtained in testing and calibration laboratories as well as in manufacturing organizations plays an important part in ensuring the quality of end products. The quality of a measurement result is reflected in its uncertainty with reference to its value and its traceability to the International System of Units through various national and international standards. A measurement result without a statement of its uncertainty is incomplete, and it cannot be effectively used for making decisions based on an analysis of measurement data.

The subject of measurement uncertainty is therefore important mainly for the following reasons:

1. Testing reports and calibration certificates have better acceptability within and beyond national boundaries if measurement results are accompanied by statements of their uncertainty.
2. There is an increased awareness among quality professionals in testing and calibration laboratories and also in manufacturing and service organizations of the importance and impact of measurement uncertainty.
3. There is a growing realization globally among analytical chemists about the importance and need for quantifying uncertainty in analytical measurements.
4. International standards published by the International Organization for Standardization (ISO)—ISO/IEC 17025:1999 for testing and calibration laboratories and ISO 9001:2000 for manufacturing and service organizations—demand that the uncertainty of measurement results be estimated and known.

There was a lack of consensus among metrologists on the methods of estimating measurement uncertainty. This had been causing difficulty in intercomparisons of measurement results, including those associated with the intercomparison of national standards by the International Bureau of Weights and Measures (BIPM). The need for a harmonized approach on this issue was therefore felt. Thus, the *Guide to the Expression of Uncertainty in Measurement* was published in 1993 due to the initiative of the BIPM and a number of national metrology laboratories. Now there is global consensus on this subject in all areas of measurement and at all levels, from industrial measurements to measurements made in research and development laboratories to those made in national metrology laboratories. In view of this, there is the need for a book covering all aspects of measurement uncertainty. Such a book would be of use to scientists in testing and calibration laboratories as well as to those working in manufacturing and service organizations.

I am happy that a book entitled *The Uncertainty of Measurements: Physical and Chemical Metrology—Impact and Analysis* has been written by Shri Krishna Kimothi, who has long practical experience in this field. The book is being published by the American Society for Quality (ASQ), an association of quality professionals of international repute. The book covers the various aspects of uncertainty in measurement in testing and calibration laboratories and in industry. In addition to measurement uncertainty, the book also covers other attributes of measurement such as calibration and traceability in physical and chemical metrology and statistical control of the measurement process.

I am sure this book will prove useful to metrologists and engineers in industry, quality system workers, and those involved in product certification and laboratory accreditation.

Professor Shri Krishna Joshi
Former director general, Council of Scientific and Industrial Research;
and Secretary to Government of India, Department of Scientific
and Industrial Research and former director,
National Physical Laboratory, New Delhi, India
October 2001

Preface

Managing uncertainty is a real challenge faced by today's managers. It appears that uncertainties have become all pervasive: uncertainties of markets, uncertainties of economy, uncertainties of technological changes, political uncertainties, and so on. The problem of uncertainty arises when there are a number of possible outcomes and a decision based on the probabilities of these outcomes has to be made. The theory of probability facilitates decision making in the face of uncertainties and thus minimizes the risks associated with these uncertainties.

The uncertainty of measurement results is one uncertainty element which is drawing the attention of not only managers, metrologists, and quality professionals but also ordinary consumers. Measurements and measurement results influence human lives in many ways. For individuals as consumers and common citizens, measurements affect trade and commerce, safety and healthcare, and environmental protection. Other entities for which measurements play an important role in decision making are manufacturing and service organizations and testing and calibration laboratories. In industry, measurements facilitate the monitoring of the quality of products and services at various stages of their life cycles.

Similarly, measurements are used in calibration laboratories to assess the operational integrity of measuring equipment, whereas in testing laboratories the characteristics of products are measured in order to determine their compliance with specifications. Measurement results also need to be compared among themselves in many situations.

A measurement result is incomplete without a statement of its uncertainty. The acceptability of a measurement system for a specific measurement application also depends upon the uncertainty of the measurement results produced by it. A measurement result is a means of communication among persons and organizations. For this communication to be meaningful, it is essential that a consensus approach be established in the evaluation and reporting of measurement uncertainty on a global basis. This has been achieved with the publication of the *Guide to the Expression of Uncertainty in Measurement (GUM)* in 1993 by the International Organization for Standardization (ISO) and other international organizations having an interest in metrology. This document has achieved global acceptability and is applicable to uncertainty evaluation and reporting in physical as well as chemical metrology at all levels of measurement, from the shop floor of industry to fundamental research.

Uncertainty of measurement is an important issue for decision makers in industries and in testing and calibration laboratories. It decides not only the acceptability of a measurement system but also the cost involved in taking measurements. A clear understanding of measurement uncertainty is therefore essential for persons who are involved in the design of measurement systems, who engage in actual measurement, and who make decisions based on measurement results. A great deal of background material is available on the subject of measurement uncertainty in various journals and publications issued by laboratory accreditation bodies and national metrology laboratories. However, there are not many books available which deal with the why and how part of the subject. The author has felt the necessity for such a book based on his experience as a quality professional and a trainer. This book is aimed at fulfilling this necessity and providing necessary information inputs to professionals working in testing and calibration laboratories and in the metrology functions of industries. In addition, the book will be useful for quality auditors working with quality system and product certification bodies and with laboratory accreditation bodies. A number of organizations and institutions conduct academic progams in which general metrology forms part of the curriculum. The book will also be useful for persons associated with these programs.

The subject matter of the book is developed in a logical sequence and can be divided into five parts:

Part 1 (chapters 1 and 2) defines the need for reliable measurements and the inputs needed to obtain measurement results of the desired quality.

Part 2 (chapters 3 and 4) defines various metrological terms, including *measurement uncertainty,* and statistical concepts necessary to understand and evaluate measurement uncertainty.

Part 3 (chapters 5 and 6) defines the compatibility of measurement results achieved through national and international measurement systems and the role of calibration in achieving traceability of measurement results.

Part 4 (chapters 7 and 8) defines the evolving concepts of measurement uncertainty evaluation before the publication of *GUM* and leads to the harmonized approach as achieved with the publication of *GUM.*

Part 5 (chapters 9, 10, 11, and 12) is application oriented. It starts with measurement assurance concepts and goes on to deal with measurement-uncertainty-related issues in industries, testing laboratories, and calibration laboratories.

As the title suggests, the book is about uncertainty in measurements in physical and chemical metrology. The thrust is to provide understanding of the concepts as well as the requirements for measurement uncertainty stipulated in various standards. The procedures for uncertainty evaluation using older approaches as well as the *GUM* approach are illustrated with a number of practical examples. With a subject like this, statistics cannot be altogether avoided; however, efforts have been made to keep it to the minimum possible. The book is primarily on an important aspect of general metrology, and efforts have been made to provide examples from various metrological areas.

It is hoped that metrologists and general quality professionals alike will find the book useful and interesting.

Shri Krishna Kimothi

Acknowledgments

I have the privilege of having professional interaction with a number of persons associated with metrology. These persons belong to industries and testing as well as calibration laboratories. The structure and contents of this book have evolved based on these interactions. My own organization, the Standardization, Testing, and Quality Certification (STQC) Directorate under the Ministry of Communications and Information Technology, Government of India, provides testing and calibration services through a network of laboratories. Discussions at various forums with my colleagues working in these laboratories and the input provided by them have helped me in conceptualizing the contents of the book. My association with the Metrology Society of India gave me the opportunity to interact with scientists working at the National Physical Laboratory, New Delhi, which also helped me in conceptualizing the contents. I am thankful to all my colleagues and friends in this branch of the scientific community for their input and their wholehearted support for the book.

I have been extremely encouraged by the support and guidance I received during the preparation of the book from some distinguished persons and well-wishers. Professor Shri Krishna Joshi is an eminent Indian scientist who has held important positions such as director of the National Physical Laboratory, New Delhi; director general of the Council of Scientific and Industrial Research; and secretary to the Government of India Department of Scientific and Industrial Research. I am extremely thankful to him for writing the foreword of the book and providing valuable input.

Dr. Shanti Lal Sarnot, director general of the STQC, Ministry of Communications and Information Technology, always supports intellectual activities. He has given me a great deal of inspiration, encouragement, and support at various stages of preparation of the book, for which I am especially grateful to him.

Dr. Bhupendra S. Mathur, scientist emeritus of the National Physical Laboratory, New Delhi; Atul Krishna Datta, director general of the National Test House, Calcutta; and Harbans Lal, director general of the Federation of Indian Chambers of Commerce and Industry (FICCI) Quality Forum, New Delhi, have given me important feedback about the book based on their reviews. My special thanks go to all of them. B. Basaviah, director of the Electronics Regional Test Laboratory (North) STQC, New Delhi, has supported me in the copyediting of the manuscript for editorial and technical contents. I am extremely thankful to him for this support. I also thank Dr. R. N. Chowdhury, counselor of the Confederation of Indian Industry (CII) New Delhi, for providing input.

Vital feedback has been provided by Philip Stein, an eminent U.S. metrologist and quality consultant, and T. M. Kubiak, also an eminent U.S. quality professional, based on their review of the book. I am extremely grateful to them for their comments and suggestions. My special thanks go also to the American Society for Quality (ASQ) for publishing the book and to Craig Powell, Annemieke Koudstaal, and Denise Cawley of ASQ Quality Press for the effective and prompt communications I have received from them.

My thanks are due to Anita Negi and Alka Chauhan for their valuable help in the preparation of the manuscript and to Shyama Madan for her support and secretarial assistance.

I also acknowledge with thanks the support provided by my son Anurag Kimothi, a student of the Indian Institute of Technology, Delhi, in the preparation of the figures, tables, and mathematical part of the text.

Shri Krishna Kimothi

1

Metrology and the Need for Reliable Measurements

Thus every sort of confusion is revealed within us, and the art of measuring, numbering and weighing come into the rescue of human understanding—there is the beauty of them—the apparent greater or less or more or heavier no longer have mystery over us, but give way to measurement and calculation and weight. . . .

Then that part of the soul which has an opinion contrary to measure is not the same with that which has an opinion in accordance with measure. . . .

And better part of the soul is likely to be with that which trusts to measure and calculation. . . .

And that which is opposed to them is one of the inferior principles of the soul.

Socrates as quoted in Plato's, *Republic*

A measurement is a kind of language. It gives everyone something to agree upon so that effective communication and commerce can take place. Good measurements must be defined by their accuracy, precision and need. Of course, measurement capabilities should fit the need, but new measurement capabilities can often redefine those needs.

David Layden, cited in *Quality Progress,* February 2001

An almost infinite number of measurements are made every day in trade, industry, and scientific laboratories throughout the world. The measurement results are used to make decisions which could be crucial at times. These decisions affect our lives as professionals and also as consumers. In order for us to have adequate confidence in the correctness of decisions made based on measurement results, it is essential that the measurement results are reliable. This chapter explains the critical role played by measurements in various activities associated with human beings. The importance of reliable measurements in decision-making and the need for correct understanding and interpretation of measurement results are also explained.

THE UNIVERSAL NATURE OF MEASUREMENTS

The false diagnosis of an ailment may have serious consequences for an individual. Sometimes it can also be very embarrassing for decision makers at the senior level. One such case was reported by the *Hindustan Times,* a leading English-language daily newspaper published in New Delhi, in its edition of 15 January 2002. A soldier was tested for HIV along with other troops in his regiment when the regiment was selected for duty with a UN peace mission abroad. The soldier was diagnosed as HIV positive, and his medical category was downgraded. This decision was made based on the HIV test carried out in the Armed Forces Transfusion Centre. Feeling humiliated, the soldier had his HIV test conducted in a civil hospital, where he tested negative. The soldier then sued the army authorities for false diagnosis in a court of law. The court issued notice to the army for the facts of the case, and the army authorities filed an affidavit before the court.

In defending their case, the army authorities based their arguments on the laws of probability and on medical literature to explain how the army had diagnosed a soldier as HIV positive when he was in fact HIV negative. They quoted an article by an expert which stated that HIV antibody tests are 99.5% accurate. Cases which are actually HIV negative but test HIV positive on some occasion are termed "false positive," and they are not unknown in the medical literature. The army therefore deduced that there is a probability of 1 in 200 persons' testing false positive. The army authorities also stated that the soldier had again been found HIV positive in the same army center where he had tested HIV positive earlier, even after he tested negative in the civil hospital. For confirmation, the soldier was again tested in another reputed civil hospital and was found HIV negative. The army informed the court that based on the test the soldier was declared HIV negative. The soldier has, however, sought compensation for the humiliation and disgrace that he faced due to the false diagnosis.

Such cases of false diagnosis do occur, though very rarely. It is not uncommon for two laboratories to report different results. In such a situation, it is important to decide whether the difference is because of the inherent variability of the analytical processes at the two laboratories or because of the bias in the analytical process of one of the laboratories. Due to variability, analytical results are probabilistic in nature. There are risks of making a wrong decision based on these results, though the probability of this may be quite low. The preceding example, however, shows what the consequences of such a wrong decision could be.

Effective decision making requires the appropriate use of available measurement data. In this context, the author had an interesting encounter with a colleague about two decades back. The colleague had been working as a technician in a repair and maintenance establishment. One of the job functions assigned to him was measuring certain technical parameters of purchased items and submitting the results to his supervisor for a decision on the acceptance or rejection of the items. He had a son studying in a secondary school. While teaching his son about the Pythagorean theorem in geometry, he had the noble idea of experimentally verifying the statement of the theorem. The next day, while narrating his teaching experience to some of his colleagues, he informed them about his newfound amusing discovery that the statement of Pythagoras's theorem is not correct. In response to a specific query about the basis of his bold statement, he offered an interesting explanation. In the process of verifying the statement of the theorem, he had drawn a right-angled triangle and measured the lengths of the hypotenuse and the other two sides, and he had observed that the square of the hypotenuse was not exactly equal to the sum of the squares of the other two sides. He repeated the exercise a number of times and drew the same conclusion. It took a lot of effort to make him understand that the exact equivalence could not be achieved due to the presence of a number of factors that contribute to errors in measurement. Some of the important factors in this case could have been the limited resolution of the measuring scale and goniometer, their inherent inaccuracies, and his own level of measuring skill.

Things have not changed too much in the last two decades. One can always find persons who have this type of conceptual misunderstanding about measurement results. They conceive of measurement results as exact and unambiguous. They are ignorant of the fact that measurement results are probabilistic in nature. This implies that another set of measurements of the same parameter could result in values which are not exactly the same as but are close to earlier reported values. It is therefore essential that such misconceptions about measurement results should be guarded against.

Measurements are extensively in demand but often go unnoticed. P. H. Sydenham has reported, based on a survey by Huntoon, that in 1967 about 20 billion measurements were performed each day in the United States. In 1965, the U.S. industrial sector invested around 3% of GNP in measurement. Measurement affects human lives in many ways. We are concerned about day-to-day measurements in the areas of trade, healthcare, and environmental pollution. As individuals, we are faced with measurements taken on a day-to-day basis by market scales; gas, electricity, and water meters; gasoline pumps; taxi meters; and so on. We are also concerned with the reliability of measurements in healthcare, such as the measurement of the temperature of the human body by clinical thermometers, the monitoring of blood pressures and heartbeats, and the clinical analysis of body fluids in pathology laboratories. As a society, we are concerned with the reliability of the measurement results of environmental parameters, such as sulfur dioxide, carbon monoxide, and carbon dioxide concentrations in the atmosphere and the density of suspended particles in the atmosphere.

Following are some examples of how measurement affects our day-to-day lives.

- **Trade and consumer protection:** A well-defined and accepted measurement system facilitates trade at the local, national, and international levels and protects consumers from unethical trade practices.
- **Safety and healthcare:** Human beings need to be assured of their safety when they use industrial products. This requires the evaluation of risks due to the products' use as well as the design and manufacture of safe products. At various stages of the manufacturing and service processes, measurements of product parameters that have bearing on human safety are made. Similarly, healthcare involves analytical and clinical measurements for diagnostic purposes in order to decide the course of treatment.
- **Environmental protection:** The environment around us has a direct bearing on the quality of our lives. Because of this, the quality of the air, water, and soil must be monitored regularly. To save the environment, the quality of industrial waste has to be controlled. There are environmental protection laws in most of countries which make it obligatory for individuals and organizations not to pollute the air, water, and soil to dangerous levels. Measurements help in achieving the desired quality level of the environment.
- **Law and order:** Police radar is used to measure the speed of motor vehicles, which helps in maintaining orderly traffic on the roads. The analytical methods used in forensic science laboratories help in solving crimes. The results obtained are presented in legal proceedings as evidence in a court of law. This ensures fair legal judgments.

Measurements are also crucially important for quality professionals working in manufacturing and service organizations. In industrial manufacturing and service organizations, measurements are required to be made at various stages of the life cycle of a product. A number of quality management standards contain specific requirements for the quality of measurement results and their evaluation. These include the ISO 9000 standards for quality management systems, the ISO 14000 standards for environmental management systems, the QS-9000 quality system standards for automotive parts, and the TL 9000 quality standards for the telecommunications industry.

In laboratories, measurements are made of various products to test their characteristics. At the international level, mutual acceptability of the measurements and test results of products is a matter of concern for eliminating technical barriers to trade.

Metrology, the science of measurement, encompasses both the theoretical and the practical aspects of measurement. In spite of the all-pervasive nature of measurement, we often do not realize the important role being played by the science of metrology and metrological institutions in our day-to-day lives. Metrological systems play an important part in regulating our lives. For example, the regulatory aspects of weights and measures are part of a metrology system; the international metrology system makes it possible for one kilogram of a substance to be weighed identically in any part of the world; and the dissemination of time by metrology institutions helps us in planning and regulating schedules in our lives. International trade and commerce would not be possible in the absence of globally accepted measurement. Thus, measurements are not only important for metrologists working in industries and scientific laboratories but are also a matter of concern to all of us as consumers and citizens. There is a strong worldwide interest in the science of measurement and measurement results not only among metrologists but also among ordinary citizens. A general awareness is emerging of the need for reliable measurement results and associated issues.

METROLOGY AND TECHNOLOGICAL GROWTH

Socrates was an ancient Greek philosopher and thinker. His thoughts on philosophy, politics and the state, economics, and social systems have paved the way for subsequent thinkers to evolve new concepts, thoughts, and theories in these areas of learning. The quotation by Socrates at the beginning of this chapter states that measurements and numbering have come to the rescue of human understanding of nature. Before this, human beings were satisfied with comparisons, such as less or more, lighter or heavier, and longer or shorter. In historical perspective, this could be considered the beginning of the science of metrology in a systematic way, when human beings started measuring things and expressing the measurements in numbers. In the second quotation, David Layden defines measurement as a kind of language which ensures effective communication. He also defines the current concept of good measurement and the relationship between new measurement capabilities and the redefinition of the need for measurement. Between these two quotations lies the history of the evolution of metrology and metrological concepts.

The history of scientific advancement is closely linked with human capability to measure and measure accurately. The need for more and more accurate measurements increased with advances in various areas of science and technology. Scientists invented newer techniques of measurement to keep pace with these advances. This led to a better understanding of the nature and scientific principles associated with measurement. After the end of the Second World War, research in the basic sciences advanced at a rapid pace. This resulted in increased growth in high-technology industrial activity. The increasing complexity and speed of many modern processes and machines made automatic control essential, and such control was not possible without satisfactory means of taking measurements. The requirement for measurement capability increased in two ways. On the one hand, there has been an ever-increasing demand for measurement of new parameters, and on the other the need for more accurate measurements has increased, which was undreamed-of in the past. Spectacular advances in the technology of the measurement of physical quantities have been witnessed in the last few decades. The development of electrical sensors and the signal-processing capabilities of electronic circuits have extended the range of what can be measured. Thus, advances in metrology helped in the rapid growth of technology, and vice versa.

The harnessing of technology brought new types of devices and machines which affect human lives directly. It improved the quality of human lives but subjected them to certain types of technological risks. The discovery of electricity revolutionized the lifestyle of human beings but with an added risk of electric shock. Ecological imbalance and environmental pollution are direct results of the uncontrolled and unplanned use of technology. X-rays have useful applications in healthcare, but excessive exposure to them has a damaging effect on human beings. The microwave oven is a very useful domestic appliance, but overexposure to microwave radiation is a health hazard.

This harnessing of technology also brought another aspect of measurement into focus. The devices and machines developed as a result of advances in technology had to be safe for human use. Their harmful effects on human beings needed to be investigated and measured in quantitative terms. The safety criteria for these products had to be decided. The products were required to be evaluated so that the risk in using them was within acceptable limits of safety. This necessitated the development of reliable measurement techniques to study the effect of new technologies on human beings and to evaluate the safety of the products.

THE INCREASING NEED FOR RELIABLE MEASUREMENTS

A measurement can be considered reliable if the characteristics of the measurement result meet the objective for which the measurement is being made. Advances in technology necessitated matching advances in metrology. In the engineering industry, the tolerance for variation in manufactured parts had been decreasing by a factor of three each decade during the last few decades. As a result of these close tolerances, a marked improvement was observed in the efficiency and reliability of internal combustion engines for motor vehicles and gas turbines for aircraft. Such improvements depended on corresponding improvements in industrial metrology. In high-technology semiconductor production, the manufacturing tolerances are such that the required metrology is at the frontiers of physics. On the other hand, technological advances in metrology resulted in the development of sophisticated devices. With the invention of the atomic clock, the accuracy and resolution of time and frequency measurements were enhanced to an unbelievable degree. This helped in the development of very accurate navigational systems and long-distance measurements in radio astronomy.

Another area in which the need for accurate measurement is being felt is environmental monitoring. The adverse effect of industrial activity on the environment is becoming a matter of concern to humankind. Dangers are posed to human health and safety by environmental degradation and by the pesticide and heavy metal residues in the food we consume. Every new technology is resulting in new types of hazards. The qualitative and quantitative measurement of the hazards is a big challenge for analytical chemists.

Besides advances in technology and environmental monitoring, there are a number of other areas in which there is a need for reliable measurement. These include various facets of human endeavor in trade, industry, biological testing, forensic analysis, analysis of food and agricultural products, and technological testing of other products.

In the area of international trade, lack of acceptance of laboratory test data across national borders has been identified as a significant technical barrier to trade. Technical requirements for products are becoming more stringent. In order to have their products accepted internationally, manufacturers must be able to prove that their goods and services meet specified requirements. They must be able to show that laboratory test reports, inspection reports, product certificates, and system certificates conform to international standards. To achieve global acceptability of products and services, the metrological institutions in various countries have to ensure that the measurements they take are consistent and accurate to the desired level, as part of a global as well as national conformity assessment system.

METROLOGY CLASSIFICATIONS

An infinite number of measurements are made every day in trade, industry, and scientific laboratories throughout the world. However, the field of measurement can be classified in two ways. The first type of classification is based on the nature of the parameter being measured, and the second type is based on the application of the measurement result. A parameter under measurement can be either a physical parameter or an analytical or chemical parameter. This classification leads to two branches of metrology: physical metrology and chemical metrology. The application-based classification leads to three branches of metrology: legal metrology, industrial metrology, and scientific metrology. The classifications are shown in Figure 1.1.

Physical and Chemical Metrology

The science of physical measurement is called physical metrology, whereas the science of analytical and chemical measurement is called chemical metrology. Physical metrology deals with the measurement of physical parameters such as mass, length, electric current, velocity, and viscosity, while chemical metrology deals with the qualitative and quantitative analysis of substances used in the chemical, biological, medical, and environmental fields. The measurements made in metallurgical testing laboratories, pharmaceutical laboratories, and pathology laboratories fall under the purview of chemical metrology. Three broad areas—forensic analysis, environmental analysis, and analysis of food and agricultural products—are also part of chemical metrology.

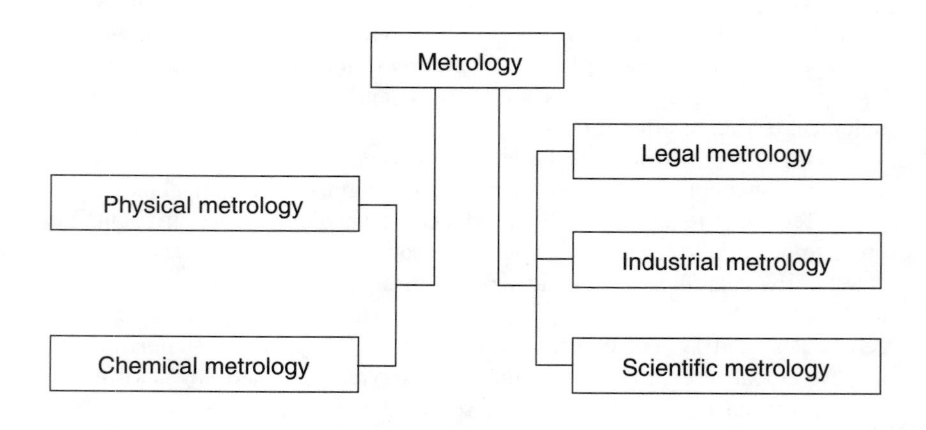

Figure 1.1 Metrology classification.

Legal, Industrial, and Scientific Metrology

Irrespective of the physical or chemical nature of the measurements taken, the discipline of metrology may be classified into legal, industrial, and scientific branches depending upon the application of the measurement result.

Legal metrology

A credible measurement system is vital for trade in any society. All measurements related to trade and consumer protection come under the purview of legal metrology. Trade is a fundamental human activity, and it operates on the principle of fair exchange of products between two parties, which may be persons or organizations. Legal metrology ensures that all measurements made for the purpose of exchanging products as part of trade are fair and credible. This type of metrology is concerned with legal issues and activities such as the evaluation and certification of measuring instruments to ensure the accuracy, consistency, and credibility of measurements. Enforcing these issues is a function of the state. The instruments that are covered under legal metrology are regularly used to satisfy the multiple needs of society. Some representative examples of these instruments and related areas concerning human lives are given in Table 1.1.

Industrial metrology

Measurements are made in industries and service organizations to check the conformance of products and services to their specifications at various stages

Table 1.1 Examples of instruments covered under legal metrology.

Area	Instruments
Trade and consumer protection	Market scales; gasoline pumps; gas, electricity, and water meters; taxi meters
Safety and healthcare	Medical scales, syringes, and thermometers; audiometers; radiotherapy equipment; tyre pressure gages; equipment used for toxic chemical measurements
Environmental protection	Instruments used for measuring noise levels and air and water pollution; equipment used for measuring vehicle exhaust emissions
Law and order	Radar equipment for determining vehicle speed; breath analyzers for testing the alcohol levels of motorists

of product development, manufacturing, and servicing. These measurements come under the purview of industrial metrology.

Today's mass production of quality goods and provision of quality services could not exist without an adequate measurement system. The requirements of the system for product and process monitoring have been stipulated in international standards such as the ISO 9000 series of quality management standards. Thus, an appropriate measurement system is an important resource for any manufacturing or service organization. In an industrial organization, measurements are required to be made at various stages in the life cycles of products, which are as follows:

- **Product development:** This stage includes the design and development of the product as well as pilot manufacturing. The prototypes of products are required to be evaluated and tested to ensure conformance to their specifications. The specifications are drawn up based on the requirements of customers.
- **Inspection of raw material and components:** Measurements are made with a view to checking the conformance of the raw material and components to their specifications by testing and evaluation. This is called incoming inspection, and it ensures that the end product meets the requirements of its specifications.
- **Product manufacturing:** At various stages of the manufacturing process, the product is inspected before it goes to the next stage. This is called in-process inspection. Again, the end product is tested to check conformance to its specifications. This is also called final inspection. At various inspection stages, measurements are made to determine the product's characteristics.
- **Process monitoring:** Statistical process control (SPC) techniques are extensively used by the industry to control the manufacturing process and to ensure that product quality is within the stipulated specification limits. Measurements at specified regular intervals are made of certain characteristics of the product drawn from the process, and key parameters derived based on the measurement results are plotted on a control chart. Average and range charts are very popular control charts used by the industry. After the results are plotted on the control chart, a decision is made as to whether the manufacturing process is under statistical control or not. If the process is found to be out of control, corrective measures are taken to bring the process under statistical control to ensure that the process is capable of manufacturing products of the desired quality.

- **Product maintenance:** During the process of corrective maintenance, the measurement results serve as diagnostic tools for finding the causes of product unserviceability. Based on the diagnosis, corrective maintenance action is completed. Measurement results are also used in preventive maintenance.

Scientific metrology

Measurements made in scientific laboratories are categorized as scientific metrology. Following are some examples of such measurements.

- **Measurements made in testing laboratories:** A product is tested in a testing laboratory with a view to determining its characteristics and deciding whether the product complies with its specifications. The product may be an industrial, agricultural, or natural product. The pathological and analytical laboratories used for diagnosis of human ailments are one type of testing laboratory. Forensic science laboratories for the solving of crimes are another type. Today, with advancements in technology, new types of products and materials are being designed and manufactured. The need to test these products is resulting in a need for newer types of test methods and testing techniques. A testing laboratory can be a "captive" laboratory set up by an organization to meet its testing needs. It can also be a third-party independent test laboratory that provides testing services to its clients. Testing activities are always associated with physical or chemical measurements.
- **Measurements made in research and development (R&D) laboratories:** Research activity is linked with the innovative capability of the human mind. Certain theories and hypotheses are proposed in the areas of the basic sciences and technology. Measurements are made in R&D laboratories to verify these theories. The correctness or otherwise of a proposed theory is decided based upon analysis of the measurement results obtained. For R&D laboratories involved in product research and development, measurement results help the scientists conclude whether the desired characteristics of a product have been achieved.

There is an interesting anecdote about the experimental verification of Einstein's general theory of relativity. Einstein predicted that light should be bent by gravitational fields. Thus, when the light from a distant star passes close to the sun, it should be deflected due to the gravitational field of the sun, causing the star to appear in a different position to an observer on earth. It is difficult to see this effect, as the light from the sun makes it impossible to observe stars that appear near the sun in the sky. However, it is possible to do so during a period of total solar eclipse. To verify the postulates of Einstein's theory, experiments were conducted by British scientists in 1919 in West Africa during a period of total solar eclipse. The experiments showed that light was deflected by the sun as predicted by Einstein's theory. This proved the correctness of the theory. However, later examination of the photographs taken during the experiment showed that the errors were as great as the effect the scientists were trying to measure. The success of the experiment was thus sheer luck. The light deflection was in fact accurately measured in a number of experiments carried out later.

- **Measurements made in calibration laboratories:** All measurements are made using some type of testing and measuring instruments. Analytical measurements also involve measuring instruments, standard chemicals, and certified reference materials. A measurement result is valid only if it has been made with a measuring instrument which has been calibrated. *Calibration* is the process of ensuring the integrity of measuring instruments. The behavior of a measuring instrument is compared against a duly calibrated laboratory reference standard. Based on a comparison of the measurement results obtained by the instrument being calibrated with those obtained by a duly calibrated reference standard, the operational integrity of the measuring instrument is ascertained and certified. This service is provided by calibration laboratories. The process of calibration also ensures the traceability of measurement results to national standards. The concepts pertaining to calibration and traceability are explained in detail in subsequent chapters.

THE PROBABILISTIC NATURE OF MEASUREMENT RESULTS

A measurement result is a quantitative statement. It is always expressed as a number followed by the unit of measurement. The unit should be the commonly accepted unit of the parameter being measured. Though the measurement result is expressed as a number, its nature is very different from that of an arithmetic number. An arithmetic number is an exact number. If we refer to the arithmetic number 3.85, there is no ambiguity in anybody's mind about this number's being 3.85. On the other hand, if we express a measurement result as 3.85, many questions may come into the minds of the person making the measurement and the persons using the measurement result. For example, if a measurement result indicates that a particular device consumes an electric current of 3.85 amperes, the following questions may come to the mind of the person making the measurement:

- If the measuring device had lower resolution, the measurement result could have been 3.8 amperes or 3.9 amperes

- If the measuring device had better resolution, the result could have been 3.854, 3.853, or another similar number
- If another person had made the measurement, a slightly different result could have been obtained

The consequence of all these "ifs" is that the measurement result might not have been the same. At the same time, they would also not have been very different from 3.85 amperes. It is thus an accepted situation that repeat measurement results will not be the same but will be similar, that is, very close to 3.85 amperes. These repeat measurements have a common origin, and hence there is a commonality among them. As repeat measurements yield slightly different results, there is always a shadow of a doubt concerning the credibility of a measurement result. The greater the spread of repeat measurement results, the greater the shadow of a doubt. It is therefore an accepted fact that repeat measurements of the same parameter obtained from any measurement system will have a certain amount of variability. The level of variability will be different depending upon the technological sophistication of the measurement system. This variability among different values of repeat measurement results leads to confusion as to what the actual value of the parameter being measured is. We therefore conclude that there is always error associated with a measurement result. As regards the reason for this variability, it is considered to be the cumulative effect of the variations in a number of factors that are influencing the measurements. Some of the factors are the inherent limitation of the measurement method, variation in the repeat observations of measuring instruments in the measurement setup, and the technical competence of the person making the measurements.

Thus, repeat measurements are spread around a specific value. This spread represents an interval. The measurement result is expected to lie within this interval most of the time. This spread characterizes the uncertainty of measurement results, the so-called shadow of a doubt. The less the measurement uncertainty, the less the shadow of a doubt, and the more the uncertainty, the greater the doubt. It should therefore be clear that a measurement result of a specific parameter obtained from a measurement system is one of the numbers contained in the interval. It is therefore not an exact number, or *the* number, as an arithmetic number is. In pure mathematics, the value of the square root of 2 or of π can be given accurately. The value can be exactly predicted to any number of decimal points. Occurrences which can be predicted accurately in this way are called deterministic occurrences. The concept of determinism implies that once an occurrence is known, successive occurrences can be predicted. Measurement systems are governed by scientific laws which are not deterministic in nature. It is never possible to predict accurately in advance successive measurement results from previous experience, but one can roughly predict a range of values between which measurement results can be expected to lie. These concepts have been clearly defined by the Nobel laureate Max Planck in the following words:

> Simple as we make the conditions and precise as our measuring instruments may be, we shall never succeed in calculating in advance the results of actual measurements with an absolute accuracy, in contrast to calculations in pure mathematics as in the case of the square root of 2 or of π, which can be given accurately to any number of decimal places. (*Scientific Autobiography,* 1947)

Thus measurement results are indeterministic rather than deterministic. This implies that measurement results are not exact values. They follow certain statistical laws and are probabilistic in nature. There are a large number of values within an interval that can claim to represent the value of the parameter. These values are not the same but are similar. There is a certain probability that the repeat measurement results will lie in this interval most of the time. This spread of measurement results is a measure of the indecisiveness of the measurement system. This indecisiveness is called measurement uncertainty and is the subject matter of this book.

WHY TAKE MEASUREMENTS?

There is a purpose for every measurement. Measurement results are used as tools for decision making. Irrespective of the nature of a measurement, it is always followed by a decision. The decision can be a conscious or a subconscious one. With measurements that are involved in trade, the trading partners are ensuring that they are both treated fairly. In an industrial scenario, measurement results are used to make decisions on product acceptance. If a product conforms to its specifications, it is accepted; otherwise, it is rejected. Measurement results are also used to improve the quality of products or processes. SPC techniques are used in industry with a view to controlling manufacturing processes. Control charts are drawn based on the measurement results of sample items picked from the production line at specified intervals. The study of these charts tells us whether a process is under statistical control or not. The determination to be made here is whether the manufacturing process is under statistical control and should be left undisturbed or whether it is out of control and needs corrective action.

Similarly, in a testing laboratory, measurement results help us conclude whether the product characteristics are as required. In a calibration laboratory, measurement results help us decide whether measuring equipment is capable of measuring a particular parameter to the desired accuracy. In an R&D laboratory, measurement results help us verify the correctness of a proposed scientific concept. The analytical results obtained in a forensic science laboratory help in deciding cases in a court of law. Similarly, the analytical results obtained in a pathology laboratory helps doctors decide whether an individual is suffering from a suspected disease.

Thus, the purpose of measurement is to obtain information on entities and make decisions based on the information contained in the measurement result. Decisions are made based on comparison of the numerical value of the measurement result with specified criteria. The criteria are also called specification limits. Specification limits are determined based on the intended end use of an item. For example, the specification limits for the dimensions of a manufactured part and for the lower limit of voltage that an electric cable should withstand fall into this category. In certain situations, specification limits are also decided based on experience or on the characteristics of similar high-quality products. The decision criteria for body fluid composition are an example of this category.

There are two ways of stipulating specification limits, depending upon the requirements and the decisions to be made:

- **One-way specification limits:** The decision criterion is a single value of the parameter. A decision is made based on whether the measurement result is smaller or larger than the specified value. One example is the decision on the quality of a microwave oven based on measurement of the leakage of microwave radiation from the oven. If the measured value is less than a specified value, it is decided that the design of the microwave oven is acceptable and that it is safe to use the oven. The specification limit is derived based on expert studies of the effects of microwave radiation on the human body. Similarly, if the tensile strength of an iron wire is greater than a specified value, a decision is made that it can be used in armoring electric cables for underground installation. A person can be charged for a traffic offense if the speed of the vehicle driven by the person as measured by police radar is greater than a specified value. The specification limit in this case depends upon factors such as traffic density and the type of road.
- **Two-way specification limits:** In this case, a decision is made based on two specified values. They are called the upper specification limit (USL) and the lower specification limit (LSL). The measurement result is compared against these two limits, and a decision is made accordingly. For example, if the measured dimensions of a manufactured automobile part are within the two specification limits, a decision to accept the part is made; otherwise, a decision to reject it is made. In calibration laboratories, measuring equipment is categorized as acceptable if measurement results show that the equipment reads the value of the reference standard within acceptable limits of accuracy. The decision criteria are shown in Figure 1.2.

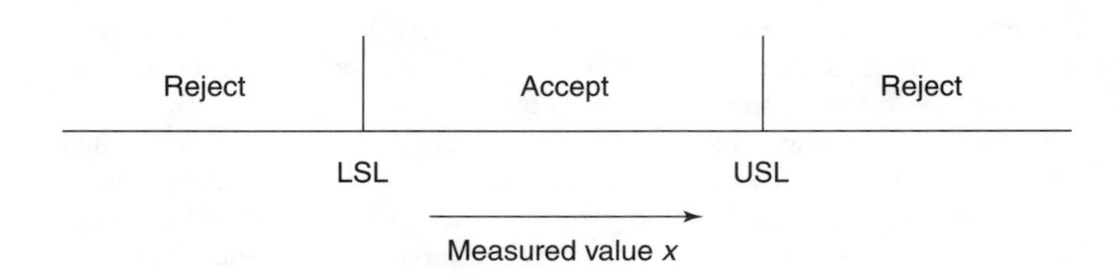

LSL < x < USL Decision to accept
x < LSL or x > USL Decision to reject

Figure 1.2 Decision criteria for two-way specification limits.

RISKS ASSOCIATED WITH STATISTICAL DECISION MAKING

Due to the probabilistic nature of measurement results, there is always uncertainty regarding the measured value of a parameter. This uncertainty is reflected in the correctness of the decision made based on a measurement result. This is especially true when the measurement result is close to the specification limits, where there is more likelihood of the decision's being incorrect. This is illustrated by the following example.

Product: Electric cable

Parameter being measured: Insulation resistance

Specification limits: Greater than 10 megaohms at 500 volts DC

Decision criterion: Measure the insulation resistance of the electric cable at 500 volts DC; cable to be accepted if insulation resistance is greater than 10 megaohms, otherwise rejected

Suppose the value of the parameter—the insulation resistance of the cable—is 10.6 megaohms: the cable should be accepted, since the value of the parameter is greater than the specification limit of 10 megaohms. This is an ideal situation. However, the actual measurement result will not always be 10.6 megaohms. Because of the probabilistic nature of repeat measurements, the measured values will lie within an interval around 10.6 megaohms most of the time. If the interval is indicated by 10.6 ±1.0 megaohms, the measurement results will lie between 9.6 megaohms and 11.6 megaohms. If repeat measurements are made on the same cable, the cable will be rejected when the measured value lies between 9.6 and 10.0 megaohms. It will be accepted when the measured value lies between 10.0 and 11.6 megaohms. If the measurements are spread uniformly within the above interval, 20% of the time the measured value will be between 9.6 and 10.0 megaohms, and 80% of the time between 10.0 and 11.6 megaohms. Thus, 20% of the time the cable will be wrongly rejected because of the uncertainty of the measurement result. This situation is illustrated in Figure 1.3.

Thus, because of measurement uncertainty, it may happen that a measured value is observed to be outside the specification limits whereas the value of the parameter is actually within the specification limits. It could also be that the actual value of the parameter is outside the specification limits but the measured value is within the specification limits. The rejection of a conforming item or acceptance of a nonconforming item could happen in both situations, whether the decision criteria are based on one-way or two-way specification limits.

The above anomaly results in two types of errors associated with statistical decision making, the so-called type I and type II errors.

- **Type I error:** The true value of the parameter is within the specification limits, but the measurement result indicates that it is outside the specification limits. This type of error is often designated as producer's risk, because a conforming item has been rejected. It is also called α-error.
- **Type II error:** The true value of the parameter is outside the specification limits, but the measurement result indicates that it is within the specification limits. This type of error is often designated as consumer's risk, as a nonconforming item has been accepted. It is also called β-error.

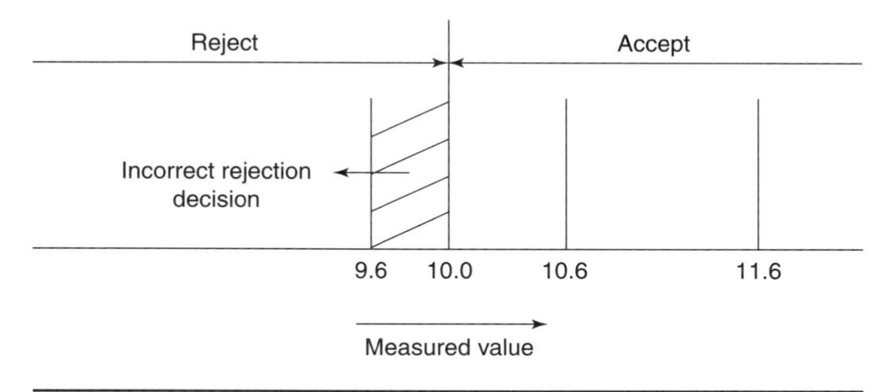

Figure 1.3 The effect of measurement uncertainty on decision making.

There is always some risk, however minimal it may be, that a decision made based on measurement results may be wrong due to the inherent uncertainty of measurement results. The origin of type I and type II errors is shown in Table 1.2.

Examples of Risk Involved and Risk Analysis

There are a number of situations in which risks due to type I or type II errors need to be analyzed very carefully when making a decision based on measurement results. Following are some examples:

- **Industrial measurements:** In an industrial scenario, the risks associated with accepting nonconforming products, or type II error, are not the same as those associated with rejecting a product which conforms to specifications, or type I error. During manufacturing, products are inspected at various stages: incoming, in-process, and final. The rejection of a conforming product at any stage may affect the delivery schedule of the end product, but the acceptance of a nonconforming product affects the quality of the end product, which may include product safety. This may sometimes result in losses due to warranty and product liability laws, in addition to the loss of reputation of an organization. Therefore, the risks need to be weighed very carefully depending upon the nature and criticality of the product and its end use.
- **Measurements related to healthcare:** Risk analysis becomes critical in measurements relating to healthcare and the medical diagnosis of patients. The risk of encountering type I and type II errors has to be minimal. One interesting example involves the use of radiation therapy for treatment of cancer. Radiation therapy is provided by using units of cobalt, which emits gamma rays. This radiation originates from the decay of natural radioactive sources. Measurements pertaining to cobalt radiation are critical. Too much radiation destroys the healthy tissues surrounding the cancer-affected part. If too little radiation is

Table 1.2 The origin of type I and type II errors.

	Parameter value within specification limits	Parameter value outside specification limits
Measured value within specification limits	No error	Type II error
Measured value outside specification limits	Type I error	No error

applied, the tumor is not destroyed and the ailment continues. The treatment is effective only if an optimum dose of radiation is administered. Measurement uncertainty considerations in such situations are crucial.

- **Measurements related to trade:** The uncertainty in trade-related measurement also needs to be considered seriously. For example, a person expecting a profit margin of 5% in his or her business has to ensure that the maximum permissible error of measurement is much less than 5%, and preferably less than 1%.
- **Measurements related to law:** Chemical measurements which are presented in legal proceedings include those involving suspected drug abuse and pesticides in fruit as well as more critical issues such as analysis carried out during postmortem. Another interesting case is that of drinking and driving. The result of measuring the amount of alcohol in blood is allowable as evidence in a court of law. In the United Kingdom, the legal limit beyond which one is found guilty is 80 milligrams of alcohol in 100 milliliters of blood (80 mg/100 ml). When this measurement is made, there is uncertainty in the observed value. At the level of the laboratory of the government chemist, this uncertainty is of the order of 2 mg/100 ml. A measured value of, say, 81 mg/100 ml with an uncertainty of 2 mg/100 ml implies that the true value lies between 79 mg/100 ml and 83 mg/100 ml. Thus, failing to consider the impact of type I and type II errors and pronouncing a guilty verdict based on the value of 81 mg/100 ml might amount to a miscarriage of justice. In such cases, the accepted convention in the United Kingdom is to subtract 6 mg/100 ml rather than 2 mg/100 ml from the measured value to make a generous allowance for any uncertainty in the measurement. The measurement result is presented as, for example, "not more than 75 mg/100 ml of alcohol," and thus care is taken that defendants are not adversely penalized by the uncertainty inherent in the analysis. A person is pronounced guilty only if the measured value is found to be 86 mg/100 ml or more. With measured values between 80 mg/100 ml and 86 mg/100 ml, one gets the benefit of the doubt due to the uncertainty of the analysis process.

MEASUREMENT RESULTS: FITNESS FOR PURPOSE

Measurement results are used for decision making. There is also a certain amount of uncertainty associated with them. This influences our faith in the correctness of a decision made based on measurement results. For example, a person being convicted for driving a car at high speed as well as the judge trying the case would like to be sure that the speed measured by the police radar is reasonably accurate. The term *reasonably accurate* itself implies that some error is inherent in the measurement result. Thus, a question arises as to how accurate a measurement result needs to be. The answer to this question depends upon the following factors:

- The end use of the measurement result; that is, the purpose of the measurement
- The criticality of the decision to be made based on the measurement; that is, the risk involved in the decision making
- The cost of taking the measurement; that is, whether the economics of operation justifies a costly measurement setup that ensures very accurate measurements

Gage blocks are used as a reference standard for the calibration of calipers and dial gages which are used for dimensional measurements on the shop floor of industries. These gage blocks are further calibrated using high-accuracy gage blocks as reference standards for dimensional comparison. Therefore, the accuracy requirement is different for measurements involving the calibration of gage blocks in a dimensional calibration laboratory than it is for the calibration of dial gages using the same gage blocks on the shop floor.

The accuracy requirement for a balance used to measure the mass of precious metals such as gold and silver is very different from that for a balance used for general-purpose measurements. The definition of *reasonably accurate* is different in each case. For example, the International Organization of Legal Metrology (OIML), in stipulating metrological requirements for nonautomatic weighing instruments, has specified four accuracy classes: special accuracy, high accuracy, medium accuracy, and ordinary accuracy. These recommendations were made in the 1992 edition of International Recommendations Document no. OIML R76-1. The requirement for "maximum permissible error on initial verification" is stipulated differently for weighing instruments of the various accuracy classes in that document. Compared with a weighing instrument of ordinary accuracy, the accuracy requirement is 10 times higher for medium-accuracy, 100 times for high-accuracy, and 1000 times for special-accuracy-class machines. Depending upon the criticality of the measurement accuracy, the user can select a balance of the appropriate accuracy class.

For a measurement result to be meaningful, it should be fit for the purpose for which the measurement was carried out. The fitness of an entity for its intended purpose is defined as its quality. Thus, its level of being reasonably accurate for its specific measurement application indicates the quality of a measurement result. This attribute of a measurement result is the uncertainty of measurement. The aspect of how much accuracy can be considered adequate for a specific measurement application is explained in subsequent chapters of this book.

A measurement result must be acceptable to all concerned with it. This is important for measurements in the regulatory and legal areas. Expressing a measurement result in commonly acceptable units is essential for its acceptability. These internationally accepted units, called SI units, make up the International System of Units. All measurement results should be linked by common roots. The common roots are the SI units which are embodied in national standards. Measurement results should be linked to national standards by a calibration process. This property of linkage of a measurement result to national standards is called traceability of measurement. Traceability and uncertainty are the quality attributes of a measurement result. For a measurement result to be meaningful, therefore, its uncertainty with reference to national standards should be known in quantitative terms and should be fit for the intended purpose. It should also have established traceability to the International System of Units. The traceability of measurement results and its achievement through calibration are explained in subsequent chapters of this book.

Metrology thus plays a crucial role in all facets of human endeavor, from fundamental research to our day-to-day needs in trade, industry, and scientific laboratories. It is also critical for regulatory aspects pertaining to the safety of human beings and to the environment around us. It is therefore important that a general awareness of the correct understanding of measurement results and associated issues emerge not only among professionals but also among consumers.

REFERENCES

Adler, Mortimer J., ed. 1993. *Great books of the Western world.* Vol. 6, *Plato.* Chicago: Encyclopaedia Britannica.

———. Scientific autobiography contained in ed. 1993. *Great books of the Western world.* Vol. 56, *National science selection from the twentieth century.* Chicago: Encyclopaedia Britannica.

Baigent, Heather. 1996. Role of testing and laboratory accreditation in international trade. *NABL News,* issue no. 4, October 1996:3–6 (published by the National Accreditation Board for Testing and Calibration Laboratories, Department of Science and Technology, Government of India).

Finkelstein, L. 1973. Fundamental concepts of measurement. In *Proceedings of the Sixth Congress of the International Measurement Conference,* 11–16. Amsterdam and London: Holland Publishing.

Hawking, Stephen. 1988. *A brief history of time.* London: Transworld Publishers.

International Organization of Legal Metrology (OIML). *Brochure of harmonisation in legal metrology through international consensus.* Paris, France: OIML.

———. 1992. Non-automatic weighing instruments, part I: Metrological and technical requirements. In Doc. R76-I, *International recommendations,* 19–24. Paris, France: OIML.

Kimothi, Shri Krishna. 1993. Measurements and measurement results: Some conceptual aspects. *MAPAN (Journal of the Metrology Society of India)* 8, no. 4:76–85.

Ku, Harry H. 2001. Statistical concepts in metrology. In *Handbook of industrial metrology,* American Society of Tool and Manufacturing Engineers, 20–50. New York: Prentice Hall.

National Bureau of Standards (NBS). 1984. *Measurement assurance program, part I: General introduction.* Special publication 676-I, 10–14. Washington: National Institute of Standards and Technology (NIST).

Pearson, Thomas A. 2001. Measure for Six Sigma success. *ASQ Quality Progress,* February, 35–39.

Quinn, T. J. 1999. Mutual recognition agreement for national measurement standards and calibration certificates issued by national metrology institutions. *Cal Lab: The International Journal of Metrology* 6, no. 2:21–23.

Stein, Philip. 1999. Measure for measure. *ASQ Quality Progress,* September, 74–75.

Sydenham, P. H., ed. 1982. *Handbook of measurement science.* Vol. 1, *Standardization of measurement fundamentals and practices.* John Wiley and Sons.

Treble, Ric. 1999. Analytical measurements and the law. *NABL News* 1, no. 13:19–20 (published by the National Accreditation Board for Testing and Calibration Laboratories, Department of Science and Technology, Government of India).

Verma, Ajit Ram. 1987. *National measurements and standards.* New Delhi: National Physical Laboratory.

2

Measurement Systems and the Quality of Measurement Results

The concept of measurement favoured by Einstein, Bohm, Putnam, et al., in the context of Quantum Theory is a realist concept. It encapsulates the realistic idea that in measurement we are about the task of reporting an independently existing objective reality. Moreover, for Einstein, Bohm, Putnam, et al. this objective reality is to be characterized classically in terms of physical quantities possessing particular value. Thus an "ideal measurement" is characterized as a process in which a measured value for some physical quantity is produced and the measured value reflects the value possessed by the physical quantity.

Henry Krips, *Quantum Measurement and Bell's Theorem*

The importance of reliable measurement in trade, industry, and scientific laboratories was explained in the previous chapter. This chapter covers various factors that need to be controlled to ensure reliable measurement. The importance of standardized measurement methods, the need for statistical control of measurement processes, and the information content which defines the quality of a measurement result have been discussed. The necessity that measurement results be relevant and adequate for their intended purpose has also been explained.

MEASUREMENT RESULTS AND MEASURANDS

Measurement has been defined as the set of operations having the objective of determining the value of a quantity. A *measurement result* is defined as the value attributed to a measurand obtained by measurement, and a measurement system is used to obtain it. One always measures a characteristic of an object and not the object itself. The physician measures the temperature of the human body and may also measure its weight but does not measure the human body itself. In a material testing laboratory, one may measure the tensile strength of an iron rod. In a food testing laboratory, one may be interested in measuring the nickel content in chocolates. Thus one always measures particular characteristics of a product or an object. The characteristics measured can be either physical or chemical.

A *measurand* is a specific quantity subject to measurement. The first step in measurement, therefore, is defining the measurand. The amount of information needed to define the measurand depends upon the desired accuracy of the measurement result and the factors influencing the measurement process. Some examples of the information to be given in defining a measurand are as follows:

- If the measurand is the length of a gage block used as a reference standard for the calibration of dial gages and calipers, the temperature at which the measurement is to be made is important information. The measurand in this case could be defined as the length of the gage block at 20°C and 50% relative humidity.
- If the measurand is the length of a rope, the tension of the rope needs to be defined, as it affects the measured length of the rope.
- The insulation resistance of an electric cable depends upon the temperature and voltage at which measurements are made, so the measurand could be defined as the insulation resistance of an electric cable at 500 volts DC and 30°C.
- If the length of an iron bar is to be measured to the micron level accurately, the temperature at which the measurement is to be made is important information in defining the measurand.

MEASUREMENT METHODS AND MEASUREMENT PROCEDURES

A *measurement method* is defined as a logical sequence of operations described generically and used in the performance of measurement, whereas a *measurement procedure* describes the operations required to perform a specific measurement according to the given method. For a measurement result to be meaningful, it is essential that the measurement is carried out using standard measurement methods. When one carries out physical or analytical measurements, a set of operations are performed in a specific sequence. For example, in analytical laboratories, analysis is carried out based on a standard method which contains instructions on the preparation of samples, the conditioning of samples, the quality of reagents needed, equipment to be used, and various other steps to be conducted in a specific sequence. Conducting tests according to the standard test method is important in all situations, but it is crucial in certain areas such as pathology laboratories. Depending upon the situation, a measurement method can also be called a test method, a calibration method, or an analytical method. Irrespective of the complexity of the measurements to be made, it is always advisable to have a documented measurement method. A measurement method has to be designed and validated by experts in the specific area of measurement. Only then can meaningful measurement results be expected. The reliability of a measurement result also depends upon the technical authenticity of the measurement method used. A measurement method is assumed to have technical authenticity if it is designed and validated by persons who are acknowledged as experts in that area of measurement. Such experts are generally associated with professional bodies. A number of professional bodies have standardized measurement methods and published them as standards. A large number of standards on methods of measurement have been published by the American Society for Testing and Materials (ASTM), the International Organization for Standardization (ISO), the International Electrotechnical Commission (IEC), and the International Special Committee on Radio Interference (CISPR). Such standards have also been published by various other national and international standardization and professional bodies.

The Effects of Deviation from a Standard Test Method: An Example

Test results are significantly influenced by the test method used. There is a saying that if one is given the freedom of choosing a test method one can get the result one desires. While testing a product, one should be interested in determining the actual value of the parameter which is under measurement and not the value one desires to be the test result. Sometimes even minor deviation in standard test methods leads to very erroneous test results. The implications of such a situation can vary from minor irritation to serious consequences in terms of loss of business. A conforming product could be judged as nonconforming or vice versa, resulting in type I or type II errors, as explained in chapter 1.

The author had an interesting encounter with such a situation in a test laboratory. For short-distance communication, very-high-frequency frequency modulated (VHF-FM) transceivers are used by the police. The contract governing the procurement of these transceiver sets stipulated that some of the sets, picked at random, would be tested in an independent test laboratory as part of the acceptance inspection. In the absence of a standard test method specific to these transceiver sets, it was agreed with the client that standard test methods for normal radio communication equipment would be used. The technical specifications of the transceiver stipulated that the maximum frequency deviation be 5 kilohertz. One of the parameters to be measured was the audio frequency response of the sets. For this measurement, a test signal of a specific frequency deviation was to be received by the transceiver. It was agreed with the client that the measurement would be carried out using a test signal with a frequency deviation of 5 kilohertz, the maximum frequency deviation allowed by the technical specifications. The measurement results of this parameter revealed that the majority of the transceivers did not conform to the specification limits. The designers of the sets opined that the design was perfect and that the problem lay with the testing.

Subsequently, a standard test method published by a professional body was located. This test method required that the audio frequency response be measured using a test signal with a frequency deviation of 3 kilohertz. When the measurements were repeated using the specified test signal, it was found that most of the sets did conform to the specifications. Thus, the majority of the sets which had appeared not to conform to specifications were found to conform to specifications when the standard test method was followed. Without this corrective approach, the acceptance process could have resulted in a type I error situation.

Measurement Method Requirements for Technical Competence

International standard ISO/IEC 17025:1999 stipulates the requirements for the competence of testing and calibration laboratories. This standard has identified "test and calibration methods and method validation" as one of the factors that determines the correctness and reliability of test and calibration results. The standard has therefore stipulated the use of standard test or calibration methods by the laboratories. In case a laboratory uses a method other than one that is standard, the standard also requires that the method shall be validated before use.

Validation of Measurement Methods

Validation is the process of confirmation that the specific requirements for which a measurement method was designed and intended to be used have been fulfilled. A measurement method is intended to measure a physical or chemical parameter to a specified level of accuracy. Validation is therefore performed to ensure that the parameter is being measured to the required level of accuracy. Some of the techniques used for measurement method validation are as follows:

- **Use of reference standards or reference materials:** Reference standards for measurement purposes are supplied by reputed manufacturers. Their values are known accurately. Similarly, reference materials are materials whose characteristics are known very accurately. These values are certified by the manufacturers of certified reference materials. The value of a reference standard or reference material is measured using the laboratory-developed method. If the measurement result is in agreement with the reference value assigned to the standard or reference material, it is accepted that the measurement method has been validated.
- **Comparison of results with those obtained by other methods:** In this technique, measurements are carried out using the laboratory-developed test method and a different established test method on the same material. If the results obtained by the two methods are in agreement, the validity of the laboratory-developed test method is established.
- **Interlaboratory comparisons:** The appropriateness of a measurement method used for testing or characterization of materials may be validated through interlaboratory comparison. The material is sent to various laboratories, which carry out tests according to the proposed method under a planned interlaboratory testing program. The laboratories chosen for this exercise are required to be technically competent. The entire exercise is performed by qualified and competent professionals who are knowledgeable about the measurement process and other critical issues involved in measurement.

Measurement Methods for Industrial Testing and Calibration

Generally, standard methods of measurement are available for testing products in testing laboratories. In addition to testing laboratories, many kinds of measurements are carried out on the shop floor of industries. These measurements are performed mainly during the inspection of purchased products, in-process monitoring and inspection, and the testing of end products. In view of the importance of measurement methods and in order to ensure objectivity in measurement results, the ISO 9000 series of quality management standards requires that organizations formulate test procedures as part of their quality system documentation. Similarly, calibration laboratories that maintain quality management systems in their operations formulate calibration procedures as part of their quality system documentation.

THE CONTENTS OF A MEASUREMENT METHOD OR PROCEDURE

The contents of a measurement method need not be either very elaborate or very short. They should, however, have details enough to ensure that a person of average competence in the area of measurement is able to carry out measurements and derive measurement results without any additional information. The contents of a measurement method may include the following:

- A brief description of the scientific principles involved in measurement
- An experimental setup giving details such as the technical requirements of the instruments and other accessories used
- Initial adjustments and warm-up requirements
- A step-by-step procedure detailing how measurement operations are to be carried out and how the measurement result is to be arrived at
- Precautions to be observed during measurement
- The purity of the chemicals and reagents to be used
- The accuracy of reference standards used for calibration
- Control of environmental conditions such as temperature, humidity, biological sterility, and electromagnetic interference
- The manner in which uncertainty is to be estimated and reported, if this is required

The contents of a measurement method thus depend upon the complexity of the measurement process and the end use for which the measurements are being made. It is important that a validated measurement method ensures that the variability of the measurement process due to the measurement method is optimized.

Test Methods of the American Society for Testing and Materials

ASTM is a professional body which has published a large number of standards on test methods for the testing of products and materials. These standards have been written with a view to harmonizing the testing process so that the test results can be compared without any ambiguity. ASTM regulations define a test method as "a definitive procedure for the identification, measurement and evaluation of one or more qualities, characteristics or properties of materials, products, systems or services that produces a test result."

It is essential that a standard test method establish a proper degree of control over all influence factors that may affect test results. To achieve this, a test method is subjected to a "ruggedness" test during its development phase.

ASTM regulations require that a well-written test method specify a mechanism for control over factors which influence test results. These factors are as follows:

- The clarity of interpretation and completeness of the test method from the operator's point of view
- The measurement accuracies/tolerances and calibration requirements of test equipment
- The temperature, humidity, atmospheric pressure, contamination, and so on of the environment
- The sampling procedure to be used to decide on the number of test specimens
- The preparation of test items
- The manner in which measurement results are to be combined to provide a test result

ASTM standards have identified three stages of a quantitative test method:

- Direct measurements or observations of a dimension or another characteristic
- The arithmetic combination of observed values to obtain a single determination based on the observations
- The arithmetic combination of a number of determinations to obtain a test result for the test method

For example, the test result for the tensile strength of a metal wire is obtained as follows:

- Measure the diameter of the wire and determine its radius and cross-sectional area
- Using a tensile testing machine, measure the force in newtons at which the wire breaks
- Calculate the tensile strength as the breaking force divided by the area of the wire's cross section

The Contents of a Standard ASTM Test Method

Generally, a standard ASTM test method gives the following information to enable testing laboratories to use the methods effectively and unambiguously:

- **Scope:** This item defines the purpose of the test method.
- **Reference documents:** This item gives reference to various related ASTM standards.
- **Terminology:** The definitions of various technical terms used in the standard are given under this heading.

- **Summary of test method:** The underlying principle of the test method is described here.
- **Significance:** This item gives the criticality of the measurements.
- **Interference:** Factors affecting measurement results are described in this item.
- **Apparatus:** Details of test equipment and test setup are given under this heading.
- **Test specimen:** This item covers how a test specimen is to be selected and prepared, including preconditioning, if any.
- **Calibration:** The calibration requirements for the various pieces of test equipment are given in this item.
- **Procedure:** A step-by-step procedure for taking measurements and observations and conducting the test is given under this heading.
- **Calculation:** This item gives information on how determinations are obtained from the observations and how the test result is obtained from these determinations.
- **Report:** Information on how the test result is to be reported in the test report issued by the testing laboratory is given here.
- **Precision and bias:** This item specifies the closeness of the agreement between repeat observations and the maximum permissible value of the error likely to be obtained by the standard test method. If the observed values of precision and bias are found to exceed the values given in the standard test method repeatedly, this implies that influence factors are not under control or that the test method is not being followed correctly.
- **Key words:** Key words which can be associated with the standard test method are given under this heading.

THE STEPS INVOLVED IN MEASUREMENTS

The following steps are involved in making measurements:

- **Definition of the measurand:** The measurand is first defined for its characteristics and the environmental conditions under which measurements are to be made.
- **Realization of the measurand:** The measurand and environmental conditions according to the definition must be realized. Sometimes the measurand is not realized according to the definition of the measurand, and this introduces error into the measurement results. For example, consider a case in which the measurand is the electrical resistance of a

nichrome wire at 20°C. The resistance is measured at 30°C, and its value at 20°C is derived from this measured value at 30°C and the known value of the temperature coefficient of the resistance of nichrome. This will introduce error on two counts: the accuracy of the temperature measurement of 30°C and the accuracy of the known value of the temperature coefficient of resistance.

- **Design and validation of the measurement method:** If a standard measurement method is not available, it has to be designed and validated.
- **Measurement operation:** This involves carrying out measurements according to the approved method. All the precautions recommended by the approved method and the equipment manufacturer's instructions are to be observed. Initial adjustments and conditioning of the measurement objects must be carried out as required by the method. The measurements are then performed according to the measurement method.
- **Derivation of the measurement result:** Based on the observations recorded during measurement, the measurement result is derived. Sometimes the observation itself may be the measurement result; at other times the result is derived based on observations. For example, the tensile strength of an iron wire is derived based on the measurement of the tensile force at which it breaks and the cross-sectional area of the wire. The cross-sectional area of the wire is derived based on the measurement of the diameter of the wire. The result must be reported and recorded as stipulated in the method. If required by the method, the uncertainty of the measurement result must be estimated and reported.

MEASUREMENT SYSTEMS AND MEASUREMENT PROCESSES

International standard ANSI/ISO/ASQ Q9000-2000, *Quality management systems—Fundamentals and vocabulary,* defines a *system* as a "set of interrelated and interacting elements." Thus, a measurement system consists of all the interrelated and interacting elements that are used in obtaining a measurement result of the desired quality. In the *Measurement System Analysis Reference Manual,* a measurement system is defined as "the collection of operations, procedures, gages and other equipment, software and personnel used to assign a number to the characteristic being measured; the complete process used to obtain a measurement."

The system aspect of a measurement process is called its measurement system. Some of the entities which can be considered part of a measurement system are as follows:

- Measurement/test method
- Measurement equipment and measurement setup
- Personnel involved in the maintenance and operation of measuring equipment and the measurement process
- Organizational structure and responsibilities
- Quality control tools and techniques
- Preventive and corrective mechanisms to eliminate defects in operations

ANSI/ISO/ASQ Q9000-2000 defines a *process* as a "set of interrelated or interacting activities which transforms input into an output," and a *product* is defined as "the result of a process." In the case of a measurement process, the product is the measurement result.

Measurements are carried out using a *measurement process,* which has been defined as aset of operations to determine the value of a quantity. The inputs to a measurement process are as follows:

- The measurement object through which the measurand is realized
- The measuring instruments and equipment or measurement setup
- The person taking the measurement
- The environment in which the measurements are made
- The laboratory or other place of measurement
- The measurement method

The output of a measurement process is the measurement result, which is an intangible entity. The sets of interrelated or interacting activities are the activities that are carried out while making measurements; these activities are defined in the measurement method. A measurement process is in many ways analogous to a manufacturing process. The inputs to a manufacturing process are the material, machine, method, person, and environment in which the manufacturing operation is undertaken, and the output is a finished product. A manufacturing process is influenced by chance causes of variation inherent in the process, resulting in variability in the quality characteristics of the end product. So too a measurement process is influenced by chance causes of variation inherent in the measurement process which result in variability of the measurement observations. Walter A. Shewhart developed statistical quality control techniques in 1939 with a view to controlling the quality of manufactured products. These techniques of controlling manufacturing processes found wide applications in industry and are still being used extensively. The same principles of statistical process control (SPC) can be used in the control of a measurement process. Metrologists are finding SPC to be a very powerful tool in testing and calibration laboratories and are using these techniques to control the quality of measurement results.

THE MEASUREMENT PROCESS: REQUIREMENT FOR STATISTICAL CONTROL

A manufacturing process manufactures end products of acceptable quality if and only if it is in a state of statistical control. A certain amount of variability is always observed in the quality characteristics of end products. The variability is inherent in the process and is due to variations in input quantities individually and collectively. Examples of these are variations in the characteristics of the raw material going into production; variations in the behavior of machinery or other equipment, which may be due to variations in the characteristics of parts forming the machinery, to aging, or to other factors; variations in the performance of the operator due to differences in skill, training, attitude, and so on; variations in the method employed; and, finally, variations in environmental conditions such as temperature, humidity, and pressure. It is impossible to control these factors of variability to the extent that there is no variation in the end product characteristics. This variability is accepted as part of the process, and these variations are referred to as variations due to chance causes.

The amount of variability that can be accepted depends upon the specifications for the end product. If the end product specifications have close tolerances, the input parameters need to be controlled to reduce the variation. This may require the use of more sophisticated machines; tight control of input raw material characteristics; highly skilled, qualified, and trained operators; and tight controls on environmental conditions. These tight controls will necessitate more investment and higher costs of operation. Thus, optimizing manufacturing process variability in the context of the requirements of end product specifications involves the economics of business. Such decisions are made at the senior management level. Irrespective of the amount of variation in a manufacturing process, if the process is influenced only by chance causes, it is said to be in a state of statistical control. This variation is random in nature and is called chance variation. Sometimes variations are observed in the process output which cannot be assigned to chance causes alone but can be attributed to causes such as faulty adjustment of a machine, an incompetent or untrained operator, uncontrolled environmental conditions, or nonconforming raw materials. These are called assignable causes of variation. The presence of an assignable cause takes the process out of statistical control. Under these circumstances, the variation in output characteristics is large and is not acceptable. The variation is not random in nature and shows some trends.

In an analogous way, a measurement process also has to be in a state of statistical control if it is to lead to meaningful results. Being in a state of statistical control implies that variability in the measurement process is due to chance causes only. The chance causes of variability in a measurement process are variations due to the measuring equipment, the skill of the person taking the measurements, environmental conditions, the object being measured, and so on. If there are assignable causes of variation present, the measurement process is said to be out of statistical control. The measurement result of such a process is not fit for its intended purpose. The assignable causes could include the use of un calibrated equipment, an improper or nonstandard test method, incompetence on the part of the person taking the measurements, and inadequate control of the environment under which the measurements are made.

The need for statistical control of measurement process has been defined by Churchill Eisenhart (1963) in the following words:

> A measurement process is the realization of a method of measurement in terms of particular apparatus and equipment of the prescribed kinds, particular conditions that at best only approximate the conditions prescribed and particular persons as operators and observers. It has long been recognized that, in undertaking to apply a particular method of measurement, a degree of consistency among repeated measurements of a single quantity needs to be attained before the method of measurement concerned can be regarded as meaningfully realized, that is, before a measurement process can be said to have been established that is a realization of the method of measurement concerned. Indeed, consistency or statistical stability of a very special kind is required to qualify as a measurement process a measurement operation must have attained, what is known in industrial quality control language as a state of statistical control. Until a measurement operation has been "debugged" to the extent that it has attained a state of statistical control it cannot be regarded in any logical sense as measuring anything at all. And when it has attained a state of statistical control there may still remain the question of whether it is faithful to the method of measurement of which it is intended to be a realization.

THE QUANTITATIVE AND QUALITATIVE ASPECTS OF MEASUREMENT RESULTS

Lord Kelvin, the great experimental physicist, emphasized the importance of numbers in measurement in the following words:

> I often say that when you can measure what you are speaking about and express it in numbers, you know something about it, but when you cannot measure it,

> when you cannot express it in numbers, your knowledge is of a meager and unsatisfactory kind.

A measurement result is represented by a numerical value of the characteristic represented by the result. In his paper, Churchill Eisenhart (1963) stated:

> Measurement is assignment of numbers to material things to represent the relation existing among them with respect to particular properties. The number assigned to some particular property serves to represent the relative amount of this property associated with the object concerned.

The numerical value of a measurement result, or "assignment of numbers to material things," is its quantitative aspect. But this is not a measurement result's totality. There is also a qualitative aspect of a measurement result. Walter A. Shewhart, who pioneered the concept of control charts as a statistical tool to control manufacturing processes, developed the concept of the qualitative and quantitative aspects of measurement. Eisenhart (1963) quoted Shewhart as follows:

> It is important to realize that there are two aspects of an operation of measurement one is quantitative and the other qualitative. One consists of numbers or pointer readings such as the observed lengths in measurements of the length of a line, and the other consists of the physical manipulations of physical things by someone in accord with instructions that we shall assume to be describable in words constituting a text.

According to Shewhart, the qualitative aspect is linked to the perfection of the documented measurement method and the way it has been realized during the carrying out of the measurements. The quality of the quantitative information will be different, even when the same measurement method is used, if it is implemented by different persons at different times using different equipment. Eisenhart (1963) described this aspect in the following words:

> More specifically, the qualitative facts involved in the measurement of quantity are the apparatus and auxiliary equipment (e.g. reagents, batteries or other sources of electrical energy etc.) employed, the operators and observers, if any, involved in the operations performed, together with the sequence in which and the conditions under which they are respectively carried out.

The statistical control of a measurement process is an essential requirement in order for the process to give meaningful results. If we were to measure the same parameter using two measurement processes, both in statistical control, the variability in repeat observations could be different. For less variability, the input parameters need to have better control, so one quality parameter is the closeness of the repeat measurement data. This characteristic of the measurement process is called precision. Eisenhart has identified two parameters—precision and systematic error—that represent the quality of measurement results. The quality of a measurement result is characterized by the measurement uncertainty, which depends upon the precision and accuracy of the measurements.

QUALIFYING QUANTITATIVE INFORMATION AND THE NUSAP NOTATIONAL SCHEME

The need for uniformity in the content of quantitative information has been felt for quite some time. Policy makers need quantitative information to decide policies pertaining to technology, economics, and social affairs. The technological issues may pertain to the effects of new technologies and industrial growth on the environment and health of people, such as the effects of radioactive pollution, acid rain, agricultural chemicals, and so on. The economic and social issues may pertain to uneven distribution of wealth, standards of living, the literacy and education levels of various communities, social customs having bearing on social harmony, and so on. Policy-related research is carried out by technologists and social scientists to obtain quantitative information on these issues in order to find solutions and decide on policy matters. The quantitative information may relate to, for example, safe levels of pesticides or radioactive pollution. Sometimes policy makers have to make decisions on matters of such public importance without delay. They need authentic information on these matters from the research community. At times the research community does not possess the knowledge and skill required to come up with an immediate and effective solution. The quantitative information provided to the policy makers by research scientists is a single number and rarely includes precise confidence limits of the classical statistical form; however, the quantitative information is considered to be robust. A policy decision based on just a quantitative number then becomes a subject of public debate. The problem of the communication of quantitative information for purposes of policy making is well known. Solving the problem of representing and evaluating technical information and identifying meaningless quantitative expression has thus become a matter of great importance for the proper accomplishment of public policy.

S. O. Funtowicz and J. R. Ravetz (1987) have precisely explained the above dilemma in their essay "Qualified Quantities: Towards an Arithmetic of Real Experience" in the following words:

> Thus the traditional assumption of the robustness and certainty of all quantitative information has become unrealistic and counterproductive. The various sorts of uncertainty, including inexactness, unreliability and ignorance, must be capable of

> representation. A quantitative form of assertion is not merely considered necessary for a subject to be scientific, it is also generally believed to be sufficient. Thus the problems discussed here are not only related to the inherent uncertainties of the subject matter (as for example in risk and environmental pollutants,) they originate in an inappropriate conception of the power and meaning of numbers in relation to the natural and social worlds. But this image must be corrected and enriched if we are to grow out of the reliance on magic numbers only in that way can we hope to provide usable knowledge for policy decisions including those for science, technology and the environment.

Such questions are quite often faced while dealing with decisions related to the environment and its effect on ecology. Funtowicz and Ravetz (1987) note that W. D. Ruckelshaus, once the top administrator at the U.S. Environmental Protection Agency, has expressed similar thoughts in the following words:

> First we must insist on risk calculations' being expressed as distributions of estimates and not as magic numbers that can be manipulated without regard to what they really mean. We must try to display more realistic estimates of risk to show a range of probabilities. To help to do this we need tools for quantifying and ordering sources of uncertainty and for putting them in perspective.

Thus, the meaning of a quantitative statement is contained in its mode of qualification as much as in its quantifying part. It is to be understood that a bare statement of quantity in the absence of its qualifying statements is scientifically meaningless.

The NUSAP Notational Scheme

Quantitative information is communicated among scientists, technologists, and social scientists for exchange of information. Quantitative information also needs to be communicated to policy makers so they can make policy decisions. Generally, the information is just a number, making it "half information," and making decisions based on it may lead to problems. In order for the quantitative information to be complete, its qualifying attributes need to be defined. For this purpose, Funtowicz and Ravetz have proposed a notational scheme called NUSAP, which is an acronym of the words *numeral, unit, spread, assessment,* and *pedigree.* These five attributes define the quantitative as well as qualitative characteristics of quantitative information. Generally, there are three types of uncertainty which influence every piece of quantitative information:

- **Inexactness of measurement and representation:** A piece of quantitative information is generally obtained based on a measurement carried out using instruments or collection of field data. The collection of information on the emission of gaseous discharge by an automobile is an example of the use of instruments, whereas the gathering of information on the socioeconomic conditions of persons living in a specific region is an example of the collection of field data. There is always some uncertainty in measurement and collection of data. There may also be some uncertainty in the representation of quantitative information.

- **Unreliability of methods, models, and theories:** The use of a different measurement method may lead to quantitative information which is not exactly the same as that obtained with an earlier method. Similarly, a different model for the collection of data may result in different quantitative information. Thus, the methods, models, and theories used for obtaining quantitative information may also lead to uncertainty in the information.
- **Revealing of the border between knowledge and ignorance in the history of the quantity:** Another source of uncertainty in quantitative information is the limitation in our knowledge about the scientific phenomena on which the information obtained is based. This is especially relevant in information related to scientific research, where different values of quantitative information are obtained as knowledge on the subject becomes greater.

In the NUSAP notational scheme these uncertainties have been taken into account and five elements have been identified which define a piece of quantitative information in its totality. The elements are given in Table 2.1.

Example of the NUSAP Notational Scheme

In order to make a policy decision on certain welfare measures, information on the educational profile of persons living in a particular city is needed. The city has a population of about 500,000. The information needed is the number of persons having an education level of secondary school or lower. The rest of the population has an education level of graduation from university or higher. The task is assigned to a social scientist, who collects the data pertaining to the number of persons with an educational level of secondary school or lower. The data are collected for a small part of the city and extrapolated to obtain information for the total population of the city.

The scientist issues the quantitative information that there are approximately 300,000 persons in the city with an educational background of secondary school or lower. The rest of the persons in the city are of university-graduate

Table 2.1 The NUSAP notational scheme.

Notation	Description
Numeral (N)	This is the principal attribute of a piece of quantitative information. It can be a whole number, a decimal, or a fraction.
Unit (U)	This represents a multiplier or standard unit.
Spread (S)	This is the generalized traditional concept of error and is expressed in an arithmetical form, such as ± or %. It is the strong qualitative element of a piece of quantitative information.
Assessment (A)	The spread is generally an estimate and cannot be given precisely. It is generally calculated as a statistical measure. The assessment qualifies the spread and can be seen as a generalization of the confidence limits used in statistical practice.
Pedigree (P)	Since *pedigree* refers to family history or parentage, this implies an evaluative history of the quantity. This is especially applicable to research fields in which early pioneering efforts have been superseded by stronger work. It indicates the border between what is currently feasible and accepted as known and what is not feasible and unknown.

level or higher. This number does not carry enough information on which to base a policy decision. If the NUSAP notational system had been followed, the quantitative information could have been presented as follows:

> The city has approximately 300,000 persons with an educational background of secondary school or lower, which constitutes about 60% of the population. The error in this estimation is not likely to exceed 5%. Thus, the actual number of such persons is likely to be between 285,000 and 315,000. This estimate is given at a confidence level of 95%. Similar studies were conducted 10 years ago, and it was observed that 65% of the population had an educational background of secondary school or lower.

The NUSAP notations for the preceding statement are given below:

Numeral (N): 300

Unit (U): Thousands

Spread (S): ±5%

Assessment (A): 95% confidence level

Pedigree (P): The last sentence of the statement gives the history of similar information collected in the past. Such complete information is very helpful for making decisions.

The quantitative information is not confined to just one number but also gives the uncertainty associated with the finding and that the number of persons with a university degree or higher has increased by 5%.

NUSAP NOTATION AND MEASUREMENT RESULTS AS QUANTITATIVE INFORMATION

The NUSAP notational system was evolved to express the various components of uncertainty contained in all quantitative information. The quantitative information may pertain to any field, including policy-related research, scientific research, economic research, and social sciences research. As a measurement result is also quantitative information, we can define the various elements of the NUSAP system in the context of a measurement result as follows:

Numeral (N): The result of a measurement is defined as the value attributed to a measurand obtained by measurement. It is a numerical value followed by a unit. The notation *N* represents the number which indicates the measurement result.

Unit (U): The notation *U* represents the unit in which the physical or chemical characteristics are measured and represented.

Spread (S): The notation *S* represents the spread or inexactness of the measurement result and hence represents the uncertainty of the measurement result. This notation is derived from the statistical variance of the data.

Assessment (A): The notation *A* qualifies the notation spread. This represents the confidence level of the measurement uncertainty and is expressed as probability.

Pedigree (P): The pedigree is also important for measurements made in testing and calibration laboratories as well as in industry. In calibration laboratories, past measurement results obtained on reference standards are used to assign uncertainty to the reference value as well as to study any pattern of drifts in their values. Testing and calibration laboratories are accredited for their technical competence by laboratory accreditation bodies. The accreditation process involves judging the capability of a laboratory to undertake specific types of tests or calibration by technical experts. This exercise is repeated at periodic intervals to ensure that the accredited laboratory continues to remain competent. Thus, an accredited laboratory has a history of taking specific types of measurements. In industry, gage repeatability and reproducibility studies are carried out before a measurement system is validated for a specific application. The known history of a specific type of measurements enhances our confidence in those measurements. This aspect of measurement is represented by the pedigree notation *P*.

Following is an example of the application of the NUSAP notational scheme to a measurement result: The measured value of the mass of a reference standard weight held by a legal metrology laboratory is 1.0005 ±0.0007 kilogram. The uncertainty of ±0.0007 kilogram has been quoted at a confidence level of 95%. The value has been measured every year and has been found to be within the above interval. The NUSAP notations are as follows.

Numeral (N): 1.0005, as above

Unit (U): Kilogram

Spread (S): ±0.0007

Assessment (A): 95% confidence interval

Pedigree (P): The parameter has been measured every year and the measured values have been found to be within the above interval.

THE RELEVANCE AND ADEQUACY OF MEASUREMENTS: THE MANAGEMENT PERSPECTIVE

A measurement result without a statement of its uncertainty lacks credibility and cannot be effectively used. The management of an industrial organization has to be very cautious in deciding its metrology needs, as the costs involved may include both a liability element and a savings element. The proper identification of the measurement needs of an organization is important. The systematic measurement of identified parameters is essential for the overall productivity and prosperity of an organization. In fact, it is as important to the overall economics of an organization as any other managerial or technical activity. It is therefore essential that the highest level of management recognize the importance of relevant measurements and adequate quality of measurement results. Too high an accuracy of measurement may result in overdesigning, thus increasing the cost. It may also result in excess testing and unnecessary rework of the product, leading to higher production costs and lower profitability. A measurement accuracy that is too low will result in a product that is not acceptable to the customer.

The relevance of accurate measurements has been described by H. L. Daneman and W. C. Wiley (1972) as follows:

> When we measure what we should measure, we make measurements which are relevant. Measurements applicable to the end objective of our institutions, measurements which raise up the value of goods and services, measurements which are germane to our national priorities.

Robert D. Huntoon of the U.S. National Bureau of Standards, now the NIST in NBS publication no. 359 defined adequacy of measurement with these words: "The widely acknowledged purpose of measurements is to arrive at product-compatible measurements by a process involving instruments, people and procedure."

The measurement system has to be adequate for the intended purpose of the measurement. The relevance and adequacy requirements of measurement systems are equally valid in testing and calibration laboratories. In testing laboratories, measurements are made to determine the numerical values of product characteristics. The requirement for a "reasonably accurate" measurement setup in a testing laboratory is also influenced by the variability of the product characteristics being measured. In this situation, a measurement system can be considered adequate if the measurement system's variation due to various sources of error is small compared with product characteristic variations. The accuracy requirement for a measurement setup is generally specified in the standard measurement method. If the accuracy requirement is not met, one cannot totally rely on the correctness of the decision made. In a calibration laboratory, the integrity of measuring instruments is compared against reference standards, which are of very high accuracy. The calibration process can be considered adequate only if the accuracy of the reference standard used for calibration is much better than the accuracy of the measuring instrument that is being calibrated.

A measurement system has to be relevant to and adequate for a specific application. On this, Thomas A. Pearson (2001) has quoted Norman Belecki, the longtime chief of the Electrical Reference Standards Division of the U.S. National Bureau of Standards, now the NIST, in the following lines:

> The first step to making a good measurement is gaining a good understanding and definition of the real need. Why are we making the measurement? We often settle for measuring what we can measure—what we can measure within time and bud-

get constraints—rather than defining the real measurement need. Perhaps we do not know the real requirements or how they relate to the things we can measure.

The second step is meeting the technical requirements for measurement uncertainty, standards, test equipment traceability, system stability and measurement process control. To complete the process, employ confirmation experiment using a second, independent measurement technique whenever possible. This is critical especially when uncertainties required to support the process come to near to specification of the available instrumentation. If this is not possible, measure with better methods and uncertainties than required if available and review all results for reasonableness. Primary factors that can influence the measurements should be determined by experimentation, and their impact on overall accuracy minimized and controlled as necessary in future measurements.

Thus, for making good measurements, it is essential that the following requirements be met:

- Real measurement needs are clearly understood and defined.
- Confirmation experiments are carried out to ensure the adequacy of measurement uncertainty according to the defined needs.
- Traceability of measurement results is ensured through calibration.
- If measurement uncertainty is not adequate, a better method with the required capability of measurement is employed.
- The effects of influence factors such as environmental conditions are studied to ensure that measurement uncertainty still meets the needs.
- The measurement process is controlled on a continuing basis using process control techniques.

All of the above aspects are covered in subsequent chapters of this book.

SUMMARY

There are various activities that must be completed before a measurement system can be put into service and used as a routine for taking measurements. These start with understanding the measurement objectives and then defining and realizing a measurand. The next activity is designing and validating a measurement method if a standard method is not available. Validation consists of verification that the measurement method meets the measurement objectives. The measurement system can then be put into service. Its capability should be verified at periodic intervals to ensure that it continues to meet the measurement objectives. This process is shown in Figure 2.1.

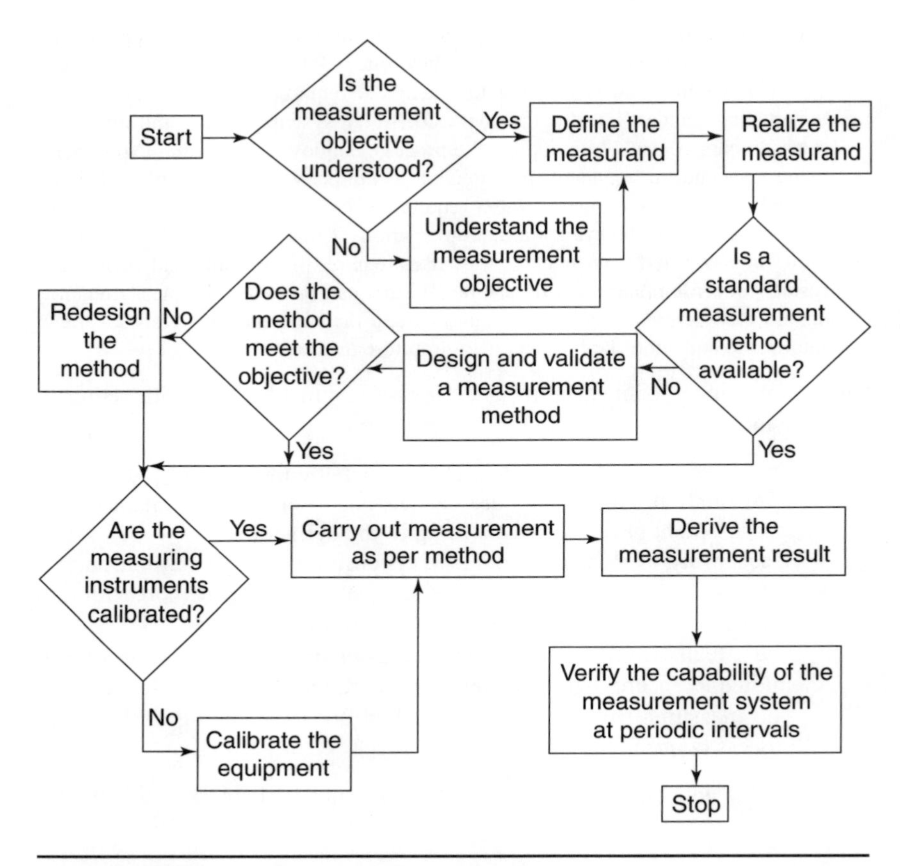

Figure 2.1 Parameter measurement flowchart.

REFERENCES

American Society for Testing and Materials (ASTM). 1990 (reapproved 1996). ASTM E177-90, *Use of terms precision and bias in ASTM test methods.* ASTM.

ANSI/ISO/ASQ Q9000-2000, Quality management system—Fundamentals and vocabulary. 2000. Milwaukee, WI: ASQ Quality Press.

Chrysler Corporation, Ford Motor Company, and General Motors Corporation. 1995. *Measurement systems analysis: A reference manual.* Southfield Michigan, Automotive Industry Action Group.

Daneman, H. L., and W. C. Wiley. 1972. Measuring what we should measure. In *Proceedings of the 1972 Joint Measurement Conference 25–29, National Bureau of Standards.* Boulder, CO.

Eisenhart, Churchill. 1963. Realistic evaluation of the precision and accuracy of instrument calibration systems. *Journal of Research of the National Bureau of Standards: Engineering and Instrumentation* 67C, no. 2:161–85.

Funtowicz, S. O., and J. R. Ravetz. 1987. Qualified quantities towards an arithmetic of real experience. In *Measurement, realism and objectivity: Essays on measurement in the social and physical sciences,* 59–88. Boston and Tokyo: D. Reidel Publishing.

Grant, Eugene L., and Richard S. Leavenworth. 1988. *Statistical quality control.* New York: McGraw-Hill.

Huntoon, Robert D. 1971. *Metrology and standardization in less developed countries: The role of a national capability for industrializing economies.* NBS publication no. 359: NBS. 265–76.

International Organization for Standardization and International Electrotechnical Commission. 1999. ISO/IEC 17025:1999, *General requirements for the competence of testing and calibration laboratories.* Geneva, Switzerland: ISO.

Krips, Henry. 1987. Quantum measurement and Bell's theorem. In *Measurement, realism and objectivity: Essays on measurement in the social and physical sciences,* 45–58. Boston and Tokyo: D. Reidel Publishing.

Pearson, Thomas A. 2001. Measure for Six Sigma success. *ASQ Quality Progress,* February, 35–40.

3

Statistical Methods in Metrology

The theory of errors—We are thus brought to consider the theory of errors which is directly connected with the problem of the probability of causes. Here again we find effect—to wit, a certain number of irreconcilable observations, and we try to find the causes which are, on the one hand, the true value of the quantity to be measured, and, on the other, the error made on each isolated observation. We must calculate the probable a posteriori value of each error, and therefore the probable value of the quantity to be measured. But, as I have just explained, we cannot undertake this calculation unless we admit a priori—i.e. before any observations are made—that there is a law of the probability of errors. Is there a law of errors? The law to which all calculators assent is Gauss's law, that is represented by a certain transcendental curve known as the bell.

Henri Poincaré, *Science and Hypothesis,* 1903

Measurement uncertainty is evaluated using statistical methods. Therefore, an understanding of statistical principles is desirable for metrologists. This chapter aims at providing a working knowledge of the statistical methods needed for the evaluation of uncertainty. The chapter explains various characteristics of data, analysis of random variables, and various frequency distributions having applications in metrology. It also explains estimation methods employed in uncertainty evaluation. This background will help the reader properly understand the concepts used in the *Guide to the Expression of Uncertainty in Measurement* and their applications.

STATISTICAL METHODS AND STATISTICAL THINKING

A new term, *statistical thinking,* is now being used quite often. It has been defined as a mind-set which accepts the following principles and allows them to direct one's actions:

- All work is accomplished through interconnected processes. A process transforms an input into an output by adding value. A process can be studied and improved, thereby improving the output.
- Variation exists in all processes. This process variation is reflected as variation in the characteristics of the output.
- Understanding and reducing variation is the key to success.

On the other hand, *statistical methods* are used for the analysis of data. The genesis of statistical methods lies in data, which have been defined as series of observations, measurements, observed facts, or numbers. The origin of data lies in the existence of variation in all processes. Statistical tools are used on data to extract information about a process or process output. Thus, statistical methods consist of the use of statistical tools to extract information from data, whereas statistical thinking is a thought process. One should have statistical thinking for successful application of statistical methods.

The concepts of statistical thinking and statistical methods have direct relevance in metrology. As explained in earlier chapters, measurements are carried out through measurement processes. There are many factors, such as the behavior of the measuring equipment, the competence level of the person making the measurements, and the environment under which the measurements are being made, that influence the outcome of a measurement process. The variations in these influence factors lead to variation in the measurement results, thereby introducing uncertainty into measurements. This uncertainty is evaluated using statistical methods. The uncertainty in measurements should be small enough that the measurements meet the specific needs for which they are being made. If necessary, the uncertainty can be reduced by analyzing the reasons for the variations in the measurement process and taking corrective action to reduce them. This can be done either by improving an existing measurement process or by using a new measurement process until the measurement system is fit enough for a specific measurement application.

Thus, a metrologist who has a mind-set compatible with statistical thinking will have more motivation to apply statistical methods in the analysis of measurement data. In fact, a professional metrologist must have this attitude.

THE USE OF STATISTICAL TECHNIQUES IN METROLOGY

Statistical techniques have a wide range of applications in metrology. Some of their important applications follow:

- **Estimation of measurement uncertainty:** Measurement results and the uncertainty associated with them are estimated partly based on statistical analysis of the measurement data. During this process, the various sources of error which contribute to the variability of a measurement process are identified. The contribution made by each factor is estimated, and ultimately the combined uncertainty of the measurement is evaluated. Statistical techniques are used for this purpose.
- **Trend analysis of measurement standards:** The long-term behavior of measurement standards maintained by calibration laboratories is analyzed using statistical techniques. The value assigned to a standard and the uncertainty associated with it are plotted against time to reveal the stability of the measured value. Thus, drift in the value of the standard with the passage of time can be identified. The more we know about the history of a standard, the more valuable an artifact it becomes. The charts used in this process are called trend charts.
- **Statistical process control of a measurement process:** To ensure that a measurement process continues to measure a parameter to acceptable uncertainty limits, statistical process control (SPC) techniques are intensively used in metrology laboratories. The application of SPC techniques enhances user confidence in the measurement process and is part of a measurement assurance program.
- **Gage repeatability and reproducibility (R&R) studies:** The suitability of a measurement system for a specific application needs to be verified initially and reverified subsequently at appropriate intervals. This reduces the chance of type I and type II errors associated with statistical decision making. The importance of this requirement is increasingly being felt by the users of measurement results. These verifications, called gage R&R studies, are used intensively in manufacturing and service organizations. The objective of R&R studies is to examine the variability of a measurement process that is due to gages, operators, and variations in the characteristics of parts that may arise from manufacturing process variability. The measurement setup is considered acceptable for a specific application if the measurement process variation is significantly small compared with the part-to-part variation. Experiments are carried out by having various operators

make repeat measurements using the various gages used in the measurement system. The exercise is repeated on a number of parts that represent the complete manufacturing tolerance. Thus, the total measured variability consists of two components, one due to the manufacturing process and the other to the measurement process. For a measurement system to be considered acceptable, the measurement process variability has to be less than 10% of the total variability. If this ratio is between 10% and 30%, the measurement system may be accepted, depending upon factors such as the criticality of the measurements and the gage costs. In no case should a measurement system be used if this ratio is higher than 30%. This topic is covered in detail in chapter 10.

- **Interlaboratory comparison studies:** Interlaboratory comparison studies are conducted with a view to judging the technical competence of testing and calibration laboratories. These are also called proficiency testing programs. Such studies are also carried out as collaborative studies for the verification and validation of measurement methods. In interlaboratory comparison studies, the measurement data obtained from various laboratories are statistically analyzed, and based on the statistical analysis the technical competence of a laboratory to generate technically valid measurement results is ascertained. Similarly, based on the statistical analysis of measurement data from various laboratories, the validity of a measurement method can be ascertained.
- **Determination of calibration intervals:** Measuring instruments and measurement standards are required to be calibrated at regular intervals to ensure that they are measuring correctly. The periodicity at which a measuring instrument needs to be calibrated depends upon the frequency of its use and the cost of its calibration. This cost includes any cost incurred due to nonavailability of the equipment during the time it is under calibration. The establishment of the appropriate periodicity depends upon the risks associated with not calibrating the equipment and the probability of the measuring equipment's not reading correctly. The measurement data from past calibrations of the measuring equipment thus help in deciding the periodicity of its calibration.

MEASUREMENT DATA AND STATISTICS

Repeat measurements of the characteristics of a product made under the same conditions result in a set of measurement results which are represented by numbers. These numbers constitute measurement data. Each member of the set of data is an estimate of the characteristic being measured and has an equal claim to the value being measured.

Statistics is the science of the collection, analysis, interpretation, and presentation of data. It helps us to draw valid conclusions based on data. Statistical techniques have applications in all branches of science, including metrology. Useful information can be derived from data in two ways. One is the display of the data in graphs, tables, and charts so that they can be easily understood. This application is called descriptive statistics. Another application of statistics is the use of sample data to draw inferences about a population. This application is called inferential statistics. In inferential statistics, we draw inferences about a universal set based on the study of sample data. This leads to a generalization that goes far beyond the data contained in the sample. Since generalization can never be completely valid, any inference has to be supplemented by information on the likelihood of the inference's being true. This information on likelihood is called probability.

Population and Sample

In inferential statistics, statistical techniques are applied to data which make up a relatively small part of a larger whole. Our interest generally lies in the larger whole. This larger whole is called a population. The relatively small part of the population that contains the objects on which studies are conducted constitutes a sample. Thus, a sample is a set of data selected from a population.

For example, suppose that it is desired to find out the heights of the soldiers in an army formation in order to make a decision on the sizes of uniforms. There could be thousands of soldiers in the army formation in which our interest lies. The population consists of all the soldiers in the army formation. To obtain information on heights, a relatively small proportion of the soldiers will be picked as a sample. Data will be collected on their heights, and inferences will be drawn about the heights of the soldiers in the army formation based on the study of the sample.

In metrology, when we make a large number of repeat observations of a particular characteristic under measurement, this collection of repeat observations can be called a population. Thus, the population of repeat observations consists of all observations made in the past and present, as long as the measurements have been made according to a standard measurement method. We generally make 3 to 10 repeat observations and draw our inferences about the measurand based on these observations. This set of observations is a sample. We can say that this sample has been drawn from a large population in the past and present. If the measurement method is not changed, even repeat observations to be made in the future for the same parameter will belong to this population.

Measurement Data as Random Variables

There is a certain amount of indeterminacy in the physical world. We are therefore not able to forecast the future with complete certainty. In many repetitive situations, we are able to predict in advance roughly what may happen based on experience, but not exactly what will happen. For example, when we toss a coin we cannot predict in advance exactly whether we will get heads or tails. Similarly, one cannot predict in advance based on experience the exact number of cars that will pass through a specific road in the next minute. One can predict the approximate number of cars likely to pass in the next minute as lying between two numbers. This rough prediction is made based on the study of past traffic patterns. In the same way, before actually measuring a parameter with a measurement system, one cannot predict the exact value of the measurement result based only on experience. One might be able to predict roughly that the measurement result is likely to lie between two specific values.

Occurrences which cannot be predicted exactly in advance are said to be random occurrences. The variables associated with such occurrences are called random variables. For example, the outcome of tossing a coin and the number of cars passing through a specific road in one minute are random variables. Similarly, a measurement result is a random variable.

Types of Random Variables

There are two types of random variables: discrete random variables and continuous random variables. Their characteristics are defined in Table 3.1.

Random occurrences are described by a statement of the likelihood that the occurrence will happen. Thus, the theory of probability is used in the analysis of random variables. The likelihood is represented quantitatively by the term *probability*. It is expressed as a ratio of the favorable outcome to the total number of possible outcomes of a random event over a large number of trials. For

Table 3.1 Characteristics of random variables.

Discrete random variables	These can have only discrete values. The number of cars passing through a road in one minute can have only an integer value; hence, it is a discrete random variable. Similarly, heads or tails is a discrete random variable.
Continuous random variables	These can assume all possible values between two limits. Measurement results are continuous random variables, as they can assume all possible values within a range. The resolution of the measurement system is the only limiting factor.

example, the probability of getting heads or tails during the tossing of a coin is one-half, as there is one favorable outcome against two possible outcomes. The probability of getting a specific number during the tossing of a die is one in six, as there is one favorable outcome against six possible outcomes.

Probability is expressed as a number with a value between zero and one. The minimum value of zero corresponds to an impossible event, and the maximum value of one corresponds to an event that is certain. Low probability implies that a random occurrence has a remote chance of happening, whereas high probability implies a greater chance that an occurrence will happen. Probability can also be expressed as a percentage, that is, from 0% to 100%. A probability figure expressed as a percentage implies that if the experiment is repeated a hundred times, the likely number of favorable outcomes will be equal to the percentage. A probability of 95% implies that if the experiment is repeated a hundred times favorable outcomes are likely to be obtained 95 times. Thus, a probability figure gives a numerical indication of the likely number of favorable outcomes of a random event out of the total number of outcomes.

Random variables do not have unique values, but there is a certain statistical regularity associated with them. If a random occurrence is recreated a large number of times, a definite average pattern of results is observed. For example, if a coin is tossed a large number of times, on average we will get heads 50% of the time, and another 50% of the time we will get tails. Similarly, if we collect data on the number of cars passing through a road in one minute at a specified time of day over a large number of days, the average will assume a definite value. In the case of a large number of repeat observations of the same parameter, it is observed that their average also assumes a definite value. This tendency toward the convergence of the overall averages of large numbers of repeat random variables is called statistical regularity.

Data consist of collections of related observations. In metrology, the data are groups of repeat measurements. For example, the repeat measurement observations obtained during the calibration of a standard resistor in a calibration laboratory is a set of data. In R&R studies, the data are the measurement observations taken by various operators, on various parts, using various gages. The repeat measurement observations contain complete information about the measurement result. Each observation also has an equal claim to represent the characteristic being measured.

A measurement result is a continuous random variable. It can assume any of the possible values contained in the interval between two limits. The number of values a measurement result can assume is limited only by the resolution of the measurement system. For example, assume that we have a 100-ohm standard resistor with a specified tolerance of ±0.1 ohm. Thus, its value lies between 99.9 ohms and 100.1 ohms. If repeat measurements are made on this standard resistor by a measurement system with a resolution of 0.1 ohm, the repeat observations can assume only three values: 99.9, 100.0, and 100.1 ohms. If the resolution of the measurement system is 0.01 ohm, the repeat observations will range between 99.90 and 100.10 ohms. This increases the number of possible outcomes. Thus, any value within the above range is possible subject to the resolution of the measurement system. However, resolution cannot be increased indefinitely, as at a certain stage the least significant digit will be responding only to inherent electrical noise of the measurement system rather than a change in the repeat observations.

THE CHARACTERISTICS OF STATISTICAL DATA

A number representing a random variable is a random number. The simple collection of random numbers as measurement results does not convey much information. The information is to be extracted from the data. There are two important characteristics which describe the behavior of the data. Knowledge of the value of these characteristics helps in extracting information from the data:

- **Central tendency of the data:** The central tendency of a set of data is defined by a number which represents the behavior of all the data. It is like the center of gravity, the point within a body at which the whole mass of the body may be considered to be concentrated. Similarly, the central tendency is a single number which represents all the numbers contained in the data set. The arithmetic mean is a universally accepted parameter of central tendency. Its numerical value is equal to the sum of all the numbers in the data setdivided by the number of the pieces of data. If $x_1, x_2, x_3, \ldots . x_i, \ldots\ldots\ldots\ldots x_n$ are the random variables representing the measurement data, the arithmetic mean of the data is expressed as follows:

$$\bar{x} = (x_1 + x_2 + x_3 + \text{———————} x_i + \text{————} x_n) / n$$

$$= (\Sigma\, x_i) / n$$

where Σ represents the summation of the terms.

Though the arithmetic mean is the most accepted measure of central tendency, at times two other measures, the median and the mode, are also used. If the data are arranged in ascending or descending order of magnitude, the median is the value which is in the middle; that is, half of the data are greater than and the other half are smaller than the median. If there are an odd number of data, the central value when the data are arranged in ascending or descending order of magnitude represents the median. For an even number of data, the mean of the two central values represents the median. The other measure of central tendency, the mode, is the value which has the highest frequency of occurrence. The most appropriate measure of central tendency of the three—mean, mode, and median—depends upon the probability distribution of the data. Fortunately, most of the data we deal with in metrology follows a normal probability distribution, and for this distribution the mean, mode, and median have the same value.

- **Dispersion of the data:** Dispersion is another important characteristic of data. It pertains to the variations among individual values contained in the data. To understand the pattern of a set of data its dispersion—its spread or variability—should be estimated. Knowledge of the dispersion enables us to judge the reliability of the central tendency of the data. While the numerical value of the central tendency may represent the data in its totality, this capability to represent the data is affected by its dispersion. The reliability of the mean as a representative of the data is greater when the spread of the data is small rather than large. For example, the mean of the data represented by 99, 100, and 101 is 100, and so also is the mean of 95, 100, and 105. The average of 100 is more representative of the data in the first case than in the second one, because the second set of data has more variability than the first one.

There are two important measures which characterize the variability of data:

- **Range:** This is the difference between the maximum value and the minimum value of the data. In the above example, the range of the first set of data is 2 whereas for the second set it is 10. This measure is frequently used in statistical quality control, but it has certain limitations. The range is calculated based on only the two extreme values, that is, the maximum value and the minimum value of the data. An error in either of them will lead to an incorrect assessment of the variability of the data. The information on range derived from the difference between the maximum and minimum values of the data may also be found not to be consistent. If different sets of data are obtained on the same parameter at different times, there will be wide variation in the range values obtained from different samples of the same population. Another drawback of the range parameter is that it is derived based on only two values of the data, so that the information contained in the rest of the data is not utilized in the range calculation.
- **Standard deviation:** The most widely accepted scientific measure of the variability of data is standard deviation. There is consistency among standard deviation values for various data samples drawn from the same population. Standard deviation is the square root of the average of the squared differences of the observations from the mean. It measures the scatter of the data in terms of differences from the mean value. This quantitative measure of

variability is derived based on information contained in all the data rather than on the maximum and minimum values of the data, as in the calculation of range. The unit of standard deviation is also the same as that of the data. Standard deviation is expressed as follows:

$$s = [\{(x_1 - \bar{x})^2 + (x_2 - \bar{x})^2 + \ldots\ldots\ldots\ldots\ldots + (x_n - \bar{x})^2 / (n - 1)]^{1/2}$$

$$s = [\Sigma(x_i - \bar{x})^2 / (n - 1)]^{1/2}$$

The square of standard deviation is called variance.

The Mean and the Standard Deviation of Populations and Samples

In statistical analysis of data, the words *sample* and *population* are used very frequently. Generally, μ and σ are used to represent the population mean and the population standard deviation. For a sample, the analogous quantities are called the sample mean and the sample standard deviation and are represented by $\bar{x}$ and s. If N is the population size, which is very large, and n is the sample size, the sample and population parameters are given as follows:

$$\mu = \lim_{N \to \infty} \Sigma x_i / N$$

$$\sigma^2 = \lim_{N \to \infty} \Sigma (x_i - \mu)^2 / N$$

$$\bar{x} = \Sigma x_i / n$$

$$s^2 = (x_i - \bar{x})^2 / (n - 1)$$

PROBABILITY DENSITY FUNCTIONS AND FREQUENCY DISTRIBUTION

Unlike that of exact numbers, the behavior of random numbers is analyzed by the probability density function (pdf). The pdf characterizes the behavior of random variables. The function is in the form of a mathematical equation and can assume geometrical shapes including curves. It can be represented graphically. The abscissa represents the numerical value of the random variable, and the ordinate represents the pdf. The probability that a random variable lies between two specified values on the x axis is given by the area under the pdf between the two specified values.

As explained above, if x is a random variable and $f(x)$ is its pdf, the probability that x will have a value between points a and b is given by the area under the pdf $f(x)$ between points $x = a$ and $x = b$. The pdf $f(x)$ is always a function of x. The graphical representation of the variation of $f(x)$ as a function of x is called the frequency distribution. The pdf that characterizes the behavior of a random variable is determined based on the study of a large number of data which represent the inherent variability of the random variable. The starting point is the drawing of a frequency histogram in which the variable is plotted on the x axis in various uniform intervals. The frequency, that is, the number of data observed in the interval, is plotted on the y axis. If the number of data is increased to a very large extent, the frequency histogram takes the shape of a frequency distribution.

Mathematically, the above statement can be expressed as

$$P(a < x < b) = \int_a^b f(x)dx$$

The probability density function has the following characteristics:

- $f(x)$ is always positive for all values of x
- The total area under $f(x)$ is equal to unity, as there is 100% probability that the random variable will be within $f(x)$

The Use of the Probability Density Function in Metrological Applications

Our main concern in metrology is to find the numerical value of a measurement result and the uncertainty associated with it. The uncertainty is an interval within which the true value of the parameter under measurement is expected to lie with a certain probability called the confidence level. If the frequency distribution of the behavior of the measuring equipment is known, the information is very helpful in estimating measurement uncertainty. A large number of frequency distributions are used in statistical analysis. As regards metrological applications, the following distributions are used quite frequently:

- Rectangular or uniform distribution
- Trapezoidal and triangular distribution
- Normal distribution

THE CHARACTERISTICS AND APPLICATION OF RECTANGULAR DISTRIBUTION

A random variable is characterized by rectangular distribution if the value of its distribution function is constant within a specified interval. If A is the central value of the random variable and a is the half-width of its distribution function, the pdf is represented mathematically as shown in Figure 3.1.

When $(A - a) < x < (A + a)$, $f(x) = 1/2a$, otherwise $f(x) = 0$

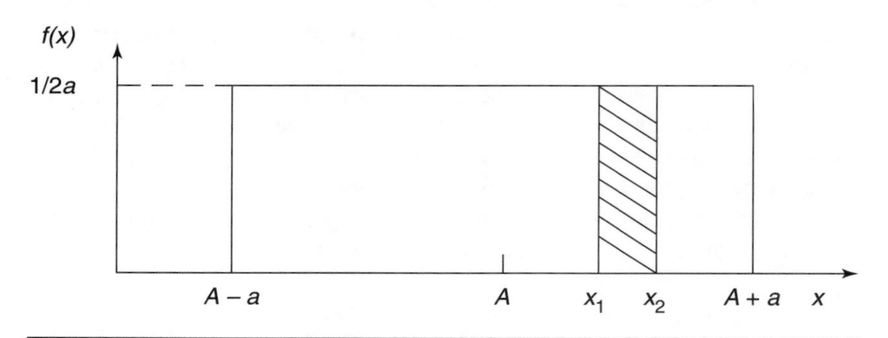

Figure 3.1 Rectangular distribution.

The probability that the random variable will assume a value between x_1 and x_2 is given by the area under the distribution function between points x_1 and x_2 and is equal to the multiplication of $(x_2 - x_1)$ and $1/2a$. The total width of the pdf is $2a$, and the value of $f(x)$ within the width is $1/2a$ so that the total area under the curve is unity. The value of $f(x)$ outside $A - a$ and $A + a$ is zero. This type of distribution is called rectangular because of its rectangular shape. In view of the uniform value of $f(x)$ within the width, rectangular distribution is also called uniform distribution. The central value or the mean of the distribution is given by A. The variance of this distribution is equal to $a^2 / 3$, and its standard deviation is equal to $a / \sqrt{3}$, where a is equal to the half-width of the interval.

The Application of Rectangular Distribution in Metrology

Rectangular distribution has many applications in metrology, particularly in estimating the measurement uncertainty due to particular types of error sources. A number of error sources can be characterized by uniform distribution. Some of them are as follows:

- **Uncertainty due to the resolution of digital meters:** One factor which contributes to measurement uncertainty is the resolution of digital meters. To overcome the parallax error inherent in analog instruments and to obtain better stability of measurement results, digital measuring instruments are used very frequently. *Resolution* is defined as the minimum value read by the measuring instrument. In the case of digital meters, it is equal to one unit of the least significant digit.

Most quantities subject to measurement are continuous random variables. They can assume all possible values within the limits contained in the measurement range. However, there are two factors that limit the possible values. The first is the limited number of digits contained in a digital display. The minimum value that can be read in a digital display is equal to the value represented by one unit of the least significant digit; below this value, numbers are not available. This limit is called the limit to resolution. Another factor contributing to the limiting of the possible values is the inherent variation in the parameter being measured. The use of a greater number of digits of resolution does not necessarily enhance our capability to measure all possible values as some of the digits may express only noise. There is a limit beyond which improving the resolution does not help. The number of digits depends upon the variability of the measurand and the measurement accuracy required.

Now let us examine how the resolution of a digital display affects measurement uncertainty. Suppose we have a three-and-a-half-digit digital voltmeter. The maximum and minimum values it can display are 1999 and 0000. If we measure a voltage of 47.84, on a 100-volt range the display will read it as either 47.8 volts or 47.9 volts, with a greater likelihood of reading 47.8 volts. Because of the digital limitation, the measured value has been truncated. A measured value of 47.8 volts leads us to believe that the value of the voltage being measured lies anywhere between 47.75 volts and 47.85 volts. Thus, we would declare the measured value to be 47.8 volts, with a statement that it could be anywhere within the range 47.80 ±0.05 volts. Similarly, a value measured as 47.9 volts implies that the value could lie anywhere within the range 47.90 ±0.05 volts. Such situations are represented by rectangular distribution; thus, representing uncertainty this way assumes that the last digit has been split. There is an equal probability of the measured voltage's being within an interval of ±0.05 volts, and the probability of its being beyond this interval is zero. In such a situation, the standard deviation is equal to the half-width of the interval divided by $3^{1/2}$. In this example, the standard deviation is equal to $0.05 / 3^{1/2}$.

The behavior described in the preceding example is common to most types of digital measuring equipment. Analog-to-digital conversion, in which an analog signal is converted into digital form, is a critical process in the operation of this type of equipment. Nonlinearity in the conversion process beyond a certain value may lead to a situation in which the displayed value is not at the center of the width of the interval. In this case, a measured value of 47.8 volts may imply that the measured value is between 47.78 and 47.86. The displayed value is not at the center of the width of the interval. Even the width of the interval may not be uniform for each value corresponding to the least significant digit. In such a situation, we may say that the last digit is not split evenly. With this type of asymmetric behavior on the part of a measuring equipment, it would be inappropriate to calculate uncertainty based on the width of the interval represented by ±0.5 digits. In such a case, the uncertainty must be calculated taking into consideration the width of the interval represented by ±1 digit so that the standard deviation will be equal to the value represented by the least significant digit divided by $3^{1/2}$. In the preceding example, the standard deviation would be equal to $0.1 / 3^{1/2}$.

There are also situations in which the resolution of the digital display does not contribute to uncertainty. This is the case when the number of digits in the display is such that the value represented by the least significant digit is of the same order as the inherent variability in the parameter being measured. In this situation, the uncertainty in the reading of the last digit is primarily due to the variability of the parameter and not to the resolution of the digital display.

- **Uncertainty of reference standards:** Reference standards are used for calibrating measuring equipment in calibration laboratories. An uncertainty statement about the measured value of a reference standard is contained in the calibration certificates of the standard. Sometimes the uncertainty statement is not qualified by a statement of its confidence level. If it can be assumed that the uncertainty is given as the maximum bounds within which all values of the reference standard are likely to lie with equal probability, uniform distribution can be used. If $\pm a$ is the uncertainty of the reference standard, the variance due to this uncertainty is given by $a^2 / 3$ and its standard deviation by $a / \sqrt{3}$.
- **Uncertainty due to the resolution of analog instruments:** As is the case with digital measuring equipment, the resolution of analog instruments also contributes to uncertainty. If the resolution is d, the measured value can be considered to lie within an interval of $\pm 0.5d$ with rectangular distribution. The standard deviation in this case would be $0.5d / 3^{1/2}$.

THE CHARACTERISTICS OF TRAPEZOIDAL AND TRIANGULAR DISTRIBUTION

In uniform distribution, the probability that the value of a random variable will lie outside the bounds is zero, and the pdf is constant within the interval. The pdf is thus like a step function. In the actual physical world, this is a rare situation. It is more realistic to expect that the values near the bounds are less likely to occur than those near the midpoint. Such a situation can be described by a trapezoidal distribution having equal sloping sides with a width of $2a$ and the top of the width $2a\beta$ $(0 < \beta < 1)$. The pdf of a trapezoidal distribution is shown in Figure 3.2.

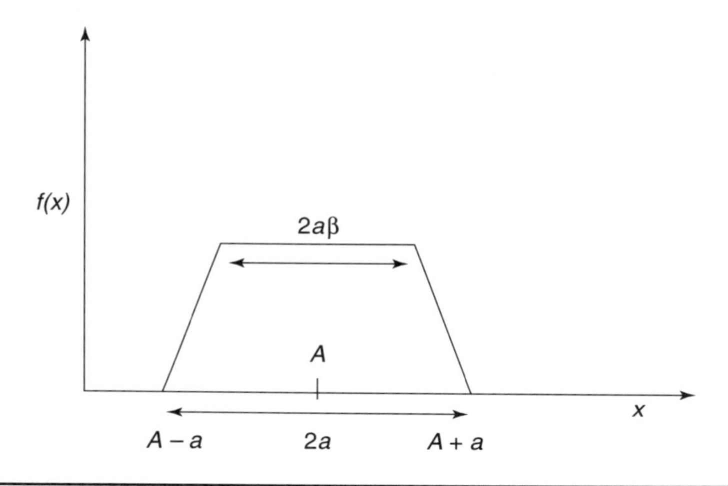

Figure 3.2 Trapezoidal distribution.

At $\beta = 1$, the distribution becomes rectangular.

At $\beta = 0$, the distribution becomes triangular. So triangular distribution is a special type of trapezoidal distribution. In this the pdf has maximum value at the mean. It decreases as we move away from the mean and ultimately its value is zero at the bounds.

The means of distribution = A

The variance of the trapezoidal distribution = $a^2(1 + \beta^2) / 6$

The standard deviation of the trapezoidal distribution = $a[(1 + \beta^2) / 6]^{1/2}$

The variance of the triangular distribution = $a^2 / 6$

The standard deviation of the triangular distribution = $a / 6^{1/2}$

Example of the Application of Trapeziodal and Triangular Distribution

Suppose that the distribution function of a measuring instrument is plotted. A quantity whose numerical value is 100.0 is measured with the instrument a large number of times. The frequency distribution is drawn, and it is observed that the quantity is never measured as less than 99.9 or more than 100.1. The probability of obtaining a value between 99.95 and 100.05 is uniform, but that of reading the value as less than 99.95 or more than 100.05 decreases. Ultimately, the probability of reading the value as less than 99.9 or more than 100.1 is zero.

The behavior of this instrument can be characterized by trapezoidal distribution with the following parameters: $A = 100$, $a = 0.1$, $2a\beta = 0.1$, and thus $\beta = 0.5$. The standard deviation for the above data is equal to $0.1\{(1 + 0.5^2) / 6\}^{1/2}$, or 0.046.

If the distribution is such that the values nearer 100.0 are more probable than the values further from 100.0 and that the probability of values less than 99.9 or more than 100.1 is zero, then it is a case of triangular distribution. The standard deviation for this case would be equal to $0.1 / 6^{1/2}$, or 0.04.

THE CHARACTERISTICS OF NORMAL DISTRIBUTION

The normal distribution is a very popular and useful distribution. The well-known mathematician Karl Gauss was instrumental in inventing this distribution, so it is also called a Gaussian distribution. This distribution comes very close to fitting the actual observed frequency distributions of many phenomena. For example, the frequency distributions of human characteristics such as weight and height follow a normal distribution. Similarly, outputs from manufacturing processes—for example, the dimensions of a manufactured part—display a normal distribution. The data generated by measurement processes also follow a normal distribution.

Certain properties are observed in the data generated as an output of a measurement process:

- The repeat results are spread roughly symmetrically about a central value
- Small deviations from the central value are observed more frequently than large deviations

The curve of a normal distribution is mathematically represented by the equation

$$f(x) = \{1 / (\sqrt{2}\pi)\sigma\} \exp[-(x - \mu)^2 / 2\sigma^2]$$

where x is the random variable, in this case the measurement data, and $f(x)$ is the normal pdf. The variable x is represented on the abscissa and $f(x)$ on the ordinate. The term *exp* represents the power of the base of the natural logarithm: $e = 2.71828$. Measurement results that follow a normal curve are said to be normally distributed. A normal probability curve is shown in Figure 3.3.

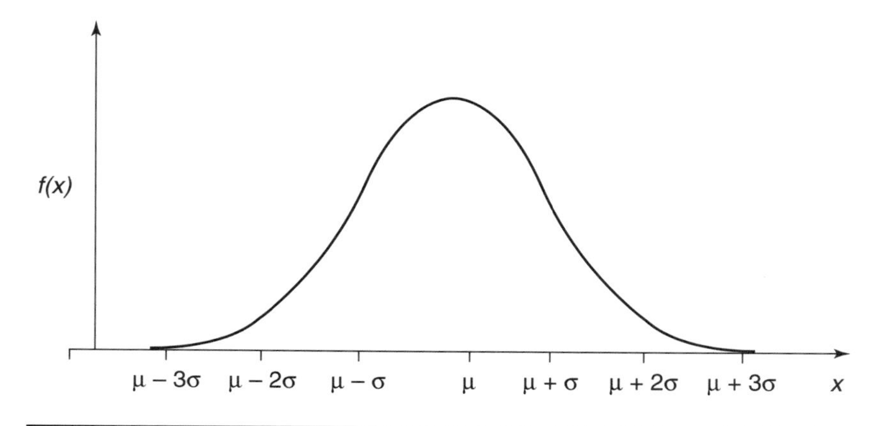

Figure 3.3 Normal probability distribution.

A normal curve has the following characteristics:

- It has a bell shape which theoretically extends from $-\infty$ to $+\infty$ and the extremities of which are asymptotic to the x axis
- It is symmetrical about the central value μ
- It is unimodal: its mean, mode, and median are the same
- The area under the curve is equal to unity
- The probability that the random variable lies between two points on the x axis is given by the area under the curve between these two points

The Characterization of a Normal Curve

Based on the equation of a normal curve, it is clear that two parameters characterize it: μ and σ. The pdf represents the behavior of the population. If we use a measurement process to take a large number of repeat observations, their behavior will be represented by the pdf which is characteristic of the population, and not of the sample.

- **Population mean (μ):** The parameter μ of the normal curve represents the central tendency of the data. It is the mean value of the population, and the curve is symmetrically distributed about the mean value. Theoretically speaking,

$$\mu = \lim_{n \to \infty} \Sigma x_i / n$$

where μ represents the limiting mean value when the number of observations approaches infinity.

• **Population standard deviation** (σ): The parameter σ represents the variability of the data. It is the standard deviation of the population. The distance between the central line and the point of inflection of the curve is also equal to σ on each side of the central line. The point of inflection is the point on a curve at which it changes its curvature. A normal curve is concave toward the center around μ, and at the point at which $x = \mu + \sigma$ or $x = \mu - \sigma$, it changes its curvature and becomes convex toward the center. The numerical value of σ is given by the equation

$$\sigma = \lim_{n \to \infty} \{\Sigma(x_i - \mu)^2 / n\}^{1/2}$$

The square of σ is called the variance of the population.

• **Area under a normal curve:** Theoretically, a normal curve extends to infinity on either side of the central value μ, implying that the random variable can assume any value. In practice, however, only values within a specific range are considered. The area under the normal curve is unity. The information needed for computing the area under the normal curve is contained in statistical tables. Generally, σ is used as a unit on the abscissa, and the area under the curve for constant multiples of σ can be computed from the tabulated values of the normal curve. If we take the interval between $\mu - k\sigma$ and $\mu + k\sigma$ for various constant values of k, the area under the normal curve is as given in Table 3.2.

The measurement observations that result from a measurement process follow a normal distribution, and hence much about the behavior of the measurement process can be established. We only need to know the population mean μ and the population standard deviation σ.

The frequency distributions described above find applications in analyzing metrology data. A summary of parameters and applications of these distributions is shown in Table 3.3.

Table 3.2 Area under the normal curve.

k	Interval	Area under the curve
0.6745	$\mu - 0.6745\sigma$ to $\mu + 0.6745\sigma$	50%
1.0	$\mu - \sigma$ to $\mu + \sigma$	68.3%
1.96	$\mu - 1.96\sigma$ to $\mu + 1.96\sigma$	95%
2.0	$\mu - 2\sigma$ to $\mu + 2\sigma$	95.5%
2.58	$\mu - 2.58\sigma$ to $\mu + 2.58\sigma$	99%
3.0	$\mu - 3\sigma$ to $\mu + 3\sigma$	99.73%

Table 3.3 Summary of frequency distributions used in metrology.

Distribution	Parameter	Application
Rectangular	Mean = A Half-width of interval = a	Applies when the limits to error are quoted between two bounds with equal probability
Triangular	Mean = A Half-width of base = a	Applies when the values are more likely to fall in the middle of the interval and the probability gradually diminishes as we move toward the extremes of the interval, beyond which it is zero
Trapezoidal	Mean = A Half-width of base = a Ratio of two parallel sides = β	Applies when the values are spread through the middle of the interval with equal probability and the probability gradually diminishes as we move away from the middle of the interval toward the extremes of the interval, beyond which it is zero
Normal	Mean = μ Standard deviation = σ	Applies widely in metrology and with other types of data

THE STANDARD NORMAL PROBABILITY DISTRIBUTION

A normal curve is characterized by two parameters: μ, the mean of the population, and σ, the standard deviation of the population. To analyze the behavior of a random variable, the values of μ and σ should be known. However, we should also have knowledge about the area under the normal curve for the various values of random variables in terms of μ and σ. If x is a random variable representing the repeat measurements of a parameter, the probability that a single observation obtained as a measurement result between a and b is given by the area under the pdf curve $f(x)$ between the points a and b on the x axis.

The location of the curve $f(x)$ is decided based on the value of μ, as the curve is symmetrically spread around the value of $x = \mu$. The area under the curve between the two points depends upon how far the points $x = a$ and $x = b$ are located from $x = \mu$. For different mean values, the normal curve would be centered at different locations on the x axis. That is why the mean of normal distribution is also called the location parameter.

The shape of the curve is determined by the value of the population standard deviation σ. Thus, keeping the same mean value but changing the standard deviation produces changes in the peak value of $f(x)$ and the width of the normal curve. A small value of σ produces a sharper and thinner curve, whereas a large value of σ produces a flatter and wider curve. This is illustrated in Figure 3.4.

Different values of the mean and of the standard deviation produce different normal curves. Thus, it appears as if we should have different statistical tables for the various values of μ and σ. However, for normal distribution to be of any practical use, we must have a means of identifying all distributions with a single normal curve. This problem has been solved by the development of a normal table based on a probability distribution called the standard normal probability distribution.

A standard normal probability distribution function is characterized by a normal curve with $\mu = 0$ and $\sigma = 1$. Let us define a new random variable z such that $z = (x - \mu) / \sigma$, where x represents the repeat measurement observations and μ represents the true value of the measurand, if the measurement process is influenced only by sources of random error. σ is the standard deviation of the measurement process.

The value $x - \mu$, or the numerator of the definition of z, is the measurement error. By dividing the error value by the constant term σ, the standard deviation of the variable, z gives us the normalized error. The normalization

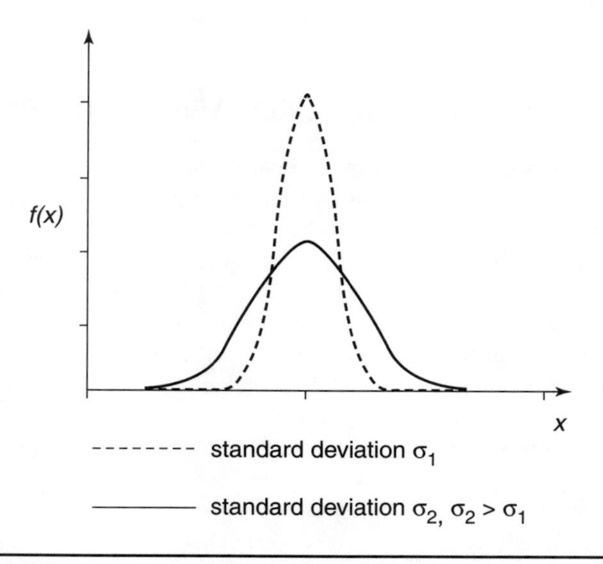

Figure 3.4 The effect of σ on the shape of a normal curve.

has been done with reference to σ, the standard deviation. So if x represents the random measurement data, z represents the normalized measurement error.

The probability density function for the normalized error random variable z is given by $f(z)$.

$$f(z) = 1 / (\sqrt{2}\pi) \exp(-z^2 / 2)$$

The pdf of standard normal distribution is shown in Figure 3.5.

This curve is also bell shaped and extends from $-\infty$ to $+\infty$. The curve is centered around $z = 0$ and has a standard deviation of unity. The probability that the normalized error will be less than a value of a is given by the equation

$$P(z < a) = \int_{-\infty}^{a} f(z)dz$$

This is equal to the area under the standard normal probability density function curve between $-\infty$ and $z = a$ of the curve as shown in Figure 3.4.

Similarly, the probability that the normalized error will be between b and c is given by the area under the curve between $z = b$ and $z = c$ as shown in Figure 3.5. This probability is given by the equation

$$P(b < z < c) = \int_{b}^{c} f(z)dz$$

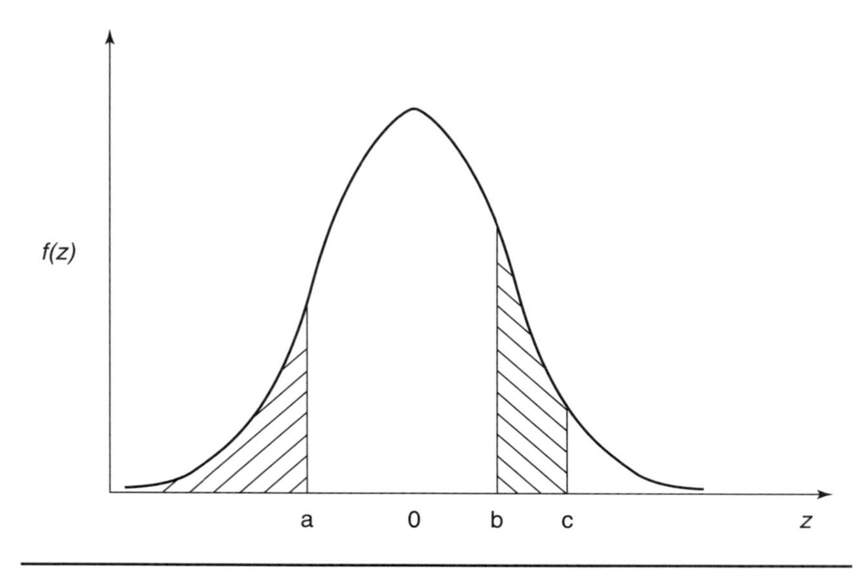

Figure 3.5 Standard normal distribution.

The theory of errors was developed by statisticians to study the phenomena of random events more than 200 years ago. Metrologists came to use these concepts in the context of measurement errors much later. The error function is a well-known function evolved by mathematicians.

The Use of Standard Normal Probability Distribution Tables

There are two types of standard normal tables available in compilations of statistical tables. One type of table gives the area under a normal curve from $z = -\infty$ to a specific value of z. The other type of normal distribution table gives the area under the standard normal probability density function between the mean and the positive values of z. One such table is shown in Appendix A. The z values generally vary from 0.00 to 3.09 in steps of 0.01. The normal curve is symmetrical about the mean corresponding to $z = 0$, and hence the value given in the table also holds good for negative values of z from $z = 0$ to $z = -3.09$. The table can thus be used to find the area under the curve for various positive as well as negative values of z.

For example, if we want to know the area under the normal curve corresponding to $z = -2.55$ to $z = 2.55$, we can refer to a normal distribution table and find the value in the standard normal table indicated under $z = 2.55$. This can be done by entering the table through the row corresponding to $z = 2.5$ under column 0.05. This value in the table is 0.4946. The same value will be applicable to $z = -2.55$. The area under the curve from $z = -2.55$ to $z = +2.55$ is equal to 2×0.4946, which equals 0.9892, or 98.92%.

Since $z = (x - \mu) / \sigma$ or $x = \mu \pm z\sigma$, this implies that 98.92% of the time a normally distributed random variable will have values between the population mean plus 2.55 times the population standard deviation and the population mean minus 2.55 times the population standard deviation.

The area under the curve corresponding to various values of z can be obtained from the table. The chosen z values need not always be symmetrical with respect to the mean. The area under the curve between $z = -1.0$ and $z = +1.5$ can be obtained as follows: The area under the curve between $z = 0$ and $z = +1.5$ is equal to 0.4332. The area under the curve between $z = 0$ and $z = -1.0$ is equal to 0.3413, the same as the area between $z = 0$ and $z = 1.0$. The total area under the curve between $z = -1.0$ and $z = +1.5$ is thus equal to 0.7745, or 77.45%.

An Example of the Use of a Standard Normal Table

The electromotive force of a standard saturated cell is measured a number of times, and the following observations in volts are noted:

1.01858	1.01850	1.01853	1.01860	1.01852	1.01856
1.01859	1.01857	1.01855	1.01857	1.01859	1.01852
1.01856	1.01856	1.01858	1.01853	1.01856	1.01853
1.01857	1.01853	1.01856	1.01852	1.01853	1.01857
1.01859	1.01854	1.01855	1.01858	1.01855	1.01858

We intend to know what percentage of the observations will lie below 1.01850 volts or above 1.01860 volts. From the data obtained, we calculate the mean and the standard deviation of the population: $\mu = 1.018556$ volts and $\sigma = 0.000025$ volts. The z value corresponding to 1.01850 is expressed by (1.01850 – 1.018556) / 0.000025, which equals –2.24. The percentage of observations lying below 1.01850 volts is given by the area under the curve between $z = -\infty$ and $z = -2.24$. The area under the curve between $z = -\infty$ and $z = 0$, based on the table, equals 0.5000. The area under the curve between $z = -2.24$ and $z = 0$ equals 0.4875. The area under the curve between $z = -\infty$ and $z = -2.24$ equals 0.0125. Thus, 1.25% of the observations will have values less than 1.01850 volts.

Similarly, the z value corresponding to 1.01860 is expressed by (1.01860 – 1.018556) / 0.000025, which equals 1.76. The percentage of observations that will lie above 1.01860 volts is given by the area under the curve between $z = 1.76$ and $z = \infty$. The area under the curve between $z = 0$ and $z = \infty$, based on the table, equals 0.5000. The area under the curve between $z = 0$ and $z = 1.76$ equals 0.4608. The area under the curve between $z = 1.76$ and $z = \infty$ equals 0.0392. Thus, 3.92% of observations will have values greater than 1.01860 volts.

SAMPLING DISTRIBUTION

Sampling distributions are used to draw inferences about a population based on the study of the data contained in the sample. The correctness of the inference depends upon how the sample has been drawn from the population. The sample must be truly representative of the population. Random sampling is the best way of drawing a sample. Random sampling implies that the items in the sample are selected in such a way that each item in the population has an equal probability of being included in the sample. Similarly, among all possible samples, each sample has an equal probability of being included. A random sampling process does not have a bias. Nonrandom sampling, on the other hand, introduces error into the estimation. This error is called sampling error.

Statistics and Parameters

A probability distribution describes a random variable just as normal distribution and rectangular distribution describe the behavior of measurement observations. A parameter is a characteristic of the population. For example, the mean μ and the standard deviation σ are the parameters for a population which follows normal distribution. The half-width of the distribution is the characteristic of a population that follows rectangular distribution. We can also draw repeat samples from a population and measure some characteristic of all the items in the sample. The characteristics of the sample are called sample statistics. A sample statistic is derived based on the measurements taken of the items contained in the sample. For example, the mean of the observations of a particular characteristic of the items forming the sample is one type of sample statistic. The standard deviation of the observations of the sample items is another type of sample statistic.

Sample statistics can be considered as another type of variable. If we draw a number of samples from the population, measure a specific characteristic of the items contained in each sample, and calculate their sample means and sample standard deviations, those means and standard deviations will have different values. Thus, a certain amount of variability is associated with sample means and sample standard deviations. The sample mean and sample standard deviation are just two examples: other sample statistics such as mode and median may also exhibit variability. The distribution of the sample statistics is called the sampling distribution, just as the distribution of random variables is called the frequency distribution. For example, a sampling distribution of the means of various samples is the distribution of the sample means. It is also called the sampling distribution of the mean.

Sampling from a Normal Population

Theoretically, the normal pdf extends from $-\infty$ to $+\infty$. This implies that there is a certain probability that the random variable can assume any value in the case of measurement observations, including an absurd value, though the probability of getting such a value is extremely low. Theoretically, repeat observations of a measurand can assume all the values in an infinite range. But in practice it never happens. We therefore accept the mathematical model that 99.73% of the observations will lie within $\mu \pm 3\sigma$. It is generally impossible to get absurd empirical values.

If the population follows a normal distribution, the sampling distribution of the mean from the normal population has some interesting characteristics. The mean of the sampling distribution is the same as the mean for the population. Now x is a random variable, and its mean and standard deviation are μ and σ, respectively. We can draw a large number of samples with n items in each sample and calculate the means of all these samples. These sample means can be taken as another random variable $\bar{x}$. If the mean of the random variable

$\bar{x}$ is represented by X (grand average) and its standard deviation by $s_{(\bar{x})}$,the relationship between the population parameters μ and σ and the parameters of the sampling distribution can be established as follows.

If x_{1i} represents the repeat observation of the first sample, its mean is, $\bar{x}_1$ and its variance s_1^2.

For the *k*th sample, the observations would be represented by $x_{k1}, x_{k2}, \ldots x_{kn}$. The mean and variance of the *k*th sample would be given by the equations

$$\bar{x}_k = \Sigma x_{ki} / n$$

$$s_k^2 = \Sigma(x_{ki} - \bar{x}_k)^2 / (n - 1)$$

The various means represented by $\bar{x}_k$ have their own sampling distribution with its own mean and variance, as illustrated here for *p* number of samples:

$$X = \Sigma \, \bar{x}_k / p$$

$$s^2_{(\bar{x}k)} = \Sigma(\bar{x}_k - X)^2 / (p - 1)$$

For a large number of samples, the grand average of the averages of the various samples, or X, converges to the population mean μ. Similarly, the average of the sample variances, or s_k^2, converges to the population variance σ^2 for large values of *p*, so $\Sigma s_k^2 / p = \sigma^2$. The variance and the standard deviation of the sample mean are given by the equations

$$s^2_{(\bar{x})} = s^2_{(x)} / n$$

$$s_{(\bar{x})} = s_{(x)} / \sqrt{n}$$

and thus the standard deviation of the sample mean is equal to the standard deviation of the original population divided by the square root of the sample size.

Thus the width of distribution of sample mean is less compared to the width of the distribution or random variable as shown in Figure 3.6.

Standard Error

The standard deviation of the distribution of a sample statistic is called the standard error. If the sample statistic is the mean of the samples, the standard deviation of the sample mean is called the standard error of the mean, which is equal to $s(x) / \sqrt{n}$. The concept of standard error is frequently used in metrology in estimating uncertainty in measurement. For a normally distributed variable, the standard error is inversely proportional to the square root of the sample size.

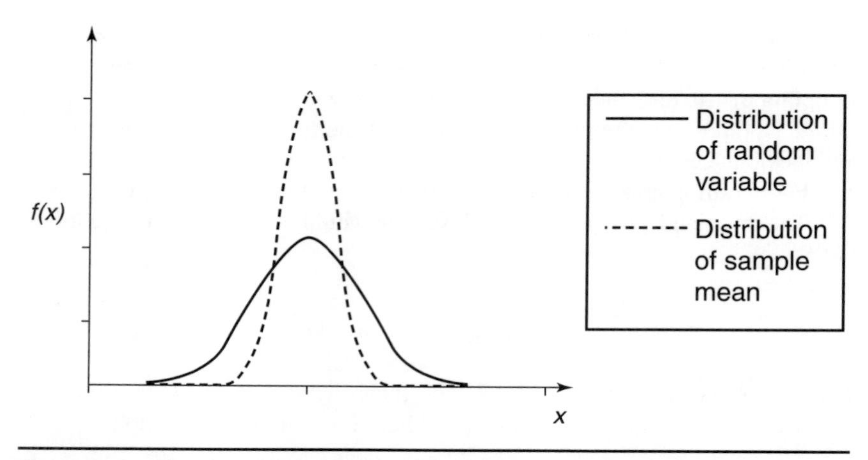

Figure 3.6 The distribution of a random variable and its sample mean.

Thus, as the sample size increases, the standard error becomes smaller in the ratio of the square root of the sample size.

The Central Limit Theorem

It has been seen that the mean of the sampling distribution is equal to the population mean. This implies that the sampling mean is also normally distributed. The central limit theorem states that, even if the distribution of a random variable is not normal, the distribution of the sample means always approaches normal if the sample size increases.

STATISTICAL ESTIMATION

In inferential statistics, we draw inferences about a population based on study of the data obtained from a sample. Statistical inference techniques use probability concepts to deal with uncertainty in decision making. Statistical estimation is one of the techniques used in inferential statistics.

We very frequently resort to estimation in our lives. If one plans to construct a house, the first thing one generally does is estimate the amount of money needed for the purpose. For this, one decides the specification details of the proposed house and then, based on the experience of people engaged in the construction industry, one draws an inference that the proposed house will cost approximately a certain amount of money. An estimation is always an approximation. Since it is an approximation, there is a certain amount of risk that the estimation may not be correct. Even when we are about to cross a road, we estimate in our minds how safe it will be to do so. We base the estimate on a quick study of the traffic pattern and traffic density. This type of estimate is based on our experience or intuition, or sometimes purely on a hunch. These estimates are not drawn scientifically, but in our day-to-day lives they work well. Many times, however, reliable information is needed in order to plan for the future. Such information can be obtained from the study of a sample to draw inferences about a population. This process of drawing inferences about the population based on studies of a sample is called statistical estimation. Statistical estimation is not based on a hunch or on the experience of others but is instead arrived at scientifically. To obtain a reasonably accurate estimation, we need to know the following:

- The pdf of the variable under study
- The information obtained from the sample data based on actual investigation

Statistical estimation is thus the process of analyzing a sample result in order to predict the corresponding value of the population parameter.

Estimator and Estimate

For normally distributed data, the sample mean and sample standard deviation are two sample statistics used to estimate the mean and standard deviation of the population. The sample mean is used to estimate the population mean; thus, the sample mean is said to be the estimator of the population mean. An *estimator* is a sample statistic used to estimate a population parameter. An *estimate* is the specific observed value of a statistic. For normally distributed data, the sample mean and sample standard deviation are the estimators of the population mean and population standard deviation.

For example, suppose that one intends to estimate the value of a laboratory standard resistor of a nominal value of 100.000 ohms. Ten repeat measurements are made on this resistor using a high-accuracy resistance bridge, and the observations in ohms are 100.001, 99.999, 100.000, 100.001, 100.001, 99.998, 99.999, 99.997, 100.000, and 100.001. The mean of these observations is a sample statistic called the sample mean. We can say that the statistic sample mean is the estimator of the population mean—the value of the laboratory standard resistor. The numerical value of the sample mean, 99.9997, is the estimate of the population parameter. Thus, the estimator is a sample statistic, whereas the estimate is the numerical value that has been estimated—99.9997 in the preceding case.

Point Estimate and Interval Estimate

A *point estimate* is a single number that is used to estimate an unknown parameter. A single measurement observation is a point estimate of the measurand. We generally quote the average of a number of repeat observations as the measurement result; this is also a point estimate of the desired quantity under measurement, as an average is also a single number. The maximum temperature of a city on a particular day reported as part of the weather report is a point estimate.

A point estimate as a measurement result gives us an indication of the value of the measurand. However, a point estimate is an insufficient estimate in the sense that it is either wrong or right. It is not known how wrong or how right the estimate is. To overcome this limitation, the concept of the interval estimate has been developed.

An *interval estimate* is a range of values used to estimate a population parameter. An estimate that the income of the persons living in a particular city is between x and y dollars is an interval estimate. Similarly, a measurement result that the temperature of an oven is 95°C ±1°C is an interval estimate. The measurement result implies that the temperature of the oven lies within an interval of 94°C to 96°C. Thus, in an interval estimate, a range of values instead of a single value is used as the estimate. Any value within the range can be considered an estimate of the parameter being measured.

The probability that the measured value lies at any particular point within the interval is not the same throughout the interval. The probability that the measured value lies in the central part of the interval is greater than the probability of its lying at either of the extremities of the interval.

In an interval estimate, error is indicated in two ways: one is the extent of the range of the interval, and the other is the probability that the true value of the parameter lies within the range. In the context of the NUSAP notation scheme discussed in an earlier chapter, the spread is the interval estimate and the assessment is the probability that the true value lies within the interval, whereas the numeral, or the number representing the measurement result, is the point estimate of the parameter under measurement.

The Characteristics of a Good Estimator

Statistical estimation is based on investigation of samples. It is therefore apparent that different samples will lead to different estimates. An estimator is considered to be a good one if it has minimum sample-to-sample variation and if the values estimated by the estimator are close to the values of the population parameter being estimated.

POINT ESTIMATE OF POPULATION MEAN AND STANDARD DEVIATION

An estimate depends upon the information contained in the sample and the known probability distribution of the population. It is an established and universally accepted fact that for normally distributed data the sample mean $\bar{x}$ is the best estimate of the population mean μ and the sample standard deviation s is the best estimate of the population standard deviation σ.

The above statement has been made without going into the details of a statistical proof. However, a simple proof of this statement can be given using the principle of least squares. Suppose that we have a sample of observations x_1, x_2, . . . x_n. We intend to estimate a value $\hat{x}$, which is a point estimate of the parameter being measured. The value of each observation deviates from the estimate $\hat{x}$. This deviation from x_i is equal to $x_i - \hat{x}$. The estimation criteria for the principle of least squares states that the sum of the squares of the deviation from the estimate should be minimum. If D is the sum of the squares of the deviation, then $D = \Sigma(x_i - \hat{x})^2$.

For D to be minimum, $(\partial D / \partial \hat{x}) = \Sigma\, 2(x_i - \hat{x}) = 0$

$\Sigma x_i - n\hat{x} = 0$

$\hat{x} = (\Sigma x_i) / n$

Thus, the mean of the sample data is the estimate.

It is to be kept in mind that for any other probability distribution this may not be the case. If the parameter being estimated is the time to failure of electronic components based on life testing studies, the estimate is not the mean of the sample data. For example, suppose that 100 components are subjected to life testing and that the times to failure of 5 of the components, t_1, t_2, t_3, t_4, and t_5, are noted. The average of these 5 values is not the estimate of the time to failure, as information contained in the life testing of the remaining 95 components has not been taken into account. Moreover, time to failure as a random variable does not follow a normal distribution but another type of distribution that is called exponential distribution. For a normally distributed data sample, however, the mean is a good estimator of the population mean.

INTERVAL ESTIMATE OF A POPULATION MEAN USING NORMAL DISTRIBUTION

The interval within which a population mean μ is expected to remain with reference to the sample mean $\bar{x}$ can be obtained using standard normal tables, if the standard deviation of the population σ is known. A sample size n is taken and the sample mean $\bar{x}$ is calculated. It is also known that the standard deviation of the sample mean is equal to $\sigma / \sqrt{n}$. To find an interval estimate, we use tabulated values from standardized normal distributions, and z is equal to $(\bar{x} - \mu) / (\sigma / \sqrt{n})$. It should be noted that a certain percentage of the area under the normal curve is bound within $\pm z$. This area, which represents the probability of random variables' being between $-z$ and $+z$, is given in the following table for various values of z.

z	±1	±1.96	±2	±3
Probability	68.3%	95%	95.5%	99.73%

This information is used to obtain interval estimates of the population mean. For example, for 95% probability,

$$-1.96 < (\bar{x} - \mu) / (\sigma / \sqrt{n}) < +1.96$$

$$\bar{x} - 1.96(\sigma / \sqrt{n}) < \mu < \bar{x} + 1.96(\sigma / \sqrt{n})$$

The interval $\bar{x} - 1.96(\sigma / \sqrt{n})$ to $\bar{x} + 1.96(\sigma / \sqrt{n})$ is the confidence interval of the population mean μ and will be correct 95% of the time in the long run. The figure of 95% is called the confidence level. For a 99.73% confidence level, the interval for the population mean would be $\bar{x} - 3(\sigma / \sqrt{n})$ to $\bar{x} + 3(\sigma / \sqrt{n})$.

In this estimation, it is assumed that the standard deviation of the population σ is known. The standardized normal tables are applicable to large sample sizes on the order of 30 or above. For a sample of this size, the standard deviation can be calculated and normal tables can still be used with this calculated value of σ. However, for small sample sizes this method cannot be used.

INTERVAL ESTIMATE USING THE *t* DISTRIBUTION

Standard normal distribution tables are used for finding interval estimates and confidence levels. However, this is valid only if the sample size is more than 30. In actual practice, it is neither possible nor desirable to use such a large sample. In metrology, it is prohibitively costly to take 30 or more repeat observations to find the interval estimate of the true value. In metrological studies the sample size is generally small, on the order of three to five.

In order to find the interval estimate and its associated confidence level for a small-sized sample, a distribution called the Student *t* distribution is used. It is also called the *t* distribution. The *t* distribution is used to find the interval estimate at a specified confidence level if the sample size is small and the standard deviation of the population is not known. This is a very popular type of distribution among metrologists, as the mean of the measurement process and its standard deviation are not known and the number of observations is small, generally around five.

The Student t distribution is a sampling distribution and has a pdf for every possible sample size. The value of t, which is a sample statistic given by $t = (\bar{x} - \mu) / (s / \sqrt{n})$, has a probability density function represented by $f(t)$. The function $f(t)$, unlike $f(z)$, is a complex mathematical function. For each value of sample size n corresponding to $(n - 1)$ degrees of freedom there is a different distribution function. The function $f(t)$ is similar to $f(z)$ and has the following characteristics:

- $f(t)$ has a value for $-\infty < t < +\infty$
- $f(t)$ has a maximum value at $t = 0$
- The curve extends asymptotically on both sides of the t axis
- $f(t)$ has two points of inflection, one on each side of the maximum

As the sample size approaches 30, the t distribution approaches the normal distribution. The t distribution is lower at the mean and higher at the tails than the normal distribution; thus, it has more area under the tails than the normal distribution does. This is why one has to go farther from the mean of the t curve in order to get the same area under the curve as in a normal distribution. Thus, for the same confidence level, the width of the interval is greater in a t distribution than in a normal distribution. In this case, the sample standard deviation is calculated based on the data in the sample and is taken as the estimate of the population standard deviation σ. The t variable is taken in a manner similar to z except that σ is replaced by s. Thus,

$$t = (\bar{x} - \mu) / (s / \sqrt{n})$$

The interval within which the true population parameter μ lies is given as

$$\bar{x} - \frac{t.s}{\sqrt{n}} < \mu < \bar{x} + \frac{t.s}{\sqrt{n}}$$

where n is the sample size, $\bar{x}$ is the mean of the sample data, s is the standard deviation of the sample, and t is a multiplier called the Student t factor.

This interval is analogous to the one obtained for a large sample size n of greater than 30:

$$\bar{x} - \frac{z\sigma}{\sqrt{n}} < \mu < \bar{x} + \frac{z\sigma}{\sqrt{n}}$$

Thus, in the interval estimate using the t distribution, z has been replaced by t.

The values of t factors are available in standard statistical tables for various confidence levels and degrees of freedom. For a higher confidence level, the value of t is higher, indicating that the assigned interval will be wider. The degree of freedom is linked to the sample size. For a higher degree of freedom, the t factor becomes smaller. Thus, for the same confidence level the interval becomes smaller. For a sample size of 30 or more, the t values are similar to the z values. The values of t factors for the values of various degrees of freedom and confidence levels are listed in Appendix B.

DEGREES OF FREEDOM

The term *degree of freedom* is extensively used in the statistical analysis of data. Data is a collection of numbers. These numbers form a sample, which is part of a population. The population has a definite mean.

The degree of freedom pertains to the number of values we can choose freely. If there are three observations in a sample, we can choose only two observations in order to get a definite mean. The third one is established automatically. The degree of freedom is always one less than the sample size. If there are n observations forming the sample, then the degree of freedom is $(n - 1)$. For example, if we use a sample of 10 to estimate the sample mean, the data has 10 minus 1, or 9, degrees of freedom. For data drawn from a population with a specific distribution, the degree of freedom is infinite.

REGRESSION AND CORRELATION

Regression and correlation are tools that have applications in metrology. At times we measure two variables and want to establish whether there is any relationship between them. One of the variables is known and is called the independent variable, and the variable that is being predicted is called the dependent variable. Regression and correlation analysis are based on the relationship between these two variables.

The relationship between the independent and dependent variables is established through scatter diagrams. The data on the two variables are obtained as ordered pairs (x, y), and the points for the various values of x and y are plotted on graph paper. The independent variable is plotted on the x axis and the dependent variable on the y axis. The plots on the scatter diagram help us to see what type of relationship exists among the independent and dependent variables. In some cases, it is possible to draw a straight line through the scatter diagram. One approximate way of doing this is to draw a straight line in such a way that an equal number of points lies on each side of the line. This relationship is called a linear relationship. If an increase in the independent variable causes an increase in the dependent variable, it is called a direct linear relationship. Conversely, if an increase in the independent variable causes a decrease in the dependent variable, it is called an inverse linear relationship. The pattern of the points on the scatter diagram may also indicate that the relationship is not linear but curvilinear. If the points on the scatter diagram have no pattern and are scattered randomly, this implies that there is no relationship between the variables. Estimating the linear relationship between two variables is done through linear regression techniques. These techniques are used, for example, in the study of the stability of a reference standard over time. In this case, time is the independent variable and the value of the measured characteristic of the reference standard is the dependent variable. Trend charts are plotted using linear regression techniques to study the drift in the value of the standard with the passage of time.

Correlation analysis is another statistical tool that describes the degree to which one variable is linearly related to another variable. The covariance of two random variables is a measure of their mutual dependence, which is defined as

$$s(y,z) = \{1 / n - 1\} \sum_{i=1}^{n} (\bar{y}_i - y)(\bar{z}_i - z)$$

Another measure of the relative mutual dependence of two variables is their correlation coefficient, which is equal to the ratio of their covariances to the positive square root of the product of their variances:

$$r(y, z) = s(y, z) / \{s(y)s(z)\}$$

REFERENCES

Adler, Mortimer J., ed. 1993. *Great books of the Western world.* Vol. 56, *National science selection from the twentieth century.* Chicago: Encyclopaedia Britannica.

Britz, Galen, Don Emerling, Lynne Hare, Roger Herl, and Janice Shade. 1995. The role of statistical thinking in management. *ASQ Quality Progress,* February, 53–59.

———. 1997. How to teach others to apply statistical thinking. *ASQ Quality Progress,* June, 67–78.

Davenport, Wilbur B., and William L. Root. 1958. *An introduction to the theory of random signal and noise.* New York: McGraw-Hill.

Eisenhart, Churchill. 1963. Realistic evaluation of the precision and accuracy of instrument calibration systems. *Journal of Research of the National Bureau of Standards: Engineering and Instrumentation* 67C, no. 2:161–85.

Grant, Eugene L., and Richard S. Leavenworth. 1988. *Statistical quality control.* New York: McGraw-Hill.

Guide to the expression of uncertainty in measurement. 1995. Geneva Switzerland: ISO

Ku, Harry H. 1967. Statistical concepts in metrology. In *Handbook of industrial metrology,* American Society of Tool and Manufacturing Engineers, 20–50. New York: Prentice Hall.

Levin, Richard I., and David S. Rubin. 1997. *Statistics for management.* New Delhi: Prentice Hall of India.

Parsons, James A. 1973. *Practical mathematical and statistical techniques for production managers.* Englewood Cliffs, NJ: Prentice Hall.

Stein, Phillip. 2001. All you ever wanted to know about resolution. *ASQ Quality Progress,* July, 141–42.

Wells, D. E., and E. J. Krakiwsky. 1978. *The method of least square:* Department of Surveying Engineering, University of New Brunswick.

4

Measurement Error and Measurement Uncertainty

Everybody believes in the law of errors, the experimenters because they think it is a mathematical theorem, the mathematicians because they think it is an experimental fact.

Lippman

Absolute certainty is a privilege of uneducated minds and fanatics. It is for scientific folk an unattainable ideal.

Cassius J. Keyser

All measurements are influenced by various factors that contribute to errors in measurement results. Errors cannot be quantified. Thus, the quality of a measurement result is quantified by its uncertainty. This chapter explains various sources of errors, error classification, and other terms frequently used in metrology, including *accuracy, error, precision,* and *bias.* The chapter also explains various concepts pertaining to the uncertainty of measurement and illustrates the terms *uncertainty* and *confidence level* with a number of examples.

MEASURAND, TRUE VALUE, AND CONVENTIONAL TRUE VALUE

Measurand

In metrology, a *measurand* is defined as a specific quantity subject to measurement. The measurement results represent the numerical value of the measurand which is generally characteristic of a product. The results are indeterministic in nature and hence are governed by statistical laws. A measurement is made through a measurement process. Because of the variability of measurement processes, repeat measurement observations are found to vary among themselves. This is true for all measurements in physical as well as chemical metrology. It means that every measurement observation contains error. It is therefore not possible to pinpoint a value among the repeat observations that represents the correct value of the measured characteristic. This idea of the correct value is generally referred to as the true value of a measurand.

True Value

True value has been defined as the value that is perfectly consistent with the definition of a given specific quantity. The true value would be that value of a measured characteristic that would be obtained by an ideal and perfect measurement system. An ideal measurement system implies that a measurement result obtained using the system would be the true value of the measurand. There would be no error in the measurement result, and repeat observations would consistently have exactly the same value. However, perfection is an idealized concept and is not achievable in the physical world. Truth is more a philosophical concept than a reality. Therefore, an ideal and perfect measurement system does not exist. However, metrologists do believe in the existence of the true value of a measurand. They also believe in its illusionary nature: howsoever one may chase the true value, one will never be able to find it. True values are by nature indeterminate.

Conventional True Value

Measurement standards are used as references for the comparison of measurement results. If there were no measurement standards, measurement results would not be meaningful or acceptable. A standard represents the numerical value of a parameter. As the true values of parameters are unknown and unknowable, the concept of *conventional true value,* or accepted reference value, has been adopted to assign values to standards. This is the value accepted by consensus among people knowledgeable on the subject. This value is not the true value but a value very close to the true value and is accepted as such. It is also referred to as assigned value, best estimate of the value, conventional value, accepted value, and reference value. Conventional true value is now a well-accepted convention. Many examples of reference value can be cited in almost every discipline of metrology. One such example is the accepted reference value of the electromotive force (emf) of a saturated Weston cadmium cell. Before the invention of Josephson-junction-based DC voltage standards, these cells were used as laboratory standards. Their emf value is universally accepted as 1.01859 volts at 20°C. Similarly, the accepted values of Rockwell hardness standards are the reference values. The values assigned to certified reference materials (CRMs) that represent some of the specific characteristics of the reference materials are examples of accepted reference values. CRMs are frequently used in the calibration and validation of measurement processes in physical and chemical metrology applications. Conventional true value has thus been defined as a value attributed to a specific quantity and accepted, sometimes by convention, as having an uncertainty appropriate for a given purpose.

MEASUREMENT ERROR

A measurement result is generally based on one observation made using measuring equipment. At times, the arithmetic mean of a number of repeat observations represents a measurement result. Since the true value of a characteristic of a product under measurement cannot be known, a measurement result always has some error associated with it. No measurement has ever been made that did not contain some error. We do not realize this at times due to our misconceptions about the perfectness of measurement systems.

Numerically, *measurement error* has been defined as the result of a measurement minus the true value of the measurand. The true value, as explained above, cannot be known; since measurement error is linked with the true value by definition, measurement error also cannot be known. Measurement error is therefore a qualitative concept only. Metrologists believe in the existence of errors, but they also believe that, like true value, the error cannot be known exactly.

The concept of measurement error has been defined by Churchill Eisenhart (1963) as follows:

> A conscious characteristic of measurement is disagreement of repeated measurements of the same quantity. Experience shows that when high accuracy is sought repeated measurements of the same quantity by a particular measurement process do not yield uniformly the same number. We explain these discordances by saying that the individual measurements are affected by errors which we interpret to be the manifestation of variations in the executions of the process of measurement resulting from the imperfection of instruments and of organs of sense and from the difficulty of achieving (or even specifying with a convenient number of words) the ideal of perfect control of conditions and procedures.

There are various sources of errors that affect the measurement process and hence the measurement observations. Some measurement errors may be attributed to human ignorance and negligence. These errors are called gross errors. They may include errors due to the improper use of measuring equipment, the wrong selection of tests and measuring equipment, the use of improper measurement methods, and many other factors such as incorrect adjustment, incorrect observations, and computational errors. For example, measuring the voltage across a high-impedance circuit using a voltmeter with comparable input impedance is an example of the improper selection of test equipment for an application. There is no scope for gross errors in a normal measurement system. Even when all the sources of gross errors are removed, there are other factors that influence the measurement process. These factors cannot be eliminated altogether. They contribute to measurement errors, which are classified as either random errors or systematic errors. This classification is based on the factors which contribute to the generation of these errors.

The nature and treatment of random and systematic errors have been a matter of deliberation among the scientific community. The great physicist Poincaré (1903) discussed them in his essay "Science and Hypothesis":

> But it is first of all necessary to recall the classic distinction between systematic and accidental errors. If the meter with which we measure a length is too long, the number we will get will be too small and it will be no use to measure several times—that is a systematic error. If we measure with an accurate meter, we may make a mistake, and find the length sometimes too large and sometimes too small, and when we take the mean of a large number of measurements, the error will tend to grow small. These are accidental errors.

Random and Systematic Error: A Mathematical Model

Random error is defined as the result of a measurement minus the mean result of a large number of repeated measurements of the same measurand. *Systematic error* is defined as the mean result of a large number of repeated measurements of the same measurand minus the true value of the measurand.

It is important to note that in the definition of random error the reference value is taken to be "the mean result of a large number of repeated measurements of the same measurand." The mean of a large number of repeat measurements represents the true value, provided the variations are caused by chance causes only and that the number is so large that it tends to infinity. Since metrologists believe that measurement processes are also influenced by factors other than chance causes, called systematic effects, metrologists term the mean result of a large number of repeat measurements the limiting mean rather than the true value. Since the definition of random error does not define how large the "large number of repeated measurements" is, their mean is also indeterminate. Random errors are thus indeterminate in nature. It is also clear from the definition that random errors are caused by chance causes of variation only. It is important to note that random error applies to a measurement result observation, which can be a single observation or the average of a limited number of observations. The random error is different for every measurement result, and it assumes positive and negative values randomly.

Systematic error is defined differently from random error. First, it is defined with reference to the true value, which is unknown and unknowable. Second, before attempting to know systematic error, we have to take a large number of repeated measurements and then find out their mean. This implies that systematic error should not be evaluated with reference to a single observation or a measurement result based on a single observation. It is also known that the average of repeated measurement observations represents the best estimate of the parameter that is being measured. Thus, the systematic error of a measurement is defined with reference to the true value. Since the true value is indeterminate and the measurement result, being an estimate itself, is not exact, systematic error is also indeterminate.

A measurement result can be expressed as a mathematical model. It has the following components:

- Estimate m of the true value, which is the average of a number of repeat measurement observations. This itself is a random variable and has its own frequency distribution.
- Systematic error B. This can be known or unknown. Systematic error may have random as well as nonrandom components. Random components can be analyzed by statistical methods.
- Random error e. This is a variable component of error with an average of zero and hence about the same variance as the parameter that is being measured. It has its own frequency distribution and can be analyzed by statistical methods.

If x is a measurement result, the mathematical relation between x, m, B, and e is given by the following equation:

$$x = m + B + e$$

The measurement result x is also a random variable and has its own probability distribution.

RANDOM MEASUREMENT ERROR

Poincaré defined random errors as accidental errors. The word *random* implies unpredictability. Random events are those events whose outcome cannot be predicted in advance. Due to their very nature, the values of repeat measurement observations are unpredictable. Thus, repeat measurement data are random variables. Statistical techniques are extensively used in the analysis of random variables and also have a wide application in the analysis of measurement data. There are a number of factors which influence measurement processes and thus contribute to random error. These factors may affect the process individually as well as collectively, through their interaction. Some of the influencing factors and their causes and effects are shown in Table 4.1.

It is difficult to establish the exact reasons for this variability. This is mainly because all the causal forces at work that influence the measurement process are not known. There is always some basic indeterminacy in the physical world. These variations are therefore categorized as variations due to chance causes. They cannot be eliminated altogether but can be minimized

Table 4.1 Factors that influence the measurement process.

Factor	Causes and effects
Equipment used	Variation in the behavior of measuring equipment or measurement apparatus contributes to variability in the measurement process. Over the long term this may be due to the aging of components, and in the short interval it may be due to variation in the environmental conditions under which the equipment and apparatus are operating.
Operator	The operational efficiency of the person making measurements varies from person to person. The variation in technical competence from one appraiser to another also causes variation in the measurement process.
Time interval	Measurement results are always found to vary when the same measurement is repeated after some time. Other conditions remaining the same, the variation over a short time interval is less than that over a longer time interval.
Place	Measurement results of the same physical or chemical characteristics are found to vary if the measurements are repeated at different places or in different laboratories.
Environment	The effects of changes in environmental conditions, such as temperature, humidity, pressure, vibration, electrical and environmental noise, air current, and optical illumination, influence the measurement setup and also the object under measurement, thereby causing variations in the measurement process.
Chemicals and reagents	In chemical metrology, the purity level and chemical characteristics of laboratory chemicals and reagents vary from lot to lot, resulting in variations in analytical results.

with proper selection of measurement setup, enhancement of the technical competence of the person involved in measurement, and better control of the environmental conditions under which measurements are being made.

Random error represents the variable component of measurement error. Random errors are characterized by frequency distribution functions. The Gaussian frequency distribution characterizes the behavior of measurement data observed in physical and chemical metrology. This probability distribution is characterized by two parameters: the mean and the standard deviation. The standard deviation, which is a measure of the variability of data, can be used to get an indication of random error.

METROLOGISTS' AND STATISTICIANS' PERCEPTIONS OF TRUE VALUE

The theory of errors was developed by statisticians long before metrologists applied statistical techniques to the analysis of metrological data. This theory was formulated for the analysis of random events, which are influenced purely by chance causes. Under the condition that an event is influenced by chance causes only, the arithmetic average of an infinite number of observations of the random variable represents the true value of the random variable. Since it is not possible to make an infinite number of observations, the arithmetic average of a large number n of observations such that n tends to infinity is considered to represent the true value of a random variable. Thus, the theory of errors accepts the existence of the true value if the process is being influenced by chance causes only. In such a situation, random errors have an equal probability of being positive or negative. During the process of averaging, these positive and negative error components cancel each other. With a large number of measurement data such that the number approaches infinity, all random error components are canceled, and the average represents the true value.

Metrologists, on the other hand, know that it is not only chance or random causes that affect the measurement process. There is also another component of error which influences the measurement process. This component is called systematic measurement error and is generally constant for a particular measurement process. Systematic error influences the measurement process but not in the same way that random error does. Due to the nonrandom nature of systematic errors, they do not cancel each other in the way that the positive and negative components of random error do when one takes the average of a large number of observations. Even with the averaging of a large number of observations, therefore, there is always some residual systematic error component embedded in the measurement result.

In the paper mentioned earlier, Eisenhart (1963) quoted N. Ernest Dorsey, who expressed this dilemma of metrologists in the following words:

> The mean of a family of measurements of a number of measurements for a given quantity carried out by the same apparatus, procedure and observer approaches a define value as the number of measurements is indefinitely increased. Otherwise they would not properly be called measurements of a given quantity. In the theory of errors this limiting mean is frequently called the "true" value, although it bears no necessary relation to the true quaesitum, the actual value of the quantity that the observer desires to measure. This has often confused the unwary. Let us call it the limiting mean. (1944)

SYSTEMATIC MEASUREMENT ERROR

Systematic measurement errors remain constant over the course of a number of repeat measurements of the same characteristics. Systematic errors may be known or may not be known. They may be random or nonrandom in nature. A known systematic error is called a bias. Systematic errors may be determined by the statistical analysis of repeat observations as is done for a random process, if they are random in nature. Nonrandom systematic error components cannot be determined by a random process. Two of the most prominent examples of systematic errors are those introduced by measuring equipment and those due to measurement methods.

Systematic Errors Introduced by Measuring Equipment

The known systematic error of a piece of measuring equipment is called its bias. This type of error is known through the process of calibration. The known systematic error is always corrected by applying a correction, which is equal to the negative of the known bias. During the process of calibration, the measurements taken on a piece of measuring equipment are compared with those measured on a reference standard. The reference standard should have valid calibration traceable to the International System of Units and should be of much higher accuracy than measuring equipment. The difference between the readings of the measuring equipment and the reference standard gives an indication of the error of measurement. From the accuracy specification of the measuring equipment, one knows the maximum permissible error for the equipment as its inherent limitation of inaccuracy. If the difference between the readings is less than the maximum permissible error, observations made with the equipment are accepted as such. If the difference is more than the maximum permissible error, a correction for the measured value needs to be applied. If the calibration certificate of a piece of measuring equipment shows that it reads x units for a reference value of y units and the difference between

x and y is not within the limits of the accuracy specification of the measuring equipment, correction for all measurements made using this equipment needs to be taken into account as per the calibration chart or calibration graph.

For example, an ammeter is used to measure electric current. According to its accuracy specification, it is an accuracy class 1.0 instrument. This implies that the measuring accuracy is ±1.0% of the full scale deflection. The full scale corresponds to 5 amperes, so the accuracy of measurement will be ±0.05 amperes. During calibration, it is observed that the ammeter reads 5.18 amperes as 5.00 amperes. Thus, there is a bias in the measurement which is equal to 5.00 minus 5.18, or -0.18, amperes. As the bias is greater than the acceptable accuracy limit of ±0.05 amperes, a correction to the measured value is to be applied. The correction is equal to the negative of the bias. When the ammeter reads 5.00 amperes, the result should be taken as 5.18 amperes.

A *correction* is defined as a value added algebraically to the uncorrected result of a measurement that compensates for an assumed systematic error.

Systematic Errors Related to Measurement Methods

A particular parameter can be measured using various methods of measurement. For example, the temperature of an oven can be measured using a mercury thermometer. It can also be measured using a thermocouple or platinum resistance thermometer as a sensor with associated processing circuitry and an indicating device. The basic principles of measurement are different for each method. With a mercury thermometer, the temperature is measured by observing the linear expansion of a column of mercury due to the absorption of heat. The linear expansion of the mercury column is calibrated to read the temperature in degrees Celsius or Fahrenheit. The thermocouple with associated circuitry and an indicating device uses the principle of the generation of a thermo-electromotive force when the two junctions of a thermocouple are at different temperatures. The thermo-emf thus generated is processed and measured on an indicating device. The voltage-measuring device is calibrated to indicate the temperature of the oven in which the thermocouple junction is placed. In the case of a platinum resistance thermometer, the principle of change in the resistance of platinum wire when its temperature is elevated is used. The parameter actually measured is the change in resistance. The indicating device is calibrated to read the temperature in proportion to the value of the change in resistance.

Another example is the measurement of a person's systolic and diastolic blood pressure by a physician. It can be measured using a mercury-type manometer. This is the most fundamental method of pressure measurement. Another method is to use a dial-type pressure gage, in which the force due to the blood pressure generates spring action and the pressure is indicated on a dial. The two methods operate on different principles, and the systematic error of each method will be different.

Each method of measurement has a different value of systematic measurement error. For a specific method of measurement and measurement setup, the systematic error is constant; however, there is a change in the systematic error if the measurement method or measurement setup is changed.

Systematic error that is introduced due to human factors, such as parallax in observations, will generally remain constant for one operator but will differ among operators. Other factors that contribute to systematic error are the various assumptions and approximations associated with measurement methods. For example, in temperature measurements using resistance thermometers, a linear relationship is assumed between the temperature and the relative resistance. The relative resistance is the ratio of the resistance of the wire element at an elevated temperature to its resistance at a reference temperature, usually 0°C. Due to the assumption of a linear relationship, higher-order terms such as $(\Delta t)^2$ are not taken into consideration. However, the relationship is not perfectly linear, and the assumption introduces a systematic error into the measurement method. In most measurement systems, transducers are very frequently used to transform a physical quantity to a proportional electrical signal. The physical quantity is the quantity under measurement. The electrical signal is processed through electronic circuitry, and the measured value is displayed on an analog or digital indicating device. This conversion process of the physical quantity to an electrical signal also introduces some systematic error. Each type of transducer generates a different amount of systematic error. The selection of an appropriate transducer is very important in designing measuring equipment of the desired accuracy. In general, systematic measurement error cannot be known completely, and hence corrections for it cannot be complete.

PRECISION, BIAS, AND TRUENESS OF A MEASUREMENT PROCESS

It is essential that parts manufactured according to the same specifications and engineering drawings be interchangeable. This should be true for all parts, whether manufactured by a single manufacturer or different manufacturers. Interchangeability is achieved only if the dimensions of the parts are identical, they are within engineering tolerances, and they are manufactured from identical material.

It is a common experience that when repeat measurements are taken of the same characteristic of a product with the same measurement method, measurement setup, and operator in one laboratory, the repeat observations are found to vary. A measurement system is used only if this variation is within acceptable limits. Similarly, if a specific characteristic of a product is measured using different measurement methods, using different measurement setups, and by a different operator in another laboratory, the measurement results will be found to vary. The results will not be the same but should be similar, with variations within the allowable tolerance of the measurement processes used by the two laboratories. If this is not achieved, then the measurement becomes meaningless. The requirement for similarity of the measurement results, whether at one place or at different places, is analogous to the interchangeability requirements for an engineering part. Consistency in measurement results can be achieved only if repeat measurement observations are consistent and there are no biases in the measurement process. The consistency of repeat observations is called repeatability and reproducibility.

Repeatability is defined as the closeness of agreement between the results of successive measurements of the same measurand carried out under the same conditions of measurement, whereas *reproducibility* is defined as the closeness of agreement between the results of measurements of the same measurand carried out under changed conditions of measurement. Thus, the consistency of repeat observations in one laboratory is called repeatability, and the consistency of measurement results obtained in different laboratories is called reproducibility.

Precision

Repeatability is also called precision. It is defined as the closeness of agreement between repeated measurements of the same quantity under the same conditions. This is a desired characteristic of measurements made in laboratories and industry. Precision depends only on the distribution of random errors and does not relate to the true value or accepted reference value. The factors that contribute to the variability of repeat measurement results were explained earlier.

The precision of a measurement process is linked with the dispersion of repeat observations among themselves. The most widely used measure of dispersion of statistical data is the standard deviation of the data. In the case of normally distributed data, the standard deviation determines the shape of the pdf. The lower the standard deviation, the more the measurement data are clustered among themselves and the more precise the measurement process is. If we are measuring a single parameter with two measurement processes, it is likely that the two processes are centered around the same mean but have different standard deviations. Thus, the measurement results will have the same value but different precision. The process with the lower standard deviation is more precise than the other. Thus, the standard deviation of measurement data is a measure of its precision. The total variability of a measurement process is obtained by multiplying the standard deviation by a coverage factor whose value depends upon the desired confidence level. The quantitative requirement for the variability of a measurement process is dictated by the end use of the measurement results.

Bias and Trueness

The average of a number of repeat observations is the estimate of a measurement result. This average is also called the expectation of the test result. The difference between the mean of repeat measurement results and the accepted reference value or true value is called the bias of the measurement process. Bias is the total systematic error. There may be more than one systematic error component contributing to the bias. The larger the systematic error, the larger is the bias of the measurement process. *Bias* has been defined as a systematic error inherent in a method or caused by some artifact or idiosyncrasy of the measurement system. An error in calibrating measuring equipment is an example of bias. In the case of analytical measurements, extraction inefficiency is another example of bias. The effect of temperature on a measurement system also contributes to bias.

Trueness is the characteristic of the measurement process which is the inverse of bias. The lower the bias, the higher the trueness. Trueness and bias represent the same characteristic of the measurement process. The only difference is that bias represents the negative aspect, whereas trueness does not have negative connotations. This is the reason that the use of the word *trueness* is being propagated among the metrology community.

THE ACCURACY OF THE MEASUREMENT PROCESS

Accuracy of measurement is defined as the closeness of agreement between the result of a measurement and the true value of the measurand. Accuracy is also defined with reference to the true value. Since the true value is unknown and unknowable, measurement accuracy, like error, is also unknown and unknowable. Accuracy is therefore a qualitative aspect of a measurement process. Sometimes, when one takes the conventional true value in place of the true value, the accuracy can be known. The definition of accuracy is also linked with the result of a measurement, which has been defined as the value attributed to a measurand obtained by measurement. A measurement result consists of two parts. The first part is a single numerical value which is an average of repeat observations and along with the units in which it is expressed, represents the measurement result. Sometimes a single observation can also be a measurement result. This single numerical value is an estimate of the quantity that is being measured. However, the complete statement of a measurement result should include information about the uncertainty of the measurement, which is the second part of the measurement result. For a measurement result to be close to the true value, the estimate of the measurand has to be close to the true value, and thus the bias or systematic error has to be as small as possible. At the same time, measurement uncertainty has to be low, so the results of repeat measurements should be close to one another.

The term *accuracy* when applied to a measurement result involves a combination of random components of error and systematic components of error. The following two factors influence the accuracy of a measurement process:

- The precision of the measurement process, characterized by the dispersion of repeated observations due to the influence of sources of random errors. The standard deviation of repeat measurement observations is an index of precision.
- The bias of the measurement process, which is due to the presence of known and unknown systematic error components.

In order for a measurement process to be accurate, it should be precise as well as unbiased. Being unbiased also means being true. Many persons use the word *precision* as a substitute for *accuracy,* as if they had the same meaning. This is incorrect and should be avoided.

In view of its indeterminate nature, the measurement accuracy of a result cannot be quantified. At the same time, the role of metrology in the quality functions of industry and laboratories is growing. Therefore, the need to quantify the quality of measurement results is also growing. To meet the increasing awareness of this need, the concept of measurement uncertainty was evolved.

The Metrologists' Dilemma and Measurement Uncertainty

Metrologists have the objective of determining the true value of a measurand which is unknown and unknowable. The larger the number of observations, the larger is the spread in the numerical values of the observations. So how do we choose the measurement result out of all these repeat observations? Do we simply quote all the repeat measurement data as the measurement result? This collection of repeat measurement data as raw data does not carry much information. So what is the answer to the above question? This can be called the metrologists' dilemma.

This issue is depicted by Eisenhart (1963) as follows:

> This cussedness of measurements brings us face to face with a fundamental question. In what sense can we say that the measurements yielded by a particular

measurement process serve to determine a unique magnitude when experience shows that repeated measurement of a single quantity by the process yields a sequence of non-identical numbers? What is the value thus determined?

Measurement as an Estimation Process

The solution to the metrologists' dilemma lies in statistical estimation techniques. We have come to understand that measurement observations are random variables which are characterized by frequency distribution functions. Most natural phenomena, and measurement observations, can be analyzed using a normal probability distribution. In metrology, we want to know the true value of a particular physical or chemical characteristic which is under measurement. We take repeat measurements of the characteristic using a measurement process. If we take a large number of observations n such that n tends to infinity, this set of a large number of observations can be called a population of observations. In order to obtain information about the measurand, we use a sample. A sample is a set of randomly chosen data from a large population. In practice, we take a limited number of observations and accept them as a sample drawn from a large population. It is neither possible nor desirable to take a large number of observations such that the number tends to infinity. We then draw inferences about the true value of the characteristic under measurement based on the data in the sample.

If μ is the true value of the characteristic under measurement, all the data in the population have a claim to represent the true value. We estimate this value through some statistic based on the sample. It has been established in statistical theory that the sample mean is the best estimate of the population mean μ if the data follows a normal distribution. Since the mean of sample data is a single value quantity, it is a point estimate of the true value. Thus, the sample mean is a point estimate of the true population parameter. This is the reason for taking the arithmetic mean of the repeat measurement observations and accepting it as a measurement result. This single value point estimate is also called the assigned value, meaning that the value has been assigned to the measurand based on sample data which is representative of the population.

Thomas Simpson in the mid-18th century, was the first to consider the repeated measurements of a single quantity by a measurement process as observed values of independent random variables having the same probability distribution. Eisenhart (1963) has quoted Thomas Simpson as follows:

> Upon the whole of which it appears that the taking of the mean of a number of observations greatly diminishes the chances for all the smaller errors and cuts off almost all possibility of any great ones: which last consideration alone seems sufficient to recommend use of the method, not only to astronomers but to all others concerned in the making of experiments of any kind to which the above reasoning

is equally applicable. The more observations or experiments there are made, the less will the conclusion be liable to err provided they admit of being repeated under the same circumstances.

MEASUREMENT UNCERTAINTY

The mean of repeat observations is an estimate of the value of the measurand. Still, a point estimate is an estimate and not a true value. It is difficult and risky to make any decision based on a single value measurement result. We do not know "how wrong or how right" we are, that is, how close we are to the true value. At best, we have been able to pinpoint a single number, which we call an estimate of the true value of the characteristic under measurement. It is known that there is a certain amount of variability associated with the mean value taken as a point estimate. Thus, the assigned value is once again not error free. One set of observations from a measurement process may lead to one estimate, another set of observations from the same measurement process may lead to a different estimate, and a third set of observations may lead to still another estimate, though all the estimates will lie within acceptable tolerance limits. This shows how vulnerable a point estimate as a measurement result can be. However, it is important to understand that the averages of various samples are much closer to one another than are individual observations. Thus, the variability is reduced considerably. If σ is the standard deviation of a random variable, the standard deviation of its mean is given by $\sigma/\sqrt{n}$, where n is the number of observations based on which the mean is calculated. Thus, the standard deviation of the sample mean is reduced in the inverse proportion of the square root of the number of observations in the sample.

Interval Estimation of Measurement Uncertainty

For effective decision making, it is important to know "how good or how bad" a single assigned value is as a measurement result. Statistical estimation techniques again help us in knowing the "how good or how bad" part of a single value as a measurement result. These techniques are called interval estimation techniques. An *interval* is a spread around a central value. There are an upper limit and a lower limit of the interval. For example, the interval between the quantities 9.95 and 10.05 has an upper limit of 10.05 and a lower limit of 9.95, and the central value is 10.00. In metrology, we estimate the interval within which the population parameter or true value of the measurand is expected to lie. This interval estimation is made based on the following:

- Information contained in the measurement data which have been used as a sample for estimating the mean

- Information known about the measurement system such as the accuracy of the measuring equipment and other factors that may influence the measurements
- Information about the probability distribution of various variables influencing the measurements

The estimated interval, which quantifies the "how good or how bad" part of the measurement result, is called the measurement uncertainty. The uncertainty of a measurement is therefore an interval between two values. The interval is around the estimated or assigned value of the measurement result. The true value is expected to lie between the minimum and maximum values. This amounts to saying that, although a measurement result has been estimated as a single value, called the assigned value, the measurement result may lie anywhere within the estimated interval. The wider the interval, the more uncertain we are about the assigned value; similarly, if the interval is small, the uncertainty of the measurement result as an assigned value is also small.

For example, suppose that a measurement result is assigned a value of 10.00 with an uncertainty interval of 9.95 to 10.05. The dispersion of the interval is 10.05 – 9.95, or 0.10. We can also say that the measurement result is 10.00 ±0.05, implying that although we assign a value of 10.00 to the measurand it could lie anywhere within 10.00 ±0.05. Similarly, a measurement result with an assigned value of 10.0 and an uncertainty interval of 9.9 to 10.1 implies that although we assign a value of 10.0 to the measurand it may lie anywhere within 10.0 ±0.1. In this case, the dispersion of the interval is 10.1 – 9.9, or 0.2. Since in the second case the measurement result is expected to lie within a wider interval than in the first case, its uncertainty is greater.

As the measurement uncertainty is a characteristic of a measurement result, it decides the suitability of a measurement system for a specific application, or its fitness for the purpose. Uncertainty is therefore the quality attribute of the measurement result.

The Definition of Measurement Uncertainty and True Value

Metrologists know that true values are unknown and unknowable. Still, the concept of true value was quite common and popular among metrologists until recently. The *Guide to the Expression of Uncertainty in Measurement (GUM)* published in 1995 has adopted the concept of uncertainty based on observable quantities—the measurement result and its variability—rather than on unknowable quantities such as true value and error. *GUM* has recommended not to use the term *true value* for two reasons. The first is that the true value is only an idealized concept. Second, it could be implied that it is included in the definition of a measurand as "a specific quantity subject to measurement."

Thus, the value of a measurand would be defined as "the value of a specific quantity subject to measurement." Since *specific quantity* means a definite, particular, or specified quantity, the use of the word *true* as a modifier of the "value of a measurand" is unnecessary; this aspect is included in the word *specific* in the definition of a measurand. Thus, the true value of a measurand would be simply the value of the measurand.

The general perception is that there is no point in harping on something which cannot be known—the so-called true value. Now we say that we intend to measure a measurand and in the process of measurement we attribute a numerical value to it. The measurand lies within an interval which can be estimated. All the values within the interval can be reasonably attributed to the measurand as measurement results.

The *International Vocabulary of Basic and General Terms in Metrology (VIM)* had earlier defined measurement uncertainty as "the interval around the estimated value between which the true value of the measurement is expected to lie with a certain confidence level." This definition has the primacy of true value. In view of the current thinking on true value and uncertainty of measurement as propagated by *GUM,* the definition is not centric on true value. The definition of uncertainty of measurement used in *GUM* has now been adopted by *VIM.* Measurement uncertainty has thus been defined as a "parameter associated with the result of measurement that characterizes the dispersion of values that could reasonably be attributed to the measurand." The dispersion is consistent with the observations and the knowledge of the measurement process.

RELIABILITY OF UNCERTAINTY ESTIMATION AND CONFIDENCE LEVEL

Having defined "how good or how bad" the measurement result is as an interval around the average of the sample measurement data, we still have one question to be answered: How much confidence can be placed in the estimated interval? The answer to this again lies in defining this confidence as a probability. This probability is called the confidence level. For example, a confidence level of 95% implies that there is a 95% probability that the measurand lies within the interval that has been estimated as the measurement uncertainty. The spread of the interval has different values for different confidence levels. The estimated value of measurement uncertainty, therefore, depends upon the confidence level at which it is to be estimated. It also depends upon the number of observations in the sample and other characteristics of the measurement process, such as its probability distribution.

Specifying a Measurement Result

A *measurement result* has been defined as a value attributed to a measurand obtained by measurement. This is a number which indicates the value of the parameter being measured. A measurement result is incomplete without information about the uncertainty of measurement.

A measurement result should therefore be specified as $\bar{x} \pm U$, where $\bar{x}$ is the best estimate of the measurand and is equal to the average of repeat observations. U is the uncertainty of the measurement. This is obtained by multiplying the combined standard deviation with a coverage factor whose value depends on the confidence level. The combined standard deviation is obtained by taking into account the effects of various influence factors that are contributing to the uncertainty of the measurement result. $\bar{x} \pm U$ implies that the value of the measurand lies between $\bar{x} - U$ and $\bar{x} + U$ at a specified confidence level.

The methods of evaluating the uncertainty of measurement are explained in detail in chapters 7 and 8.

THE RELATIONSHIP BETWEEN MEASUREMENT UNCERTAINTY AND MEASUREMENT ERROR

The terms *measurement error* and *measurement uncertainty* are sometimes used interchangeably due to ignorance; this should be avoided. As explained earlier, they represent different metrological aspects of a measurement result. At best, their relationship can be explained as one of cause and effect. Measurement uncertainty is the effect caused by various sources of error. The terms *error* and *accuracy* depict the qualitative aspect of a measurement result. They cannot be quantified as they are defined with reference to the true value, which is an indeterminate quantity. However, qualitatively, we say that the less the measurement error, the more accurate is the measurement result, and the greater the measurement error, the less accurate the result is.

Measurement uncertainty, on the other hand, consists of an interval around the measurement result. The value of the measurand is expected to have a certain probability of lying within this interval. Measurement uncertainly is therefore a well-quantified parameter of a measurement result. Once known systematic error is corrected for, all other sources of error, including unknown systematic effects, contribute to the variability in the measurement results. There are various sources of errors that influence the measurement process and cause uncertainty in the measurement result.

The difference between uncertainty and error is illustrated by the following example. A DC reference standard voltage source of a nominal value of

10.0000 volts is measured with a high-resolution digital voltmeter. The result is quoted as 9.9926 volts ±0.0185 volts at a 95% confidence level. This implies that the value of the reference voltage source lies between 9.9926 – 0.0185, or 9.9741 volts, and 9.9926 + 0.0185, or 10.0111 volts; however, a value of 9.9926 volts has been assigned to it, and the quantity ±0.0185 is the uncertainty of the measurement process. The calibration certificate of the voltage source states its value as 9.9993 ±0.0012 volts. Thus, 9.9993 volts can be considered as a reference value with an uncertainty of ±0.0012 volts. The error compared with the reference value can be given as equal to 9.9926 – 9.9993, or –0.0067 volts.

Following is another example involving a piece of equipment with bias. A power calibrator is used to calibrate a power meter. A reference power from the calibrator is fed to the wattmeter so as to indicate 2.50 watts on the power meter. The measurement result is quoted as 2.60 ±0.02 watts at a 95% confidence level. This implies that what the wattmeter is reading as 2.50 watts is actually lying within an interval between 2.58 and 2.62 watts. Thus, the uncertainty in measurement is ±0.02 watts. The bias in the measurement result is 2.50 – 2.60—that is, 0.1 watts. This example is illustrated in Figure 4.1.

Graphical Representation of Uncertainty and Error

Suppose that a parameter under measurement is represented by a normal distribution centered around μ, that is, the true value, and has a spread as shown in Figure 4.2. It is desired to measure this parameter. Repeat measurements are made of this parameter, and a probability distribution based on the histogram is drawn. This distribution is shown as a sample distribution. It is centered around $\bar{x}$, the average of the repeat observations, and has its own spread. $\bar{x}$ is the estimate of μ and is the measurement result. The error in this case is represented by the distance between $\bar{x}$ and μ, whereas the uncertainty is an interval spread around $\bar{x}$. If there is bias in the measuring equipment, the value of $\bar{x}$ will be far from μ. Bias becomes known through the calibration process wherein a reference value is measured and the difference between the measured value and the

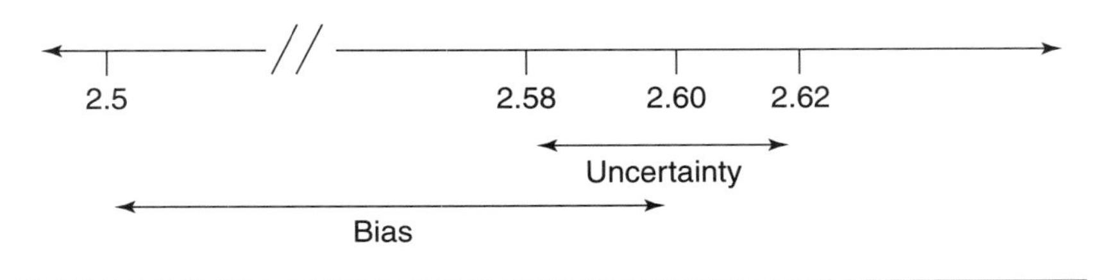

Figure 4.1 Number line representation of uncertainty and bias.

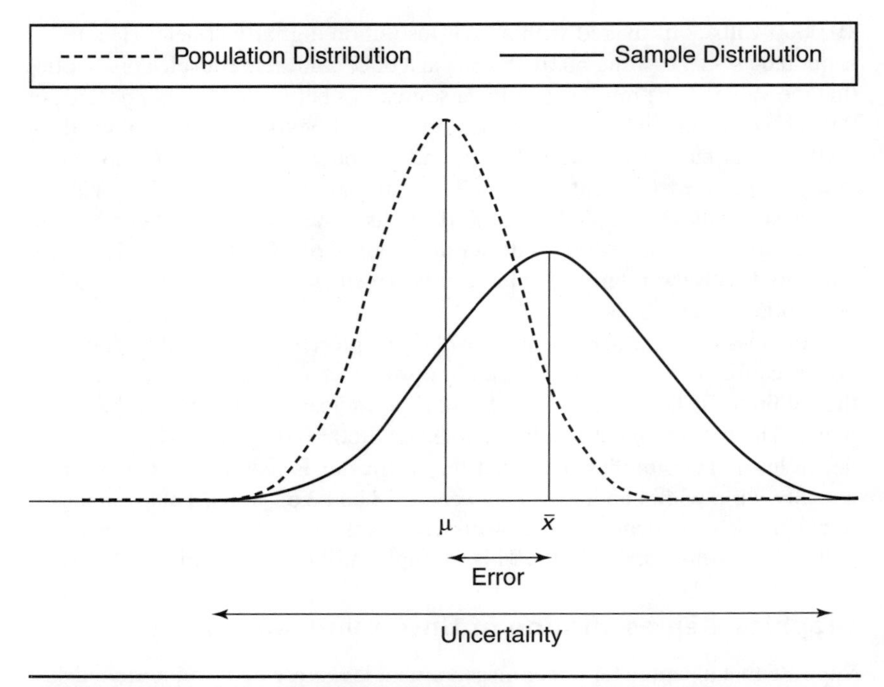

Figure 4.2 Graphical representation of error and uncertainty.

reference value is calculated. The error and the uncertainty are graphically represented in Figure 4.2.

However, error can be quantified only when the conventional true value is used instead of the true value. For example, in making measurements on a CRM, the certified value of the characteristic represented by the CRM is taken as the conventional true value. Similarly, in calibration laboratories, measurements are made and compared against reference standard values. The reference standards are assigned their values by a higher-echelon calibration laboratory, and the value so assigned can be considered the conventional true value.

ACCURACY SPECIFICATIONS OF MEASURING EQUIPMENT AND MEASUREMENT UNCERTAINTY

All testing and measuring equipment comes with information on its technical capabilities. This information is called the technical specifications of the equipment and is provided by its manufacturers. The first assessment of the suitability of measuring equipment for a specific measurement application is based on the study of its technical specifications. The specifications cover the following minimum aspects of the technical capability of the equipment:

- The measurement capability, or the range of the parameter that can be measured with the measuring equipment
- The maximum amount of uncertainty that can be contributed by the measuring equipment; categorized as measurement accuracy by many measuring equipment manufacturers
- The environmental conditions such as temperature, humidity, and electromagnetic interference under which the measuring equipment is capable of operating and the effect of those conditions on the measuring accuracy of the equipment

The accuracy specifications help in estimating the maximum likely error, or the worst-case error, when the equipment is used for measuring a specific parameter. They never give an indication of the actual error present in the measurement result. In fact, the accuracy figure indicates an estimation of the maximum uncertainty that can be attributed to the measuring equipment. In most cases, the term *accuracy* is used by the manufacturer to indicate the uncertainty in measurements contributed by the measuring equipment, which is also called systematic uncertainty. It is a common perception among metrologists that accuracy is a qualitative concept and is indeterminate. Hence, *accuracy* is a misnomer, but it is quite frequently used and accepted as an estimate of systematic uncertainty.

A set of technical specifications is applicable to a family of equipment, generally referred to by a model number or type number. A large amount of equipment of the same model or type supplied by a manufacturer to various users will have the same technical specifications. The accuracy limits as part of the specifications depend upon the capability that has been designed into the equipment and how the equipment has been manufactured. The accuracy specification is decided based on the evaluation of a number of pieces of such equipment. The accuracy figure derived from the technical specifications gives a limiting value of the uncertainty. As long as the equipment measures the value within its uncertainty limits as calculated from the specifications, it is assumed that the measuring equipment complies with its specifications and that it has no bias.

For example, suppose that a digital multimeter is made by a manufacturer as model yyy. The accuracy specifications for all model yyy equipment will be the same. There will be a large number of multimeters of model yyy, each one with a unique identity given by its serial number. Thus, a model represents a large set of equipment with the same technical specifications. If an ideal error-free voltage of 100.000 volts is read by a large number of these multimeters, the observations so obtained will have a normal distribution centered around a value of 100.000 volts. The standard deviation of these observations can be calculated. Let us say the calculated standard deviation is 0.003 volts. We know from the theory of normal distribution that 99.73% of these digital multimeters will read the error-free 100.000 volts within ±3 standard deviations, that is, within ±0.009 or ±0.01 volt. The digital multimeter can then be assigned an accuracy of 0.01 volt when used in measuring 100.000 volts. Thus, 99.73% of the multimeters will measure 100.000 volts within ±0.01 volt. This value of ±0.01 volt represents the maximum permissible error, or the systematic uncertainty. This value is also referred to as the *measurement tolerance.* It gives the worst-case error, or the limits of the error. The actual error will always be less than this value when a voltage is measured with a specific multimeter uniquely identified with a serial number of, say, xxx. The accuracy specification of a piece of measuring equipment takes into consideration the variability in the behavior of the equipment, not the variability of the parameter that is being measured. That is why we have considered an ideal error-free voltage source in this example.

Following are some examples of accuracy specifications and their interpretations:

We intend to find the maximum permissible error if 750 volts DC is measured on a $6\frac{1}{2}$-digit digital multimeter (DMM) in the 1000-voltage range.

Accuracy specification: ±(0.001% of the reading + 5 counts)

Limiting error contribution due to ±0.001% of the reading: (0.001 × 750)/100 = 0.0075 volts

On a $6\frac{1}{2}$-digit DMM, 1 digit corresponds to 0.0001% of the range.

Limiting error contribution due to ±5 counts: (5 × 0.0001 × 1000)/100 = 0.0050 volts

Maximum permissible error due to the DMM: ±0.0125 volts

This does not include the uncertainty contributed by the DC voltage source of 750 volts. If the DMM reads a voltage of 750 volts as within the above specified interval it is accepted as complying with its specifications. The accuracy limit of ±0.0125 volts is thus the uncertainty of the DMM due to its inherent limitation and is not its error. This interval can be represented on a number line. However, the probability of obtaining a reading is not the same across the entire interval. The probability of obtaining a reading within an interval near the center of the number line is greater than that of obtaining one near the end of the number line. This uncertainty component is also called systematic uncertainty. If a voltage of 750 volts with a negligible degree of uncertainty is read by the DMM beyond the interval specified above, then the DMM is considered to have a bias.

This maximum uncertainty limit is derived based on accuracy specifications. In order to evaluate the actual measurement uncertainty of a particular DMM, a number of repeat observations will have to be taken. Based on the standard deviation calculated from the data, an estimate of the uncertainty can be made. If the DMM complies with its specifications, the evaluated uncertainty will be less than the maximum uncertainty derived from the accuracy specifications of the DMM.

The accuracy specifications for analog measuring equipment are specified differently. In analog instruments, the measured value is read on a scale with the help of an indicator. The accuracy is specified as a percentage of the full scale deflection. This percentage value is called the accuracy class. An accuracy class of 1.0 indicates that the maximum permissible error that can be expected is 1% of the full scale deflection. If an analog voltmeter of accuracy class 1.0 with a full scale deflection of 300 volts is used to measure a voltage, the maximum permissible error throughout the scale will be 1% of 300 volts, or 3.0 volts. If a voltage of 30 volts is measured on the scale of 300 volts, the maximum permissible error is still 3.0 volts. Thus, the maximum permissible error will be 10% of the measured value, whereas in measuring 300 volts it will be 1% of the measured value. Therefore, analog measuring equipment is not used to measure a quantity whose value is less than half of the range of the scale.

Thus, maximum permissible error is not error in terms of its definition. It is the systematic uncertainty which is equal to the maximum permissible error and is derived from the accuracy specifications of measuring equipment.

ACCURACY SPECIFICATIONS OF MEASUREMENT STANDARDS

Measurement standards are used as the physical representation of a parameter. The accuracy specifications of standards are also applicable to a family of standards, generally referred to by a model number. Here the accuracy specifications decide a particular range, and all the standards identified with that model number will have values within this range. For example, standard resistors are available in nominal values of decimal multiples with accuracies of 0.1% and 0.01%. If a standard resistor has a nominal value of 1 ohm with an accuracy specification of ±0.01%, this implies that almost all the standard resistors identified with the same model number will have values within the interval 1.0000 ohm ±0.01%, that is, between 0.9999 ohm and 1.0001 ohms. Once a standard is procured by a laboratory, it has a unique identity given by its serial number, and it represents a resistance value within this range, say 0.99995 ohm. This is an assigned value and has its own uncertainty.

For example, suppose that a standard resistor of a nominal value of 1 ohm is used as a reference standard in a calibration laboratory. The calibration certificate of this standard resistor received from a national metrology laboratory gives the value as 1.0003 ±0.0002 ohms at a 95% confidence level. This means that the best estimate of the resistance value is 1.0003 ohms. There is an uncertainty of ±0.0002 ohm around the value of 1.0003 ohms assigned during the calibration process. This implies that there is an interval between 1.0001 (1.0003 – 0.0002) ohms and 1.0005 (1.0003 + 0.0002) ohms within which the true value of the reference standard will lie 95% of the time. In this case, 1.0003 ohms is the point estimate of the parameter, and 1.0001 to 1.0005 ohms is the interval estimate of the parameter at a 95% confidence level and represents the measurement uncertainty. Thus, 1.0003 ohms becomes the reference value and 0.0002 ohm is the measurement uncertainty.

Suppose that this standard resistance is used as a reference to calibrate a DC resistance bridge. The value measured on the bridge is 1.0004 ±0.0008 ohms at a 95% confidence level. This implies that the measured value of the reference resistance standard as obtained by the measurement process using the DC resistance bridge is 1.0004 ohms. The earlier reference value was 1.0003 ohms. The error in comparison to the reference value is then 1.0004 minus 1.0003 ohms, or 0.0001 ohm. The uncertainty of the measurement process while measuring the resistance using the DC resistance bridge is ±0.0008 ohm. This implies that as measured by the bridge the true value of the reference standard lies between 0.9996 (1.0004 – 0.0008) and 1.0012 (1.0004 + 0.0008) ohms 95% of the time.

Another example involves the identification of uncertainty as a certain percentage of the capacity of a standard. Suppose that the uncertainty specification for a proving ring used for the calibration of a ring dynamometer force measuring device has been quoted as ±0.025% of its capacity. For a proving ring of a capacity of 10,000 pounds, the uncertainty of the calibration will be 2.5 pounds. If the ring is used to calibrate a dynamometer while measuring 5,000 pounds, the uncertainty in this 5,000-pound setting will still be 2.5 pounds, as the uncertainty was quoted as a certain percentage of the ring's capacity.

SUMMARY

A measurand is a quantity that is being measured. Measurements are made using measurement systems. Measurands as well as measurement systems are under the influence of influence factors as given in the figure. There is variability in repeat measurements due to influence factors. The output of a measurement system consists of measurement results which are not error free. Statistical methods are used to derive the measurement results from the output of the measurement system. A measurement result consists of a reported value and the uncertainty associated with the reported value. Thus, a measurement is never absolute in nature. The relationship among the various factors involved in measurement is illustrated in Figure 4.3.

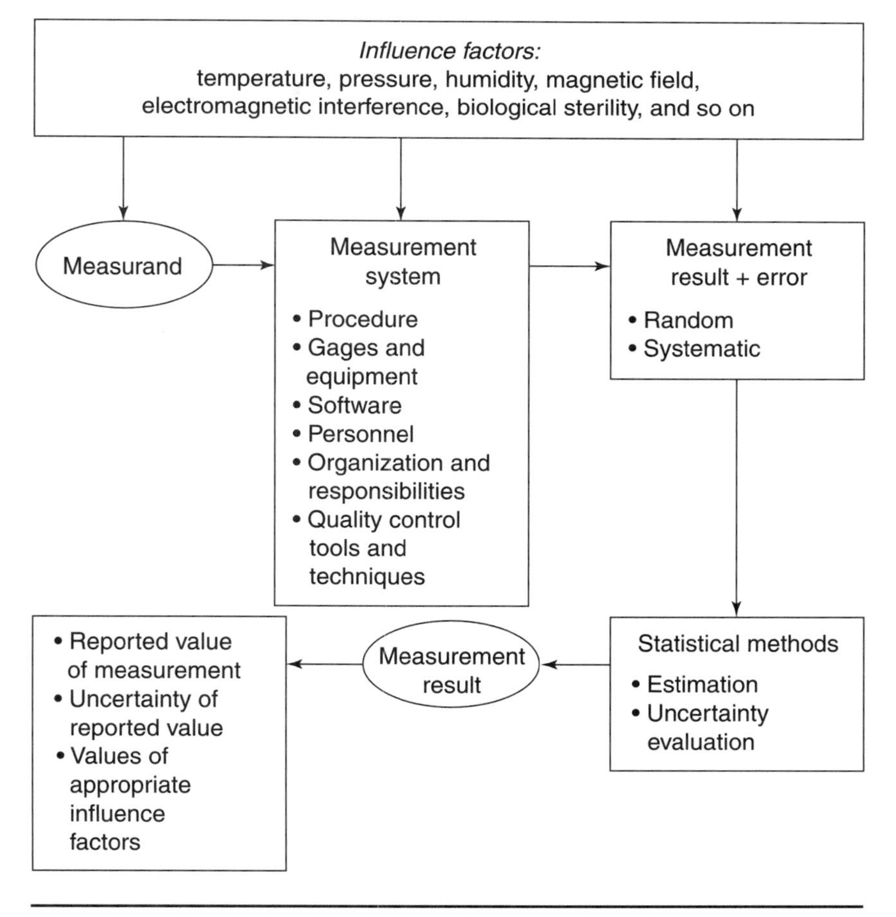

Figure 4.3 The relationship of the various metrological factors.

REFERENCES

Cooper, William David. 1978. *Electronic instrumentation and measurement techniques.* New Delhi: Prentice Hall of India.

Dorsey, N. Ernest. 1944. "The velocity of light." *Transactions of the American Philosophical Society* xxxiv, 1–110.

Eisenhart, Churchill. 1963. Realistic evaluation of the precision and accuracy of instrument calibration systems. *Journal of Research of the National Bureau of Standards: Engineering and Instrumentation* 67C, no. 2:161–85.

Guide to the expression of uncertainty in measurement. 1995. Geneva, Switzerland. International Organization for Standardization.

International Organization for Standardization (ISO). 1994. ISO 5725-1:1994-12-15, *Accuracy (trueness and precision) of measurement methods and results, part 1—General principles and definitions.* Geneva, Switzerland: ISO.

Levin, Richard I., and David S. Rubin. 1997. *Statistics for management.* New Delhi: Prentice Hall of India.

Muthu, S. K. 1986. Statistical theory of errors. *Journal of the Metrology Society of India* 1, no. 1:41–53.

National Institute of Standards and Technology. 1984. *Measurement assurance programs, part I: General introduction.* NBS Special Publication no. 676-1.Washington, DC: NIST.

Scarr, A. J. T. 1967. *Metrology and precision engineering.* New York: McGraw-Hill.

Schumacher, Rolf B. F. 1981. Systematic measurement errors. *Journal of Quality Technology* 13, no. 1:10–24.

5

National and International Measurement Systems and Measurement Standards

His majesty in his wisdom, seeing that the variety of measures was a source of inconvenience to his subjects and regarding it as subservient only to the dishonest, abolished them all and brought the medium gaz of 41 digits into general use. He named it the Ilahigaz and it is employed by the public for all purposes. For measuring the quantity of land the measure was bigha which was 60 gaz long and 60 gaz broad. 1 bigha was equal to 20 biswah, and 1 biswah was equal to 20 biswansah.

Abul Fazl-i-Allami, *Ain-i-Akbari*

It is a common experience that when the same parameter is measured in different parts of a nation the measurement results are comparable. This is true even when the same parameter is measured in different parts of the world. One kilogram of a substance measured anywhere in the world will weigh one kilogram. This unanimity on measurement results is possible because measurement systems operate within nations and also at the international level. Measurement standards are important elements of measurement systems. This chapter provides details about the roles of standards and of the various national and international metrology organizations in achieving unanimity on measurement results.

MEASUREMENT UNITS

Emperor Akbar ruled over a large part of India in the 16th century and has been considered an ideal emperor by historians. Abul Fazl-i-Allami was his courtier and court historian. He wrote a graphic account of the events and systems prevailing during Akbar's rule in his book *Ain-i-Akbari.* The epigraph at the beginning of this chapter is from this book. The narration states that various measures were used during that period for the measurement of length and area, which was causing inconvenience to the general masses and allowing dishonest people to take advantage of the situation. The emperor abolished all the existing measures and introduced a unit for the measurement of length which was called the *Ilahigaz.* This unit was used by the people throughout the empire.

The objective of measurement is to determine the value of a quantity. Useful measurements are possible only if there are defined units in which to express the measurement results. Measurement is basically a process of comparison, in which the quantity being measured is compared with another which has been selected as a unit. Thus, when we say that the length of a rod is 2 meters, it implies that the rod is twice in length in comparison with a unit of length which is one meter. Measurements have been made by human beings from time immemorial; hence, measurement units have been a historical necessity. Primitive humans used a hand span for the measurement of length. Historically, the three units of concern to human beings which have influenced their day-to-day lives have been length, mass, and time. Historical records indicate that the unit of length had an anatomical origin. Henry I of England decreed that the yard should be the distance from the tip of his own nose to the end of his own thumb when his arm was outstretched. The digit was defined as the breadth of the first finger. The Egyptian royal cubit was a unit of length decreed to be equal to the length of the forearm plus the width of the palm of the king ruling at that time. For units of mass, stones or other objects were used. Stones with weights marked on them have been found in archaeological sites dating back to 3400 B.C. Similarly, time was measured in the units of days, lunar months, and years, which were linked with the relative motion of the sun, moon, and earth.

One of the essential requirements of a meaningful measurement result is that it be understood the same way by all concerned with it. Thus, the first and foremost prerequisite for the meaningful exchange of information about measurement is that all concerned use the same definition of the unit. Persons in different parts of the world will agree that the length of a particular rod is 2 meters because the meter is well defined. In defining units, two aspects are of enormous importance and have influenced the development of units for use in metrology at various periods throughout history. These aspects are as follows:

- **Accuracy of the definition of measurement units:** Any quantity that is measured will have measurement accuracy to the extent that the unit in which it is measured has been defined. For early humans, the accuracy of this definition was less important, as their purposes, such as establishing the dimensions of a hut, would have required only crude measurements. However, in today's world of technological advancement, measurement accuracy requirements are critical, and hence the definition of the unit also has to be of that order.

- **Common understanding of measurement units:** Before the 19th century, there was hardly any trade between communities and nations. The need for exchange of goods based on measurements was mainly confined within communities, and the need for a harmonized system of units was not felt. For early humans, the size of the community would have been even smaller. Thus, a common understanding of measurement units was not difficult to achieve. Today, high-speed communication and transportation have ushered in the saying that the world is a global village. It is now all the more important that measurement units are understood the same way by all concerned around the globe.

Measurable quantities are interrelated and can always be expressed in terms of fundamental quantities. The fundamental units have been defined as those units which do not depend upon other units for their expression, whereas all other units are derived from the fundamental units. In the 18th century, French scientists proposed the metric system of units, which helped in the evolution of the centimeter, gram, and second (CGS) system and other subsequent systems of units. Mass, length, and time were considered fundamental quantities in the early French system.

The Harmonized International System of Units

We know that there was the foot, pound, and second (FPS) system of units generally prevalent in Britain. The other well-known system of units was the CGS system, which later was changed to the meter, kilogram, and second (MKS) system. In the electrotechnical area, there were the electrostatic unit (esu) and the electromagnetic unit (emu) of measurement for defining electrical quantities. The multiplicity of unit systems created a lot of confusion in technical communication and international trade. A measurement result can be understood the same way by all concerned only if the units in which it is expressed are standardized on a global basis. This was a major challenge for metrologists all over the world beginning in the 19th century.

The use of different systems of units affects trade between countries. The General Conference on Weights and Measures (CGPM) in 1954 agreed to form a uniform and comprehensive system of units based on the metric system then in use. In 1960, the CGPM established the International System of Units (SI) for global use. This system originally had six base units, but in 1971 the unit of chemical substances, the mole, was added to it. The SI is the foundation of modern metrology and is practically the sole legitimate system of measurement units in use throughout the world.

A *unit of measurement* has been defined as a particular quantity defined and adopted by convention, with which other quantities of the same kind are compared in order to express their magnitude relative to that quantity. The International System of Units has been defined as the coherent system of units adopted and recommended by the General Conference on Weights and Measures.

SI units consist of seven base units and two supplementary units. The base units are regarded as dimensionally independent, and all the other units are derived from the base units or from other derived units. The definitions of the base and supplementary SI units are given in Table 5.1.

Table 5.1 SI base and supplementary units.

Unit	Definition
Length	The SI unit of length is the meter (m). It is defined as the length of the path traveled by light in a vacuum during a time interval of 1/299,792, 458 of a second. It is no longer defined as a material artifact.
Mass	The SI unit of mass is the kilogram (kg). It is equal to the mass of an artifact cylinder of platinum iridium kept at the International Bureau of Weights and Measures.
Time	The SI unit of time is the second (s). It is defined as the duration of 9,192,631,770 cycles of radiation corresponding to the transition between two hyperfine levels of the ground state of cesium 133 atoms. It is realized using a laboratory cesium frequency standard.
Electric current	The SI unit of electric current is the ampere (A). It is defined as the electric current producing a force of 2×10^{-7} newtons per meter of length between two long wires one meter apart in free space.
Thermodynamic temperature	The SI unit of temperature is the kelvin (K). It is defined as 1/273.16 of the thermodynamic temperature of the triple point of water.
Luminous intensity	The SI unit of luminous intensity is the candela (cd). It is defined as the luminous intensity in a given direction of a source that emits monochromatic radiation at a frequency of 540×10^{12} hertz with a radiant intensity in that direction of 1/683 watts per steradian.
Amount of a substance	The SI unit of measure for the amount of a substance is the mole (mol). It is defined as the amount of a substance that contains as many elementary items as there are atoms in 0.012 kg of carbon 12.
Plane angle	The SI unit for the plane angle is the radian (rad). It is defined as a plane angle with its vertex at the center of a circle that is subtended by an arc equal in length to the radius.
Solid angle	The SI unit for the solid angle is the steradian (Sr). It is defined as a solid angle with its vertex at the center of a sphere that is subtended by an area of the spherical circle equal to that of a square with sides equal in length to the radius.

Table 5.2 SI derived units.

Parameter	Unit (symbol)	Value
Frequency	Hertz (Hz)	1/s
Force	Newton (N)	kg·m/s^2
Pressure	Pascal (Pa)	N/m^2
Energy	Joule (J)	N·m
Power	Watt (W)	J/s
Electric potential	Volt (V)	W/A
Electric resistance	Ohm (Ω)	V/A
Electric charge	Coulomb (C)	A·s
Electric capacitance	Farad (F)	C/V
Conductance	Siemens (S)	A/V
Magnetic flux	Weber (Wb)	V·S
Magnetic flux density	Tesla (T)	Wb/m^2
Inductance	Henry (H)	Wb/A
Celsius temperature	Degree (°C)	-
Luminous flux	Lumen (lm)	cd·Sr
Illuminance	Lux (lx)	lm/m^2
Activity	Becquerel (Bq)	-
Absorbed dose	Gray (Gy)	J/kg
Dose equivalent	Sievert (Sv)	J/kg

SI Derived Units

In addition to the base and supplementary units, the SI has 19 derived units which are obtained by forming various combinations of the base units, the supplementary units, and other derived units. The derived units and their symbols are given in Table 5.2.

The global acceptability of SI units has helped in the understanding and interpretation of measurement results in the same way by the various users of the measurement results even when they are in different parts of the world.

MEASUREMENT STANDARDS

SI units have been accepted internationally and are the basis of modern measurement. They are used for most legal, scientific, and technical purposes in the various parts of the world. Just defining them is not enough; they have to be made an integral part of a measurement system. The evolution of practical national and international measurement systems is achieved in four stages, as follows:

- **Definition of the unit:** The accuracy of the definition of a unit is important, as it will be reflected in the accuracy of the measurements that can be achieved.
- **Realization of the unit:** The definition of the unit has to be realized so that it can be used as a reference for measurement. This task is carried out by national metrology laboratories. The units are realized in the form of experimental setups.
- **Representation of the unit:** The realized experimental setup of the system of the unit is the physical representation of the unit. National metrology laboratories are responsible for the maintenance of this representation. They also ensure that these experimental setups continue to represent the SI and are available for reference.
- **Dissemination of the unit:** The end users of measurements are trade, industry, and laboratories. They generally do not have access to the representations of the SI units held by national metrology laboratories. The end users also need the values of the SI units for reference. This is done through the process of dissemination, wherein the units are made available to the end users of measurement results.

For example, the meter was originally defined as one 10 millionth of the distance from the North Pole to the equator measured along the meridian running through Paris. From the survey of this quarter of this great circle, the length of the meter was established and marked on a selected bar made of an alloy of platinum and iridium. This standard meter bar was preserved at the International Bureau of Weights and Measures. The bar was the representation of the meter which had been realized based on the definition. Later, more accurate geographic surveys showed that the realized meter was not exactly one 10 millionth of the length of the quarter meridian running through Paris. A new definition was therefore given to the meter, which was the distance at 0°C between two lines engraved on a bar of platinum and iridium known as the international meter and preserved at the International Bureau of Weights and Measures.

A harmonized measurement unit ensures that everybody concerned with a measurement result understands it the same way. For acceptability, it is also essential that the measurement results for the same parameter measured at different places and by different people be in agreement. This implies that measurement results should be correlated. To achieve this agreement in measurement results, it is essential that everybody draw their measurement units from a common acceptable standard. Thus, a *standard* is a physical object or a characteristic of a physical apparatus that represents the conceptual unit chosen to represent a particular measurable attribute. Like the need for measurement, the need for measurement standards has been felt by human beings from time immemorial. It is believed that the foot ruler as a measurement standard might have been started with the declaration by some king that the standard of length should be equal to the length of his foot. There is historical evidence of artifacts' being used as standards of measurement. It is believed that around 3000 B.C. Egyptians used the royal cubit as a standard of length. The royal cubit was considered a master standard and was made of granite to endure for all time. Craftspeople used cubits made of wood for measurement purposes while engaged in the construction of the pyramids. All the craftspeople were required to bring their cubits for comparison with the royal cubit on the day of each full moon. This might be an early example of the concepts of calibration and measurement traceability. Through this process of measurement, the Egyptians achieved a measurement accuracy of 0.05% in measuring the dimensions of the great pyramids.

The development of physical standards and their units has been a long-drawn process. At various times throughout history rulers have realized the need for legal metrology. Today, the whole task of evolving a measurement system is quite complex. It starts with the development and maintenance of primary physical standards, which is a specialized area of metrology. Here the tasks are the realization and representation of the units at the highest level of accuracy and reproducibility. The primary consideration is the best possible metrological performance rather than portability, size, ease of mass production, cost, time, or other such factors. Developing new standards and obtaining agreement on their use may take years. The task of developing and maintaining primary physical standards is the responsibility of the national metrology laboratory in a country. Measurement standards are used at various levels of the hierarchy of a national measurement system.

A *measurement standard* has been defined as a material measure, measuring instrument, reference material, or measuring system intended to define, realize, conserve, or reproduce a unit of one or more values of a quantity to serve as a reference. Definitions of the various categories of standards are given in Table 5.3.

Standard Reference Materials

The standards for physical parameters are used as references for measurement purposes. However, the analytical measurements covered by chemical metrology also need reference standards so that analytical data can be compared and analytical results can be understood the same way by all concerned. Examples of these are the standards for the composition of particular chemicals, pure solutions, particular grades of iron, pollution levels, and so on. These standards

Table 5.3 Definitions of the various types of standards.

Type of standard	Definition
International standard	A standard recognized by international agreement to serve internationally as the basis for fixing the value of all other standards of the quantity concerned. For example, the international prototype of the kilogram maintained at the International Bureau of Weights and Measures is an international standard of mass. Similarly, Josephson-junction-based voltage standard is maintained as an international standard of voltage.
National standard	A standard recognized by an official national decision to serve in a country as the basis for fixing the value of all other standards of the quantity concerned. Generally, the national standard in a country is also a primary standard to which other standards are traceable. For example, national prototypes of the kilogram which are identical to the international prototype of the kilogram are maintained as national standards of mass in various national metrology laboratories.
Primary standard	A standard that is designated or widely acknowledged as having the highest metrological quality and whose value is accepted without reference to other standards of the same quantity. National standards are generally primary standards. However, sometimes a standard of better metrological quality than that of the national standard is developed. It may take a long-drawn scientific and legal process before the new standard is finally accepted as an international or national standard. For example, the metrological quality of the Josephson-junction-based voltage standard is far superior to that of the standard cell. However, it could take quite some time to replace the standard cell as the national standard of voltage. . Until then it remains the primary standard.
Secondary standard	A standard whose value is based upon comparison with some primary standard. Note that a secondary standard, once its value is established, can become a primary standard for some other user. For example, the national standard of length consists of a stabilized laser source. High-accuracy gage blocks are used as a secondary standard of length. These standards are assigned values based on their comparison with national standards.
Reference standard	A standard having the highest metrological quality available at a given location from which the measurements made at that location are derived. For example, state legal metrology laboratories maintain NIST-calibrated kilogram standards. These serve as reference standards for them.

(continued)

Table 5.3 *continued*

Type of standard	Definition
Working standard	A measurement standard not specifically reserved as a reference standard which is intended to verify measuring equipment of lower accuracy. For example, multifunction calibrators are used as working standards for the calibration of testing and measuring equipment to be used in the measurement of various electrical parameters.
Transfer standard	A standard that is the same as a reference standard except that it is used to transfer a measurement parameter from one organization to another for traceability purposes. For example, standard cells are used as transfer standards for the transfer of voltage parameters from the national standard to other standards.

are materials or substances that are officially recognized as having certain attributes. The materials with known characteristics that are used for the purpose of chemical analysis are called standard reference materials. They are also called certified reference materials.

A *reference material* has been defined as a material or substance one or more of whose property values are sufficiently homogeneous and well established to be used for the calibration of an apparatus, the assessment of a measurement method, or the assigning of values to materials. *Certified reference material* has been defined as reference material accompanied by a certificate, one or more of whose property values are traceable to aprocedure which establishes traceability to an accurate realization of the unit in which property values are expressed and for which each certified value is accompanied by an uncertainty at a stated level of confidence.

MEASUREMENT COMPATIBILITY AND THE INTERNATIONAL MEASUREMENT SYSTEM

Measurement results have wide applications in various facets of human endeavor in trade, industry, and scientific laboratories. In international trade, the results of measurements should be acceptable to trading partners in different parts of the world. The acceptability of a measurement result is achieved only if all concerned have faith in the measurement system that produces the measurement result. If a quantity is measured at different times and places and the measurement results are in agreement, it is said that the measurement results are compatible. It is not enough for measurement results to be compatible within a country; they should also have global compatibility. This means that when the same quantities are measured in any part of the world and at any time the measurement results should be in agreement. There are certain conditions that need to be met to ensure the global acceptability of measurement results:

- **The harmonization of measurement units:** The units in which a measurement result is expressed have to be harmonized at the international level. This ensures common understanding about the measurement result. For example, in weather forecasting, the daily temperatures of various cities around the world are broadcast in terms of degrees Celsius. There is no confusion in the minds of the persons using this information as degrees Celsius is a harmonized unit of temperature. This has been achieved with the adoption of the SI.
- **The harmonization of measurement standards:** Measurement standards are used as physical representations of measurement units. These standards should be realized as artifacts as the physical representations of the various units of the harmonized system of units. The harmonized standards should be used as references in laboratories for measurement purposes and for propagation of the units to their users. To ensure global trade, it is not enough to keep measurement standards in one place for reference purposes. The standards should be established in the laboratories of various countries. These standards should be harmonized among themselves by intercomparison of the values assigned to the standards by national metrology laboratories in various countries.

These are necessary conditions for the mutual acceptability of measurement results to facilitate international trade. These conditions should be supplemented by a national metrological infrastructure called a national measurement system. The integration of national measurement systems with harmonized units and harmonized standards can be called an international measurement system.

National Measurement Systems

The establishment and maintenance of national standards of measurement and the dissemination of the values of these standards to their actual users in trade, industry, and scientific laboratories in a country form part of a metrology system called a national measurement system.

The requirement for measurement is increasing not only in terms of the quantity of measurements but also in terms of the quality of measurements. The end user not only needs accurate measurements but also wants to have full confidence in the accuracy of the results. Confidence in the accuracy of measurement results is ensured through national measurement systems. A national measurement system is a technical and organizational infrastructure in a country that ensures a consistent and internationally recognized basis for measurement. It has two main objectives:

- To enable individuals and organizations to make measurements to the desired accuracy and to demonstrate the validity of the measurement results
- To coordinate with the measurement systems of other countries to ensure the universal comparability of measurement results

In order to establish a national measurement system within a country, the following requirements should be met:

- **Acceptability of SI units:** The nation should accept SI units as its legal system of units.
- **Establishment of a national metrology laboratory:** A national metrology laboratory is an institution which is legally responsible for establishing and maintaining the national standards of measurement of a country. It is also called a national metrology institution. National metrology laboratories also play a major role in the dissemination of units to the actual end users through these national standards.
- **Dissemination of national standards:** The values represented by national standards need to be disseminated to the end users of measurements in trade, industry, and scientific laboratories. A mechanism should be established through which these units can be disseminated to their users through the national metrology laboratory and other intermediate calibration laboratories.

THE TREATY OF THE METER AND INTERNATIONAL METROLOGY ORGANIZATIONS

The need for international compatibility of measurement results arose in the beginning of the 19th century, when trade among different communities living in different parts of the world began. Before this, trading activities between different countries were of such a low level that issues related to international measurement units and measurement standards did not merit serious thinking. With the start of international trade, the necessity for harmonization of measurement units and measurement standards on a global scale was felt. The first success in these efforts was achieved in 1875 when the Treaty of the Meter was signed. The Treaty of the Meter is an international treaty of which most of the developed and developing countries are signatories. The treaty is concerned with establishing, maintaining, and disseminating the units and the standards of various physical parameters among the countries which are signatories.

The International Bureau of Weights and Measures

Under the Treaty of the Meter, the signatory states agreed to set up a permanent scientific organization called the International Bureau of Weights and Measures (BIPM). The BIPM has established its laboratory near Paris, France. It is a global center of excellence for all technical matters relating to scientific metrology. Its main objectives are as follows:

- To establish and maintain basic standards for physical parameters as international prototypes.
- To ensure international compatibility of measurement results. In order to do this, the BIPM must ensure that the national standards for various physical parameters maintained by national metrology laboratories are in agreement with one another. This task is accomplished through the intercomparison of the various national standards. This intercomparison activity is coordinated by the BIPM and is carried out under its aegis.
- To evolve techniques for the intercomparison of national standards.
- To determine the values of the fundamental physical constants. These constants play a large role in the development of standards for physical parameters. Their values need to be determined with absolute accuracy.

Basically, the Treaty of the Meter is a diplomatic treaty. For the execution of the various tasks required to achieve the objectives of the treaty, there are two other international organizations actively engaged in international metrology activities.

The General Conference on Weights and Measures

The General Conference on Weights and Measures (CGPM) comprises delegates from the member states of the Treaty of the Meter. This is the supreme policy-making and decision-making body under the treaty. It also controls the functioning of the BIPM. The CGPM has the following missions:

- To discuss and initiate measures necessary for the propagation and development of the International System of Units

- To ratify the results of new fundamental metrological determinations and to adopt various scientific resolutions of international importance
- To make important decisions concerning the organization and development of the BIPM

The CGPM meets every four years to make decisions on all important matters.

The International Committee for Weights and Measures

The International Committee for Weights and Measures (CIPM) is appointed by the CGPM and is responsible for planning and executing the decisions of the CGPM. The members of the CIPM are eminent scientists and metrologists of different nationalities. These persons participate in the CIPM not as official representatives of their countries but purely in a personal capacity, because of their acknowledged expertise in metrology. The CIPM works through various consultative committees. The members of the consultative committees are drawn from various national metrology laboratories around the world. These persons are known for their research and contributions to their particular fields of metrology. At present there are nine consultative committees operating under the CIPM in different disciplines of metrology, as follows:

- Electrical parameters, including radio frequency and microwave standards
- Photometric and radiometric measurements
- Temperature, including the International Temperature Scale
- Definition and realization of meter and angle measurements
- Definition and realization of the second.
- Dosimetry and radioactivity measurements
- International prototypes of the kilogram and definition of the unit of mass based on fundamental constants
- Amount of substance and international comparison to establish traceability in chemical measurements
- Development of the International System of Units

These consultative committees advise the CIPM on matters of a scientific nature, including the workings of the BIPM and coordination with national metrology laboratories. The consultative committees sometimes appoint working groups to deliberate upon specific issues.

For a measurement result to be globally acceptable, it is also necessary that the national measurement system of a country be similarly linked to international metrology organizations such as the CIPM and the BIPM. National metrology laboratories participate in the international comparison of their

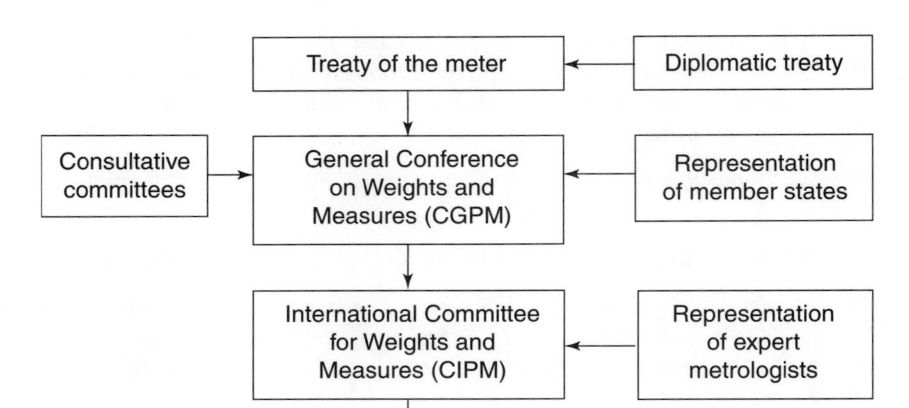

Figure 5.1 International metrology organizations.

national standards against the national standards of other member countries. This type of traceability can be called international traceability.

The concept of national and international measurement systems has been accepted globally. Nations have given their national metrology laboratories the responsibility of realizing and maintaining national standards of measurement, which are the representations of the SI units for particular parameters. These laboratories also take part in the process of dissemination of these units to the actual users. This task is carried out by the National Institute of Standards and Technology (NIST) in the United States, the National Physical Laboratory (NPL) in the United Kingdom, Physikalisch-Technische Bundesanstalt (PTB) in Germany, and the NPL in India. The traceability of measurements is achieved through the use of standards in the calibration process.

The concept of the national measurement system is working in most of the developed and some of the developing countries. However, not all countries have the capability to maintain primary physical standards. Some countries do not even have national metrology laboratories. As all nations have some share in international trade, these countries generally establish standards for trade purposes and set up legal weights and measures. For international traceability, however, they depend upon other nations in which national measurement systems are in operation.

MEASUREMENT STANDARDS BASED ON ATOMIC AND QUANTUM PHENOMENA

The first success in formulating international and national standards was the signing of the Treaty of the Meter in 1875. The first international prototype of the meter was made using a platinum iridium alloy bar with defining lines engraved on it. The distance between these lines was declared to be one meter. Similar alloy bars were maintained in various national measurement institutions for intercomparison purposes. These were the physical artifacts. Similarly, the international prototype of the kilogram, made of a platinum iridium rod, was by definition taken as the unit of mass; this rod is still considered to be the international kilogram artifact. Experience showed that in spite of best efforts these standards were not stable.

Due to the instability of these artifacts, it was decided that the standards should be made time and place invariant. To achieve this, it was considered necessary that the standards of physical quantities should be based on physical phenomena whose value does not change with place and time.

Before the advent of the atomic clock, the unit of time was derived from astronomical observations. It was derived from the time the earth took to rotate around its axis. The second was considered to be equal to 1/86,400 of the mean solar day. However, it was subsequently realized that the earth's rotation around its axis is not constant. A slight wobbling of the earth on its axis was also observed. The difference due to these phenomena was small, on the order of 1 in 10^8, but this instability was considered adequate enough not to interfere with the definition of the unit of time. Later on, the second was defined based on the time of year rather than the mean solar day. Subsequently, though, the need for more accurate time measurement in various areas of defense and space applications was felt by scientists.

Studies conducted at the NPL in the United Kingdom showed that the imperial yard made of bronze was shrinking at the rate of 1 micron per year. Even the international prototype of the kilogram is changing: in the last hundred years, an increase of about 50 micrograms in its mass has been observed.

These phenomena compelled scientists to think differently about the realization of measurement standards, as the artifacts were not stable with respect to time and place. Their accurate realization and duplication in various places was not an easy task.

Research in atomic and particle physics led to the discovery of many atomic and quantum phenomena. This research also led to the determination of physical constants of nature with absolute accuracy. Some of these physical constants are the charge of an electron, the rest mass of an electron, Planck's constant, and the speed of light. It was later realized that nature itself provides the fundamental units and that measurement units could be related to them. Successful attempts were made to realize the units for physical parameters using the physical constants of nature and other natural phenomena. This led to the development of measurement standards based on atomic and quantum phenomena. These standards based on natural phenomena are time and place invariant and have very high stability everywhere on the globe and for all time to come. Because of the importance of the absolute nature of physical constants, study of the values of these constants is one of the charters assigned to the BIPM.

Sustained and intensive studies have been carried out by scientists to define more and more units in terms of atomic or quantum phenomena. Today, all the base units of the SI have been defined in terms of atomic and quantum phenomena except the 1-kilogram artifact of mass at the BIPM. Efforts are being made to find an atomic-phenomena-based SI unit of mass, and one consultative committee at the CIPM is exclusively devoted to this task.

Before these measurement standards are realized, the base SI units need to be defined in terms of natural phenomena. The following SI units have been linked to parameters of nature:

- **Meter:** The CGPM in 1960 defined the meter in terms of the wavelength of light emitted by krypton 86 atoms, thus delinking the unit of meter from the international prototype of the meter bar, which had been in existence since 1889. One meter was redefined as equal to 1,650,763.73 wavelengths (in a vacuum) of radiation corresponding to the transition of orange light emitted by krypton atoms of mass number 86. To ensure the reproducibility of the standard of length, certain conditions for the use of the discharge lamp containing krypton 86, such as the operating temperature, current density, and direction of observation, were recommended. Through this definition, the meter was linked to the natural parameter of the wavelength of krypton 86.

  Subsequently, in 1983, after the velocity of light had become known with very high accuracy and had been agreed to by the CGPM in 1979, the meter was redefined as the length of the path traveled by light in a vacuum during the time interval of 1/299,792,458 of a second. By this definition, the meter as an SI unit of length was linked to a natural physical constant, the velocity of light. This redefinition of the meter has improved the accuracy of the realization of the meter by one order of magnitude.

- **Second:** At the time of the adoption of the SI, the second was defined based on the tropical year 1900. According to its definition, the second was 1/315,569,259,747 of the tropical year 1900. The earlier definition had been based on the mean solar day, but due to inconsistency in the earth's rotation, the base was shifted from the solar day to the year, which is the time taken by the earth to revolve around the sun.

The need for more accurate measurement of time was felt by scientists in many areas of space research and in the development of navigational aids used by ships and aircraft. Research in the area of atomic radiation resulted in the development of the atomic clock. In 1967, the CGPM decided to replace the definition of the second derived from the tropical year 1900 to one derived from atomic radiation. The second was defined as the duration of 9,192,631,770 periods of the radiation corresponding to the transition between two hyperfine levels of the ground state of cesium 133 atoms. The standards based on the atomic transition phenomenon had the highest order of accuracy. Time and frequency are interrelated parameters, as frequency is the number of cyclic variations in a unit of time. The atomic standards of frequency and time could be realized to an accuracy on the order of 1 in 10^{-11} to 10^{-12}. The new definitions relating to atomic and quantum phenomena helped improve the accuracy of the realization of these standards considerably.

One should not expect the most recent definitions to hold forever, as new scientific and technological progress may result in still better standards.

SOME IMPORTANT FUNDAMENTAL CONSTANTS

A number of fundamental physical constants have been used in deriving quantum-phenomena-based physical standards. Some important fundamental constants used in the realization of metrological standards and their values are presented in Table 5.4.

THE LIMITATIONS OF ARTIFACT STANDARDS AND THE ADVANTAGES OF INTRINSIC STANDARDS

Artifact-type measurement standards are unique. Examples of artifact-type standards are a kilogram artifact used in a calibration laboratory as a standard of mass and a 1-ohm wire-wound standard resistance maintained by an electrical calibration laboratory. These standards need to be calibrated periodically by a higher-echelon laboratory to ensure the traceability of measurements. At times, they have to be transported to other calibration laboratories for calibration.

The advantage of quantum-based measurement standards is that they need not be unique. Quantum-phenomena-related units and standards behave the same way everywhere and every time. These standards can thus be reproduced with adequate confidence about the physical characteristics they represent at any place and at any time. They do not need calibration at periodic intervals in the traditional sense, as is required for artifact-type physical standards in order to ensure measurement traceability. Quantum-phenomena-based standards are

Table 5.4 Important fundamental constants.

Constant	Remarks
Speed of light c	This constant enters into the definition of the meter. Its value is equal to 299,792,458 meters per second.
Permeability of free space μ_o	This constant enters into the equation giving the magnetic force between conductors used in the definition of the ampere. Its value is $\pi \times 10^{-7}$ $H{\cdot}m^{-1}$.
Permittivity of a vacuum $\in_o$	The value of $\in_o$ is exactly equal to that of $(\mu_o c^2)^{-1}$. According to an electrostatic theorem, the capacitance per unit length between opposing cylinders of a symmetrical four-cylinder configuration is given by $\in_o(\ln 2)/\pi$. The resulting capacitance can be used as a standard of impedance and is ultimately connected to the definition of the ohm. Its value is $8.854187818 \times 10^{-12}$ $F{\cdot}m^{-1}$.
Electronic charge e	The value of *e* is used to determine the Klitzing constant R_k and the Josephson constant Kj. Its value is 1.6021892×10^{-9} coulombs.
Planck's constant h	The value of h is used to determine the Klitzing constant Rk and the Josephson constant Kj. Its value is 6.626176×10^{-34} $J{\cdot}Hz^{-1}$.
Josephson constant Kj	This constant is theoretically equal to twice the ratio of the electronic charge to Planck's constant (2e/h). It is the basis of the realization of the volt by means of the Josephson effect. Its value is 483597.9 GHz/v.
Klitzing constant R_k	This constant is theoretically equal to the ratio of Planck's constant to the square of the electronic charge. It is the basis of the realization of the ohm by means of the quantum Hall effect. Its value is equal to 25,812.807 ohms.

actual realizations of SI units, and the traceability of measurements is a built-in property of such standards. They ensure the reproducibility of physical quantities at any place and at any time. The only condition is that the process of reproducibility has to be completed in accordance with a well-defined method. Such standards are called intrinsic standards.

Intrinsic Standards

Intrinsic standards are realized based on standard procedures. It is assumed that following the procedure correctly will generate a standard which will produce a quantity which has a low uncertainty of realization—an uncertainty within accepted limits.

The National Conference of Standards Laboratories (NCSL), a U.S. metrology organization, has defined the intrinsic standard as follows: "a standard recognized as having or realizing, under its prescribed conditions of use and intended application, as an assigned value, the basis of which is an inherent physical constant or an inherent and sufficiently stable physical property."

Another U.S. standard, ANSI/NCSL Z540-1-1994, states that intrinsic standards are "based on well-characterized laws of physics, fundamental constants of nature or invariant properties of materials and make ideal stable, precise and accurate measurement standards if properly designed, characterized, operated and maintained." Though an intrinsic standard does not need periodic calibration in the traditional sense of the term, in order to ensure that the standard has been properly realized, operated, and maintained the above standard also states that the "laboratory should demonstrate by measurement assurance techniques, inter-laboratory comparisons or other suitable means that its intrinsic standards in measurement results are correlated with those of national or international standards."

The following are examples of the intrinsic standards used by metrology organizations and calibration laboratories globally:

- Josephson-junction-based voltage standards
- Quantum-Hall-effect-based resistance standards
- Cesium atomic standards for time and frequency
- The International Temperature Scale of 1990 (ITS-90)

THE REALIZATION OF NATIONAL STANDARDS FOR SI BASE UNITS

National measurement institutions establish and maintain SI units and disseminate them within the country. The realization techniques used for the base units by the various national measurement institutions are similar. The details of some of these are given in Table 5.5.

Measurement standards are realized and maintained at various levels of metrology organizations, from the BIPM and national metrology institutions to the different levels of calibration laboratories. Even certain end users of measuring equipment maintain metrological standards. In fact, the international and national metrology systems are operational only because of the intercomparability of metrology standards. In view of the important role played by artifacts and intrinsic standards in obtaining valid measurements, much scientific effort is being put into improving the existing technology for the realization and maintenance of standards and evolving new technologies in this area.

Table 5.5 Realization techniques of SI base units.

Base unit	Realization technique
Length (meter)	Realized through a laser source as recommended by the CIPM
Mass (kilogram)	Realized through a national prototype of the kilogram
Time (second)	Realized through a cesium atomic clock
Electric current (ampere)	Realized through units of voltage (volts) and resistance (ohms)
Temperature (kelvin)	Realized through the triple point of a water cell and the ITS-90 with a number of fixed points at thermal equilibrium
Luminous intensity (candela)	Realized through a group of incandescent lamps and a calibrated radiometer

REFERENCES

Bentley, John P. 1983. *Principles of measurement systems.* New York: Longman.

British Standard Institute (BSI). 1983. *BSI Handbook 22.* Quality Assurance, BSI.

Cooper, William David. 1978. *Electronic instrumentation and measurement techniques.* New Delhi: Prentice Hall of India.

Cronin, L. B. 1999. Measurement accreditation of electrical calibration laboratories. *Cal Lab: The International Journal of Metrology* 5 (July–August), no. 4:20–44.

Fluke Corporation. 1994. *Calibration philosophy in practice.* 2nd ed. Everett, WA: Fluke Corporation.

Gopal E. S. R. 1999. Reference standards. In *Proceedings of the National Conference on Test Engineering and Metrology 1999,* 1–3. New Delhi: STQC Department of Electronics.

Gray, Dwight E., and John W. Coutts. 1958. *Man and his physical world.* 3rd ed.: D. Van Nostrand.

National Physical Laboratory (NPL) of New Delhi. *SI units.* New Delhi: NPL.

Nemeroff, Ed. 1996. NIST notes from the U.S.-Egypt bilateral workshop on metrology, standards and conformity assessment. *Cal Lab: The International Journal of Metrology* 3 (September–October), no. 5:16–17.

NRC Institute for National Measurement Standards. 2000. *Canada brochure on the International System of Units (SI).* Ottawa Canada: NRC.

Sydenham, P. H. 1982. *Handbook of measurement science.* Vol. 1, *Standardization of measurement fundamentals and practices,* 49–94. New York: John Wiley and Sons.

Verma, Ajit Ram. 1987. *National measurements and standards.* New Delhi: National Physical Laboratory.

Wallard, A. J., and T. J. Quinn. 1999. Intrinsic standards: Are they really what they claim? *Cal Lab: The International Journal of Metrology* 5 (November–December), no. 6:28–30.

Witt, Thomas J. 1995. The BIPM and the SI unit. In *Proceedings of workshop on the role of metrology in quality management and quality improvement,* 12–20. New Delhi: Commonwealth-India Metrology Centre NPL.

———. 1995. Organization of the BIPM. In *Proceedings of workshop on the role of metrology in quality management and quality improvement,* 37–75. New Delhi: Commonwealth-India Metrology Centre NPL.

6

Calibration and Measurement Traceability

Thou shall have a perfect and just weight, a perfect and just measure shall thou have: that thy days may be lengthened in the land which thy Lord thy God giveth thee.

Moses

A measurement result is considered valid only if its linkage to national measurement standards can be established. This linkage is called the traceability of a measurement result. Traceability is achieved through the calibration of measuring equipment. This chapter explains the importance of traceability in measurement and how it is achieved through calibration.

WHY CALIBRATION?

Moses, the gifted Hebrew leader who lived around 1300 B.C., is revered in the Judaic tradition as the greatest prophet and teacher. His influence continues to be felt in the religious life, moral concerns, and social ethics of Western civilization. The perfection and fairness of weights and measures in a society were a historic necessity, as is evident from the epigraph at the beginning of this chapter. In today's world, this has become all the more important in view of measurement requirements in various fields. The fairness of weights and measures is ensured through the process of traceable calibration.

The users of measuring equipment sometimes have blind faith in the capability of their equipment. They think that once equipment has been procured it will measure faithfully until it breaks down. This is a wrong conception, and at times this is learned at a large cost. The author has personal knowledge of some incidents in which this has been the case.

One manufacturer of electrical energy meters supplied a consignment to its client, which was an electric supply company. The manufacturer and the client were located 500 kilometers apart. Delivering the large consignment of energy meters entailed a lot of effort and cost. On receipt of the consignment, the client picked a few energy meters and subjected them to measuring accuracy testing in its own laboratory. The items did not meet the accuracy specifications and hence were rejected. The whole consignment was returned to the manufacturer. The meters were tested by the manufacturer in its laboratory and were found to comply with the specifications. In order to resolve this dispute, the client and the supplier availed themselves of the services of a third-party accredited laboratory. During testing in this laboratory, it was revealed that the results of the manufacturer were correct and that the client had unnecessarily rejected the meters and sent back the whole consignment. The client was unaware of bias in its measurement system and had blind faith in its measurement capability. The unnecessary cost and unpleasantness could have been avoided if the client had had a proper calibration system in place.

Another case pertained to the testing of a high-voltage electric cable. The manufacturer of the cable had obtained a quality mark for the cable from a national standards body. The government agencies who were its bulk buyers purchased industrial products with quality marking only, so the mark helped the manufacturer to obtain bulk supply orders. The manufacturer operated on a small scale but generally had orders on hand because of the strength of the mark. The quality mark scheme envisaged that a sample of the cable would be subjected to testing for all parameters according to the requirements of the country's specifications once a year in an independent laboratory. As part of the quality certification scheme, a sample of the cable was submitted to a laboratory for testing. The cable was found to comply with its specifications except in the parameter of resistance per unit length. The specifications stipulated a value of not more than a certain number of milliohms per meter. Such low-level measurements are prone to error.

The laboratory measured the parameter a number of times on a longer length of the cable using different equipment and methods and concluded that the cable did not comply with this part of the specifications. A report to that effect was submitted by the laboratory to the national standards body. Based on this report, the quality mark was withdrawn, and the ultimate consequence was the cancellation of a bulk order by a government agency. The cable had already been manufactured, and the manufacturer was awaiting clearance from the quality marking body. There was a lot of turmoil as the small company had to suffer financial losses. The quality manager was fired from his job. The reason for the failure of the cable in that particular test was investigated. It was found that the company had purchased a new Kelvin bridge which was used for measuring the resistance per unit length. At the time of purchase of the bridge, its calibration was not verified, and subsequently it was found that the bridge was out of calibration. This case shows that companies should take calibration seriously and that not doing so may cost them a lot.

There are two reasons for the occurrence of such incidents. One is the misconceived faith in the infallibility of measuring instruments, and the other is total ignorance about the factors contributing to the reliability of measurement results. The author once had an interaction with a scientist working on the design and development of sophisticated electronic products. The scientist was of the opinion that periodic calibration is necessary for industrial measuring equipment but not for equipment used in design and development laboratories, because the latter, unlike industrial equipment, is used in a controlled environment, by highly qualified persons, and only once in a while. This is a very simplistic view based on concepts of infallibility. Confidence in measurement results comes through calibration only; however, the level of calibration and its periodicity will differ depending upon the task at hand.

The author has also had an experience with a person who wanted his equipment to be calibrated instantly. The person was the owner of a small manufacturing business but had no technical background. The company was to supply instrument transformers, which were to be inspected in the factory before delivery according to the contract. The inspector arrived at the factory but refused to undertake the inspection as the equipment was not calibrated. The owner was taken aback, thinking this was an additional requirement of the contract. Once he was convinced about the necessity of calibration, he came to a calibration laboratory with the unusual request that the equipment be calibrated instantly; he didn't realize that this is a technical process and takes some time to complete. This is a classic case of ignorance about calibration.

The ISO 9000 quality system standards stipulate specific requirements for the calibration of measuring equipment. Since the inception of these standards, general awareness of the necessity of calibrating measuring equipment has increased considerably, particularly among small manufacturers.

CALIBRATION AND TRACEABILITY

Testing and measuring equipment should not only measure the desired parameter correctly but should also give evidence that it has done so. The verification of the capability of a measuring instrument to measure correctly and to generate evidence to that effect is done through the process of calibration. The designed accuracy capability of measuring equipment is reflected as its measurement accuracy in its technical specifications. Measurements are made in the International System of Units (SI) as represented in national standards. A measurement result which cannot be traced back to a national standard and ultimately to the SI cannot be relied upon. The relationship of industrial measurements and analytical measurements to the SI is shown in Figure 6.1.

Thus, calibration ensures the ability of measuring equipment to provide an accurate indication of the quantity being measured. It is defined as a set of operations that establish under specified conditions the relationship between

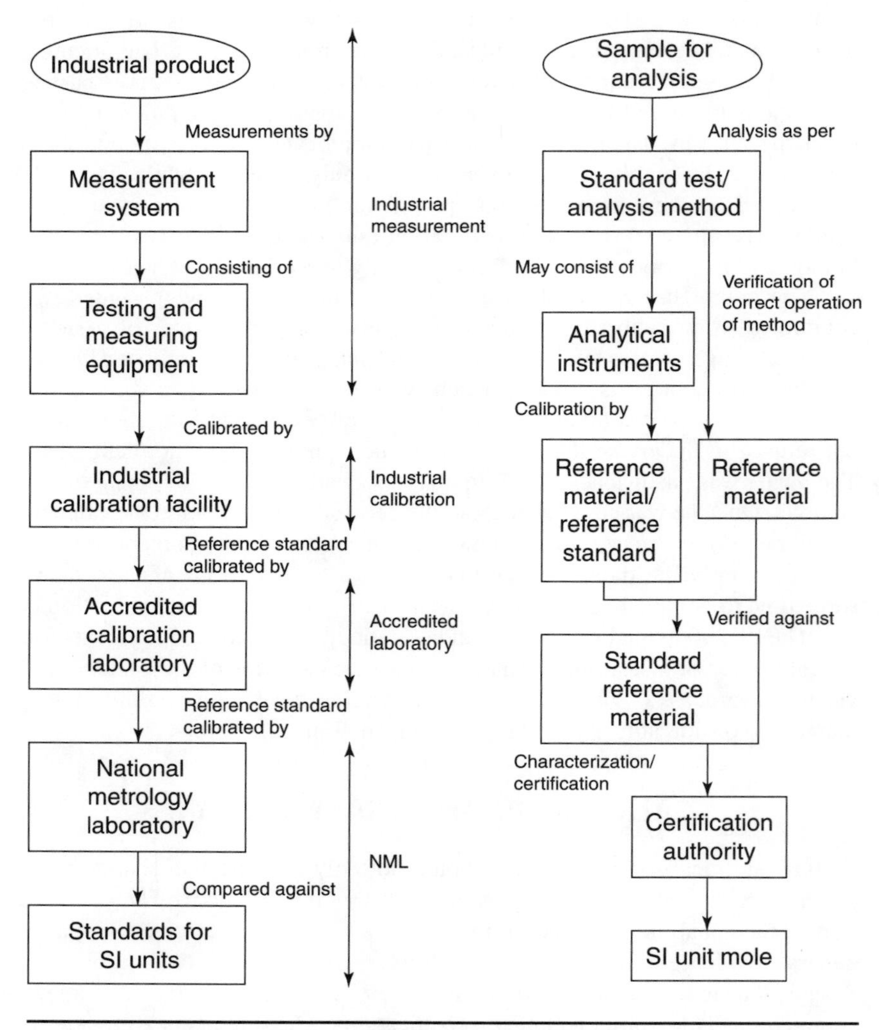

Figure 6.1 Linking measurements by end users to SI units.

values of quantities indicated by a measuring instrument or measuring system or between the value represented by a material measure or a reference material and the corresponding value realized by a standard.

Traceability is an essential requirement for a calibration process. It is defined as the property of the result of a measurement or the value of a standard whereby it can be related to stated references, usually national or international standards, through an unbroken chain of comparisons having stated uncertainties. A measurement result that meets this requirement is called a traceable measurement, and the unbroken chain of comparisons is called a traceability chain. A calibration is said to be traceable if it can be traced back along a defined line of calibration to the national standards used in the SI system of units. The calibrations are recorded in a calibration certificate or a report. Thus, a certificate of traceable calibration is a feature of the accounting aspects of a piece of equipment and serves as evidence of traceability. The validity of the values obtained by a piece of measuring equipment is not automatic but exists because a value has been assigned to the equipment by the calibration process. Measurements made with uncalibrated equipment have no means of verification of their correctness if the equipment is malfunctioning or not working. In such a situation, all measurements made by the equipment are suspect. Sydenham has emphasized the importance of traceability in his remark that ensuring the traceability of an instrument is akin to taking out insurance before a disaster occurs.

Calibration is the process by which national standards of measurement are disseminated to end users in trade, industry, and scientific laboratories. Each piece of measuring equipment used in trade, industry, or scientific laboratories is thus linked to national standards through an unbroken chain by the calibration process. As the national standards are the representations of the SI, the unbroken traceability chain joins the two entities: at one end is the SI and at the other end is the end user's measurement result. Thus, traceability can be conceived as a system for transferring SI units from the point of definition to the end user.

The calibration of measuring equipment is a necessary condition for obtaining reliable measurements. It is, however, not a sufficient condition. Traceable measuring equipment is an important element of a measurement system, but it has to be supplemented by competent persons to make the measurements, validated measurement methods, a controlled environment, and so on.

Achieving Traceability

The first step toward establishing traceability is the experimental realization and representation of SI units as national standards of measurement. This task is performed by national metrology laboratories (NMLs) as a statutory requirement. The transfer of these units to end users is accomplished through the following:

- An established calibration program that forms part of the national measurement system
- The use of intrinsic standards realized based on standard procedures
- The use of standard reference materials

TRACEABILITY THROUGH AN ESTABLISHED CALIBRATION PROGRAM

It is neither possible nor desirable to link each piece of measuring equipment in a country directly to the national standards. Persons needing to measure mass cannot borrow the primary standard held by an NML. They must use lower-level standards that can be checked against national standards. Everyday measuring devices such as scales and balances can be calibrated against working-level standards to verify their accuracy. The working-level standards are in turn calibrated against higher-level mass standards. These reference standards are ultimately calibrated against national standards.

Thus, a system of hierarchical calibration laboratories is established within a country to achieve traceability. These levels of the hierarchy are called the various echelons of the traceability chain. The NML is at the apex and is called echelon I. There are intermediate laboratories at echelon II and echelon III. At the bottom are the tests and measuring equipment used by the end users in trade, industry, and so on.

During the calibration process, testing and measuring instruments are compared with reference equipment of higher accuracy to ensure that the instruments are measuring within their claimed accuracy. The reference equipment used for calibration is calibrated using a still better reference standard in a higher-echelon laboratory. At each stage, the reference is checked for its claimed accuracy through the process of calibration. This process is repeated at each step until the process reaches the national standard for reference purposes. The transfer of a unit from an NML to another calibration laboratory at echelon II is called direct traceability to national standards. Subsequent transfer of the unit from the calibration laboratory at echelon II to intermediate laboratories produces a traceability chain, which is called indirect traceability. The units are transferred at each stage of the hierarchy chain using transfer standards. A device used as a transfer standard should have adequate metrological qualities and should have the appropriate stability, accuracy, range, and resolution for the intended purpose. The transfer standard is compared at the apex level with the national standards held at the NML. The transfer standard is then transported to the calibration laboratory at echelon II. The reference standard held by the calibration laboratory at echelon II is compared against the transfer standard which has been calibrated at the NML. Similarly, another transfer standard is used to calibrate the standard held at an echelon III laboratory. The transfer standard is physically transported from higher-echelon to lower-echelon laboratories, and the process of transferring the unit continues until the end user's measuring instrument is reached. The hierarchy of units and standards used for achieving traceability is shown in Figure 6.2.

Each calibration laboratory in the traceability chain has primary and secondary reference standards as well as working standards. Each has its own calibration program within the laboratory whereby the working standards held by the laboratory are calibrated against its reference standards. The reference standards are calibrated by a higher-echelon laboratory and are generally not used for day-to-day calibration purposes. Day-to-day calibration services are performed using the working standards.

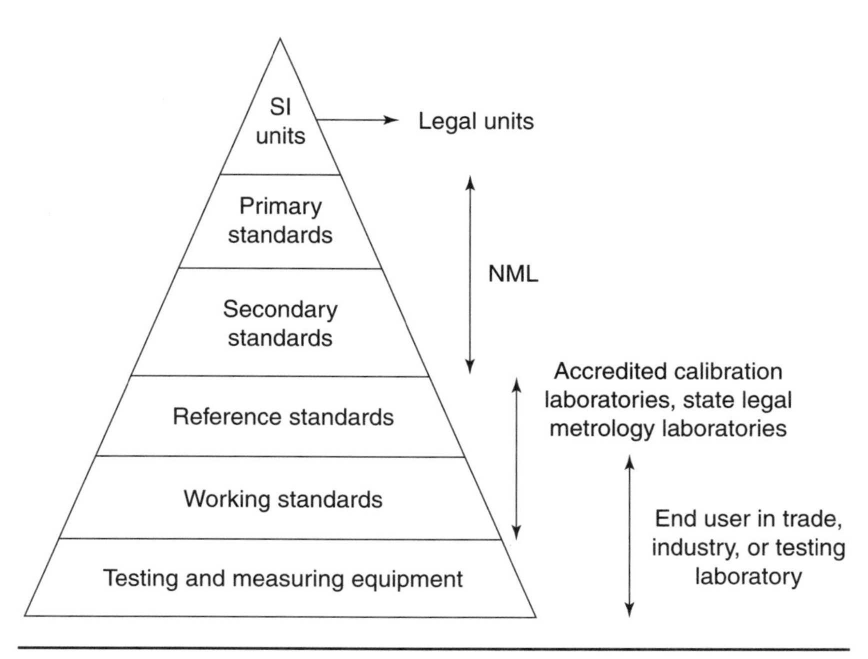

Figure 6.2 Hierarchy of units, standards, and measuring equipment.

It is important to note the following characteristics of an established calibration program:

- The system does not permit breaking of the chain. It has to start at the national standards and terminate at the end user's measuring equipment. Thus, an unbroken chain of traceability is an essential requirement.
- The uncertainties of the standards of SI units held at NMLs are extremely low. As we go down the chain, some errors creep into the system during the transfer process at each stage of transfer. Thus, uncertainty goes on increasing as we go down the traceability chain. It is at its maximum at the end user level and it decreases as we move up from echelon III to echelon II and ultimately to echelon I and national standards.
- Through this system, not every user is required to seek traceability directly from the NML. It is neither desirable nor practicable to do so. Every user has different accuracy requirements for its measurement applications. Depending upon these accuracy requirements, the user can avail the services of a calibration laboratory at any echelon level.

NATIONAL AND INTERNATIONAL TRACEABILITY

The traceability of measurements can thus be achieved through the establishment of hierarchical calibration laboratories within a country as part of its national measurement system. This ensures that the end user measurements at the bottom of the chain are linked to the national standards at the apex of the chain. This concept can be called vertical traceability or national traceability of measurement.

To facilitate international trade, the national measurement systems of various countries should also be in agreement among themselves. The national standards held by the NMLs of various countries should be compatible with the international standards maintained at the International Bureau of Weights and Measures (BIPM) and should also be traceable to each other. This task is coordinated at the international level by the BIPM with the participation of various NMLs as per the Treaty of the Meter. This concept of the traceability of national standards held by NMLs against each other may be called horizontal traceability or international traceability. The national and international traceability chain is shown in Figure 6.3.

SOME EXAMPLES OF NATIONAL TRACEABILITY

Traceability in Mass Metrology

The unit of mass is extensively used by human beings in trade, industry, and laboratories. The unit is disseminated through the use of primary, secondary,

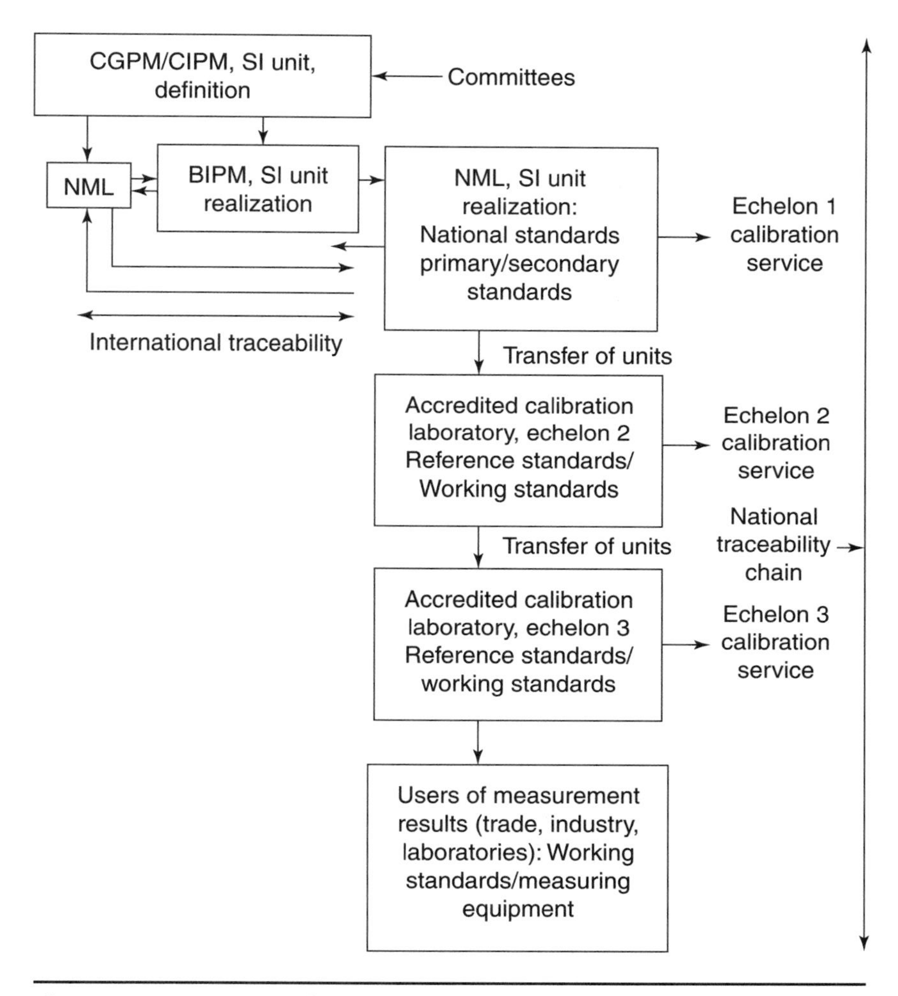

Figure 6.3 International and national measurement systems: Hierarchy of standards and traceability chain.

and working standard weights and of precision balances. In a federal setup, the national standard is maintained by national measurement institutions, and the states maintain their own primary standards in their metrology laboratories. International organizations for legal metrology have stipulated accuracy and other technical requirements for various categories of weights and weighing machines as guidelines.

In the United States, the national standard of mass is maintained by the mass group of the National Institute of Standards and Technology (NIST). The

NIST has established an office called the NIST Office of Weights and Measures (OWM) with the objective of promoting uniformity in U.S. weights and measurement laws, regulations, and standards. To help accomplish this mission, the OWM established the National Conference on Weights and Measurers (NCWM). The NCWM is a professional organization of state and local weights and measures officials that includes representatives from business, industry, consumer groups, and federal agencies. The members of the NCWM develop uniform laws, regulations, and practices which are published by the NIST. The OWM, through its state standards program for recognition of state weights and measures laboratories, provides a basis for ensuring the traceability of state weights and measures standards to the NIST.

State legal metrology laboratories maintain primary standards of mass. The unit is disseminated from the NIST to state legal metrology laboratories and to local weights and measures laboratories. The accuracy classifications of these standard weights are based on NIST handbook 105-1 or document OIML R-11. The International Organization of Legal Metrology (OIML) standard stipulates the requirement for the maximum permissible error in the weights of various nominal values and accuracies. These weights are classified as classes E1, E2, F1, F2, and M. E1-class weights are used as primary reference standards by state metrology laboratories, whereas E2-class weights are used as high-precision standards for the calibration of weights and special precision analytical balances. For weights of classes E1 and E2, periodic calibration is essential. Class-F1 weights are used for the calibration of high-accuracy balances, whereas F2-class test weights are used as working standards for the calibration of balances used for precision and analytical work. Periodic calibration of F1 and F2 weights is recommended and preferred. Weights of accuracy class M are used for industrial scales and platform scales. The OIML document also categorizes balances into accuracy classes I, II, III, and IV, implying special, high, medium, and ordinary classes of accuracy.

The periodicity of calibration requirements in the U.S. mass metrology system is as follows:

- **Legal requirements:** Each state establishes legal requirements for the periodic verification of class F test weights used for commercial applications. In most cases, this is a fixed interval of one to two years.
- **Industry/scientific requirement:** There is no fixed calibration interval for industrial and scientific applications. The interval is decided based upon application, tolerances required, historical data of weight artifacts, and so on.
- **Primary standards:** There is no fixed interval for the recalibration of primary standards in state legal metrology laboratories. The period is decided based on past verification reports and the availability of measurement assurance programs.

Each state legal metrology laboratory is required to maintain NIST-calibrated control standards to periodically verify measurement control. The traceability of the mass unit is achieved through dissemination of the standard from the NIST to state legal metrology laboratories to local weights and measures laboratories to the actual users of the unit in trade, industry, and laboratories through an unbroken chain. The traceability chain is shown in Figure 6.4.

Traceability in Dimensional Metrology

Measurements of different dimensions of various products are required to be made in trade, industry, and laboratories. In the engineering industry, dimensional measurements must be carried out very frequently. Various types of

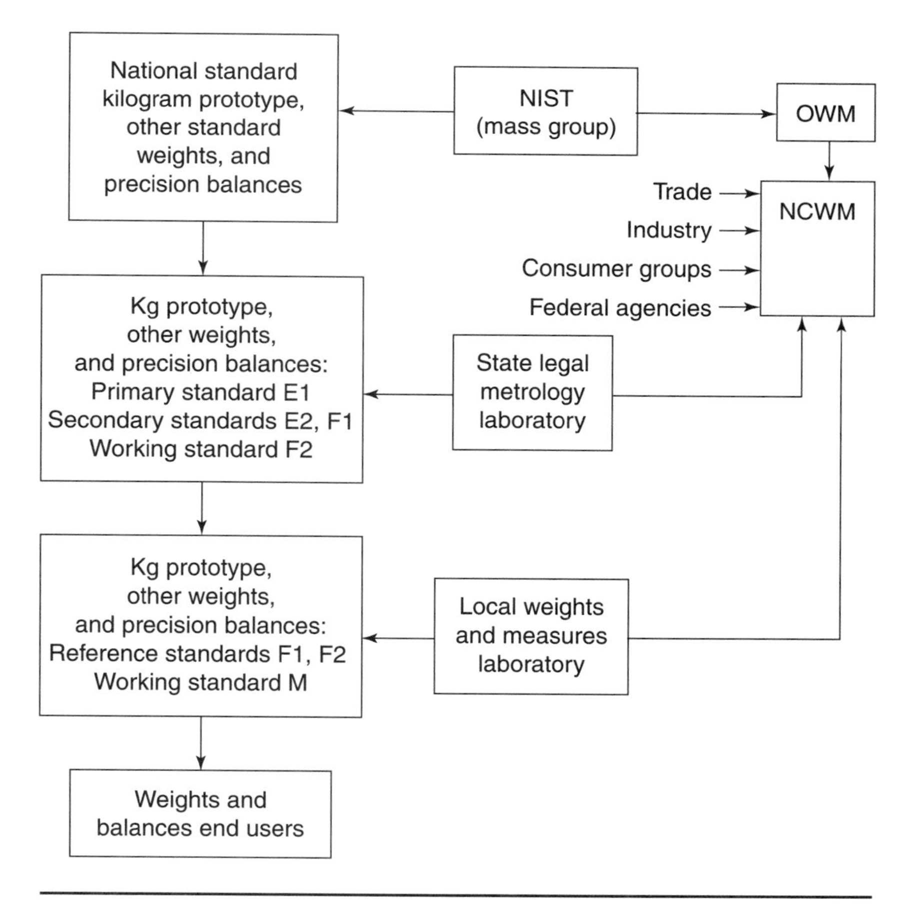

Figure 6.4 Traceability in mass metrology in the United States.

instruments and devices, such as vernier calipers, dial gages, micrometers, height gages, scales, and measuring tapes, are used to measure dimensions. NMLs also maintain stabilized laser and monochromatic wavelength sources as quantum-phenomena-based national standards.

Reference standard gage blocks are calibrated using a stabilized laser source as a reference. The calibration is done with an interferometer, using interferometric calibration techniques. A gage block calibrated by this method can further be used as a reference for calibrating gage blocks of lower accuracy. The lengths of identical gage blocks are compared using electronic comparators. The gage blocks are thus calibrated using a reference gage block of better accuracy. The dial gages, vernier calipers, micrometers, height gages and other scales and tapes are calibrated using the calibrated gage blocks as reference standards. These calibrated devices are used for the measurement of product dimensions. Through this traceability chain, the unit of length is disseminated to the actual user. The measurement results are thus linked to the national standard of length. Coordinate measuring machines are used as transfer standards in dimensional measurements, positional geometrical deviations, and contour measurements. Angle standards are also held by laboratories for the calibration of angle gages.

Traceability in Temperature Metrology

Reliable temperature measurements are needed by process and metallurgical industries that use furnaces. Scientific laboratories also need reliable temperature measurements. NMLs use the International Temperature Scale of 1990 (ITS-90) for realizing temperature standards. The ITS-90 is based on 17 thermal equilibrium states known as fixed points that cover the temperature range from 0.65 K to the highest temperature practicably measurable. The ITS-90 also stipulates requirements for standard formulas for interpolation, which are different for different temperature ranges. The fixed points are generally the vapor pressure points, triple points, gas thermometer points, melting points, or freezing points of various substances. Since the realization of these fixed points is not an easy task and not all 17 fixed points may be needed, the NMLs usually do not realize all the fixed points.

Standard platinum resistance thermometers (SPRTs) are used as reference standards for the calibration of platinum resistance thermometers used by industry. Standard thermocouples are also used as reference and working standards in calibration laboratories for the calibration of thermocouples used in industry. Liquid-in-glass-type thermometers are also used as working standards within a certain range of temperatures.

In calibration laboratories, precisely controlled temperature baths are used as temperature sources, and SPRTs, standard thermocouples, or standard thermometers are used as references for calibration purposes. Thus, traceability of the temperature measurements of actual users is achieved.

Traceability in Electrical Metrology

The ampere is the SI base unit of electric current. In addition, a large number of derived SI units are electrical in nature. Some examples are electric energy and power, electrical potential, electrical resistance, capacitance, inductance, conductance, frequency, magnetic flux, and magnetic flux density. Many of these units are interrelated. These standards cover the DC units and AC units of the SI at various frequencies.

The traceability chain for some of the SI units used in electrical metrology is as follows.

The DC Volt

The unit of the volt is defined in terms of the ampere base unit and is defined as the difference in electric potential between two points of a conducting wire carrying a constant current of 1 ampere when the power dissipated between these points is equal to 1 watt.

The realization of the primary standard of the DC volt is achieved through the Josephson-effect-based intrinsic standard. The Josephson effect provides a means of realizing very stable DC voltage in terms of the Josephson constant K_j and a precisely known frequency. This junction is kept at liquid helium temperature. The junction is irradiated by a microwave source with a frequency of 10 gigahertz. A number of Josephson junction arrays are used to obtain a very highly accurate DC voltage source.

Prior to the invention of the Josephson-junction-based intrinsic standard, saturated standard cadmium cells were used as the primary standard of the DC volt. In some laboratories, they are still used as reference standards. Zener-based solid-state DC voltage standards are also used as primary and transfer standards. These standards use temperature-compensated zener diodes and do not need critically controlled temperature conditions. They are also mechanically rugged and are easier to use and calibrate than standard cadmium cells.

DC voltage calibrators are used as working standards for voltage. Solid-state zener-diode-based devices are used as traveling standards to transfer the accuracy of primary standards to working standards.

These primary, secondary, and working standards, along with DC microvoltmeters and high-accuracy digital multimeters, help in achieving the traceability of DC volts.

DC Resistance

Since 1990, the unit of DC resistance, the ohm, has been related to the quantum Hall effect by means of the Klitzing constant. However, many laboratories still use the standard resistance artifacts. These standards are made from resistance elements enclosed in hermetically sealed double-walled metal containers.

These standards have high temperature coefficients of resistance and are maintained in temperature-controlled oil baths. Depending upon their designed accuracy, these artifact standard resistors are used as primary, secondary, or working standards. Using various types of bridges, the ohm is disseminated to the actual users of measurement results.

AC Standards Metrology

The standards for AC voltage and current are obtained through AC/DC thermal transfer standards. The focus of AC/DC metrology is the use of electrothermal devices to measure the average power produced in pure resistance by an AC voltage in order to assign it a value that is equal to that of a DC voltage which produces the same power in the resistance. This establishes an AC equivalent of DC voltage, which is measured using DC metrology techniques.

A set of multijunction thermal converters are maintained as the primary standard of AC voltage and current in the frequency range of 10 to 100 kilohertz. The absolute value of AC/DC transfer error is assigned to each converter.

AC calibrators are used as secondary and working standards of AC voltage. These AC sources generate AC voltage which can be precisely selected over wide ranges of frequencies and provide various values of voltages and currents. AC power and energy standards are also based on the principle of AC/DC transfer techniques.

Traceability in Time and Frequency Metrology

The traceability of most physical parameters is achieved through calibration at periodic intervals and involves the shipping of standards and devices to calibration laboratories. However, time and frequency metrology offers many convenient ways to establish continuous real-time traceability to NMLs.

The international standard for time and frequency metrology is the Coordinated Universal Time (UTC) scale maintained by the BIPM. The BIPM also maintains a time scale known as International Atomic Time (TAI) based on data from about 45 NMLs. The UTC scale is identical to TAI except that leap seconds are added to UTC to keep it in line with the astronomical time scale. The UTC and frequency signals are broadcast over radio, telephone, or network paths. These signals are received at the users' sites and are used for the calibration of reference standards, working standards, and measuring instruments.

The traceability chains for mass, length, and temperature are given in Table 6.1, and those for DC voltage, resistance, and time and frequency are given in Table 6.2.

Table 6.1 Equipment and standards for mass, length, and temperature: Traceability chains.

Level of metrology organization	Physical parameter		
	Mass	**Length**	**Temperature**
BIPM	International prototype of the kilogram	Stabilized laser source/krypton 86 source of light	ITS-90
NML	National prototype of the kilogram, other prototypes of primary and secondary standards	Stabilized laser source/krypton 86 source of light, gage blocks of high accuracy and length comparators as primary and secondary standards	ITS-90 SPRTs
Accredited laboratory (direct traceability)	Reference/working standard weights and balances	Gage blocks of grade 0, length comparators	SPRTs, PRTs, standard thermocouples, standard thermometers
Accredited laboratory (indirect traceability)	Reference/working standard weights and balances	Gage blocks of grades 0 and 00, length comparators	SPRTs, PRTs, standard thermocouples, standard thermometers
User of measurement results (trade, industry, or laboratory)	Standard weights and balances	Gage blocks, calipers, dial gages, scales, etc.	Thermometers, thermocouples, etc.

Table 6.2 Equipment and standards for DC voltage, resistance, and time and frequency: Traceability chains.

Level of metrology organization	Physical parameter		
	DC Voltage	**DC resistance**	**Time and Frequency**
BIPM	Josephson-array-based voltage source	Quantum-Hall-effect-based standard resistance	UTC and SI second, cesium beam frequency standards
NML	Josephson-array-based voltage source, standard cell/Zener-based DC reference standards	Quantum-Hall-effect-based standard resistance, standard resistors of high stability and accuracy	UTC and SI second, cesium beam frequency standards, broadcast of time and frequency
Accredited laboratory (direct traceability)	Standard cell/Zener-based DC reference standards, DC calibrators	Standard resistors, bridges, resistance meters	Reception of standard time and frequency signals from NML, calibration of standards, rubidium frequency standards
Accredited laboratory (indirect traceability)	DC calibrators	Standard resistors, bridges, resistance meters	Reception of standard time and frequency signals from NML, calibration of standards, rubidium frequency standards
User of measurement results (trade, industry, or laboratory)	Digital multimeters and DC voltmeters	Bridges, resistance meters	Frequency and time interval meters

ACCREDITATION OF CALIBRATION LABORATORIES

Accredited calibration laboratories play an important role in establishing the traceability of measurement results. These laboratories are part of national measurement systems. *Accreditation* is a formal recognition that a body or person is competent to carry out specific tasks. In the context of a laboratory, accreditation is a formal recognition that the laboratory has demonstrated its capability, competence, and credibility to carry out the task it claims to be able to perform. It represents formal recognition of the technical competence of a testing or calibration laboratory. Laboratory accreditation bodies accord formal recognition. These accreditation bodies periodically monitor the laboratories to ensure the validity of the tests or calibrations they carry out. The laboratory accreditation process helps in the conformity assessment of products and thus reduces technical barriers to international trade, which is one of the objectives of the World Trade Organization. Laboratory accreditation bodies operate in almost all the developed and developing countries. In some countries, there are more than one such body. In the United States, the American Association for Laboratory Accreditation (A2LA) and the National Voluntary Association of Laboratory Accreditation Programs (NAVLAP) are the laboratory accreditation bodies among other types of accreditation bodies. In the United Kingdom, the United Kingdom Accreditation Service (UKAS) performs this task. In Germany, the German Calibration Service (DKD) is the accreditation body for calibration laboratories. In India, accreditation services for testing and calibration laboratories are provided by the National Accreditation Board for Testing and Calibration Laboratories (NABL).

Before a testing or calibration laboratory is granted accreditation, it is evaluated by a group of experts. The evaluation is carried out for the effectiveness of the quality system operating in the laboratory and the technical competence of its operations. The evaluation process includes careful considerations of the following factors:

- The technical competence of the laboratory staff, supervisors, and managers
- The quality practices followed
- The technical capabilities of the equipment and facilities
- The environmental conditions in and around the laboratory
- The measurement and test procedures used
- The results of interlaboratory comparison and proficiency testing
- The traceability of measurement results
- The independence, impartiality, and integrity of the laboratory

The technical competence requirements for testing and calibration laboratories were originally stipulated in the international standard ISO/IEC Guide 25:1990, which has now been replaced by ISO/IEC 17025:1999. Calibration laboratories have to comply with the above standards in addition to specific additional technical requirements stipulated by laboratory accreditation bodies.

After having been satisfactorily evaluated, a testing or calibration laboratory is issued a certificate of accreditation by the laboratory accreditation body that defines the scope of the capabilities for which it is accredited. The accreditation status of a calibration laboratory implies that it has demonstrated its quality and technical competence in specific areas of calibration. The calibration certificate issued by such a laboratory is accepted not only in its own country but in other countries as well. Since traceability of measurements is an essential requirement for accreditation, the calibration certificate of an accredited calibration laboratory ensures traceability and acceptable limits of uncertainty of reference standards.

The International Laboratory Accreditation Cooperation (ILAC) is an informal apex organization at the international level which aims to achieve international acceptance of laboratory tests and calibration results. It also aims to provide general information on the subject of laboratory accreditation and to facilitate its international development.

THE PROPAGATION OF UNCERTAINTY IN THE TRACEABILITY CHAIN

The first and foremost task in establishing an international or national measurement system is to agree on the definitions of the SI units. Once the definitions are agreed upon, they are to be realized. At the apex level, a standard is the realization of the definition of a unit. For example, the definition of the meter as a unit of length was earlier defined as the length equivalent to a number of wavelengths of monochromatic light emitted by krypton 86. According to this definition, the length standard was to be realized through an experimental setup which consisted of a monochromatic source of light and other experimental setups. How a standard is to be realized is specified in the recommendations of the International Committee for Weights and Measures (CIPM). The procedure for realization stipulates the details of all the steps to be taken and the environmental conditions to be maintained during the realization process. Though at the apex level the task is performed by expert scientists, it is accepted that realization of the definition is not perfect. The deviation may be due to the inability to control environmental conditions very accurately or some other known limitation. This translation from the definition to the realization of a metrological standard introduces some uncertainty into the realized value compared with the defined value. Thus, the definition is perfect and has no uncertainty, but the realization process introduces some uncertainty in the realized value. This uncertainty is so insignificant that a layperson is generally not even aware of the phenomenon. For the metrologist working at the apex level, though, it is important not only to realize the existence of such uncertainty but also to estimate and quantify the value of the uncertainty in the realization of physical standards.

Consider, for example, the realization of the definition of the ampere. The SI base unit of the ampere is linked with force and the amount of current passing between two parallel wires which are 1 meter apart. Making an experimental setup to realize the unit of the ampere requires two coils spaced exactly 1 meter apart with a provision for measuring the force through the use of a mass balance. The realization of such an experimental setup is not an easy task, and even once it is accomplished there are a number of factors, such as uncertainty in measuring the spacing of the coils and the force, which contribute to the uncertainty of the realized standard. This is the reason that the ampere is realized through the units of volts and ohms.

The national standards for the various SI units are realized in NMLs in a manner similar to that used at the BIPM. These are complex experimental setups, and a lot of effort is needed to maintain them. They are generally not transportable. These national standards are the ultimate legal standards in a country. They are intercompared with similar standards held at other NMLs and also with those held at the BIPM. The intercomparison is accomplished through transfer standards.

The value of an SI unit is realized in an NML through an experimental setup. This is further transferred to the primary reference standard held by the NML. The value of the reference standard is further transferred to the standards held by the calibration laboratories at echelon II. A similar method is used to transfer the unit to the next lower echelon and ultimately to the end user in the traceability chain.

Starting with the realization of the unit in the definition stage and continuing with each transfer of the unit down the chain, uncertainty is added at each stage. Thus, the further we are down the traceability chain, the higher is the uncertainty of the measurement. The uncertainty at the end user level is greater than that at the beginning of the chain. Figure 6.5 shows the increasing uncertainty in the traceability chain for the various levels of metrology organizations. The curve is an indication of increasing uncertainty and is not drawn to scale. Table 6.3 shows some typical values of uncertainty figures at various stages of the traceability chain.

INTERNET-BASED CALIBRATIONS

In a hierarchical system of calibration laboratories forming part of a national measurement system, standards and measuring equipment are physically sent to laboratories for calibration purposes. The accuracy of the measuring equipment to be calibrated determines the level of hierarchy of the calibration laboratory. There are two factors which have to be kept in mind in this type of

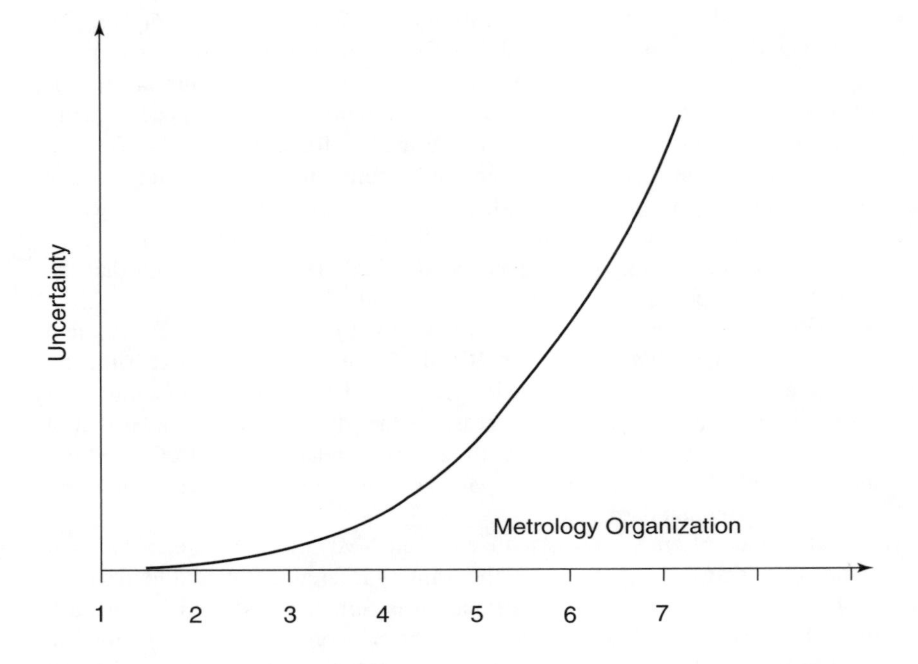

Figure 6.5 Increasing uncertainty in the traceability chain.

calibration process. The actual-use environment of the equipment will not be the same as that of the calibration laboratory, and during transportation there is the risk of damaging the equipment as well as that of affecting its calibration uncertainty.

To overcome these problems, the National Physical Laboratory of the United Kingdom (NPL UK) has undertaken research to develop technology for Internet application in the area of calibration. The Internet offers many new

Table 6.3 Uncertainty figures at various metrological levels for different parameters.

Level of metrology organization	Physical parameter		
	Mass	**Length**	**DC Voltage**
BIPM	International prototype of the kilogram, 1 part in 10^8	He-Ne laser, 1 part in 10^{10}; monochromatic wavelength, 1 part in 10^8	Josephson arrays, 0.4 ppm relative to definition, 1 part in 10^{-19} relative to other arrays
NML	International prototype of the kilogram, 1 part in 10^8; transfer standards, 2 parts in 10^8	He-Ne laser, 1 part in 10^{10}; monochromatic wavelength, 1 part in 10^8; gauge blocks, 1 part in 10^7 to 4 parts in 10^8	Josephson arrays, 0.4 ppm relative to definition, 1 part in 10^{-19} relative to other arrays; Zener-based standard sources, 0.1 to 0.8 ppm relative to Josephson arrays
Accredited calibration laboratory	Reference/working standard weights and balances, 1.5 parts in 10^5 to 1.5 parts in 10^6	Gage blocks of grades 0 and 00 and length comparators, 1 part in 10^5 to 6 parts in 10^6	Standard voltage sources and calibrators, 5 to 50 ppm
User of measurement results (trade, industry, or laboratory)	Standard/commercial weights and balances, 5 parts in 10^5	Gage blocks, calipers, dial gages, scales, etc., 1 part in 10^4 to 1 part in 10^5	Digital multimeters and DC voltmeters, 0.01% to 1%

technologies, such as electronic mail, video conferencing, and e-commerce. NPL UK has proposed the concept of Internet calibration application (ICA), wherein the NPL would provide calibration services in certain metrology areas at remote laboratory locations through the Internet. Internet application may not be applicable to all metrological areas due to the compatibility limitations of certain types of standards, but it has tremendous potential.

In ICA, the national Metrology laboratory through its server computer would be connected to the client laboratory through an Internet connection. The instrument to be calibrated would be connected to the Internet through a personal computer. The national standards laboratory would retain the details of customer equipment and artifacts on a database accessible over the Internet. Any client laboratory needing calibration traceable to national standards would log on to the national standard laboratory which would then take control of the measurement system through the Internet, operate the measurement hardware, interpret the data, correct it using the database of calibration data, and provide a measure of the uncertainty. The system would thus provide direct traceability to the national standards laboratory. This technology is just emerging and will revolutionize the whole concept of the calibration and traceability of measurement results. It is likely that in the future all standards and measuring equipment will be designed and manufactured to be compatible with ICA.

ACHIEVING INTERNATIONAL TRACEABILITY

One of the essential requirements for international compatibility of measurement results is that the various national standards maintained at the national metrology Laboratory (NML) of member countries must be in agreement among themselves. The task of achieving agreement among the national standards of various countries and their harmonization is one of the charters of the BIPM. The membership of the BIPM consists of scientists and technical experts working in the NMLs of various countries.

International traceability is achieved through intercomparison of the measurement standards held by the various NMLs. This task is carried out by the BIPM in association with the national laboratories. It is also the responsibility of the NMLs to ensure that their measurement standards and measurement capabilities are recognized worldwide. In the absence of such recognition, not even the testing and calibration services provided in a country will be acceptable in other countries, resulting in technical barriers to international trade. International trade is carried out primarily based on the assumption of equivalence of national measurement standards and reliability of the link between these standards and the services provided by the testing and calibration laboratories of the countries involved.

The equivalence of national measurement standards is achieved through the participation of NMLs in a number of international comparisons under the auspices of the BIPM and of regional metrology organizations. This activity is carried out under mutual recognition agreements signed by various NMLs related to national measurement standards. A set of key comparisons of national measurement standards are identified by the consultative committee of the CIPM and executed by the BIPM. For an international comparison, some 80 key comparisons are chosen and a small number of supplementary comparisons are identified to test the reliability of the calibration services provided in the countries. The key comparisons are chosen in order to judge the principal techniques in each field of metrology. The objective of the exercise is to establish the degree of equivalence of the national measurement standards maintained by various NMLs, thereby providing confidence to governments and trading partners regarding the mutual acceptability of measurement results. The outcome of the key comparison is a statement of the measurement capability of each NML participating in the comparison. Information on the results of the key comparison is maintained by the BIPM and is made available to all the participating laboratories.

Following is some information on a few intercomparisons that have been carried out.

Periodic verification of the national prototypes of the kilogram: This verification was carried out during 1989–92. The objective was to verify the mass of the national prototypes of the kilogram held by various countries and to gather more information on the stability of the international prototype of the kilogram. Forty-two countries participated in this intercomparison exercise. The official kilogram prototype and its copies held at the BIPM were also included in this exercise. Based on these intercomparisons, the uncertainties associated with the kilogram prototypes of various countries were reassigned. The kilogram prototypes were cleaned and washed during this intercomparison process. An interesting conclusion of the intercomparison exercise was that the masses of the prototypes appeared to have increased; this conformed to the trend observed during the previous intercomparison carried out during 1946–53. It was also concluded that over the last 100 years the average increase in the masses of the kilogram prototypes was about 100 micrograms. The masses of the oldest and best-maintained national prototypes increased at the rate of 0.25 microgram per year.

Intercomparison of the unit of time: International Atomic Time is a time scale based on atomic clock comparison data supplied by 45 national laboratories. Since the unit of time is an accurate representation of the second, the objective of atomic clock intercomparison is to provide an accurate time scale. This intercomparison of atomic clock data is also used to make corrections in the Coordinated Universal Time scale due to irregularity in the earth's rotation. This intercomparison data is compiled once every two months.

Intercomparison of the national standards of electrical SI units: In the area of electrical standards, intercomparison is generally carried out between Josephson array standards for voltage and quantum-Hall-effect resistance standards. This process helps in calibrating the voltage and resistance standards of NMLs. The uncertainties of national standards compared with the standards held at the BIPM are plotted and published by the BIPM. An intercomparison of Josephson array standards maintained by NMLs of the United States, Germany, Canada, and the United Kingdom, along with those at the BIPM, was conducted during January 1991 to March 1993. As a result of the intercomparison, it was observed that the standards demonstrated coherence among themselves to better than 1 part in 10^9. Similar studies of zener-diode-based voltage standards at 1.018 volts demonstrated coherence among themselves to better than 20 parts in 10^9.

The Role of Regional Metrology Cooperation in Achieving International Traceability

It is difficult for the BIPM to facilitate the participation of the many countries in international comparisons of measurement standards. To avoid this and at the same time to ensure the compatibility of national metrology standards, regional metrology cooperations have formed in various regions of the world. One such regional cooperation is the Commonwealth Asia Pacific Metrology Program (APMP). This program has 21 member countries from the Asia Pacific region.

The objective of such regional cooperations is to improve the measurement capabilities of member countries' national laboratories by sharing facilities and expertise. In addition to facilitating the intercomparison of national standards, these programs also engage in the activities of training, consultancy, and information exchange in the area of international metrology. The Commonwealth Science Council supports the APMP. Similar programs are operational in Commonwealth Africa and the Caribbean region.

Such programs are helping a great deal in improving the metrology capabilities of developing countries. In the cooperation programs, the NML of each country is provided an equal chance to participate and get benefits. The concept is one of self-help, wherein a laboratory makes its needs known to other member laboratories and fills its needs by itself with the help of the others' expertise and technology.

Under these metrology cooperation programs, one or two national laboratories are assigned the role of nodal laboratory. The nodal laboratory keeps in touch with the BIPM and assumes the same role in regional intercomparisons of national standards as that of the BIPM in international comparisons.

There are a number of regional metrology cooperation programs operating in various regions of the world. One such program is the Western Europe Metrology Cooperation (WEMC), which has the participation of 14 Western European nations. Similar cooperation programs are operational in North America and South America.

Regional cooperation in metrology has also been facilitated by the regional associations of laboratory accreditation bodies of various countries. The Western European Calibration Cooperation (WECC) is one such organization. Its members include the national laboratory accreditation bodies of the United Kingdom and Germany—UKAS and DKD, respectively. The Asia Pacific Laboratory Accreditation Cooperation (APLAC) is a regional metrology program of the national laboratory accreditation bodies of the Asia Pacific region.

The European Metrology Organization (EUROMET) is another important regional metrology program. With the participation of European countries, it aims at the development of national primary standards so as to provide traceability without national limitations.

The intercomparison of national standards on a regional basis is an ongoing program. The status report of one such program conducted by the APMP on pneumatic pressure comparison was presented at the Third International Conference on Metrology in the New Millennium and Global Trade, held February 8–10, 2001 at the National Physical Laboratory (NPL) of New Delhi. Ten member countries of the APMP participated in the intercomparison, and NPL India was the pilot laboratory for coordinating the intercomparison program.

In addition, bilateral comparison programs are conducted in which two national laboratories compare the values of their national standards among themselves.

MEASUREMENT TRACEABILITY IN CHEMICAL METROLOGY

Traceability in physical metrology has been a well-conceived and practiced concept. The concept has been in practice in a limited way since the signing of the Treaty of the Meter in 1875. In the last 50 years, the concept has become well accepted and is now being practiced globally. The need to meet traceability requirements as stipulated in the ISO 9000 series of quality system standards propagated the concept to all categories of industry, including small-scale industries.

The compatibility of measurement results is also an essential requirement for analytical results covered by chemical metrology. The analytical measurements made by different laboratories in the same country and by laboratories in different countries have to be comparable. Thus, the concept of the traceability of measurement results is equally valid in chemical metrology.

Traceability by definition links measurement results to SI units. The issue of traceability in chemical measurements has been gaining the attention of metrologists and chemical analysts in order to achieve comparable results.

To define and harmonize the practice of methodologies and traceability concepts in analytical measurements, the CIPM appointed a consultative committee on quantity of matter in 1993. An international cooperation program called the Cooperation on International Traceability in Analytical Chemistry (CITAC) has been formed with the participation of leading laboratories of the world. CITAC aims to foster collaboration among existing organizations to improve the international comparability of chemical measurements. At the European level, EURACHEM is a network of organizations having the objective of establishing a system for the international traceability of chemical measurements and the promotion of good quality practices. It provides a focus for analytical chemistry and quality related issues in Europe. It also provides a forum for the discussion of common problems and for the development of an informal and considered approach to both technical and policy issues.

One of the base SI units, the mole, is used for traceability purposes in national and international measurement systems in chemical metrology. The unit of the mole is linked to the amount of a substance, so traceability to the mole requires that measurements be made using a method that has a direct link to the mole. Thus, performing chemical measurements that are traceable to this unit has the advantage that they will be consistent with other measurements made with respect to other SI units.

There are three ways of achieving traceability in measurement, as traceability requires a clearly identified common point of reference.

1. **Accepted unit of measurement:** This is the most common method and is used extensively in physical measurement, where all measurements are linked to the SI through a hierarchical system of calibration laboratories.
2. **Accepted method:** If measurements are carried out according to a method which has been formulated and internationally agreed upon by experts, a certain amount of traceability is built into the results which are obtained by the accepted method. Most chemical measurements can be considered traceable to the extent that the common point of reference is the method. The realization of intrinsic standards is one example of the achievement of traceability through an accepted realization method.
3. **Use of reference material:** Certified reference materials are frequently used as tools to establish traceability. Traceability to certified reference materials held at the reference laboratory is achieved. The reference materials can be further traced to the SI unit of the mole.

REFERENCE MATERIALS AND CERTIFIED REFERENCE MATERIALS

Reference materials play an important role in establishing the traceability of analytical measurements and thus in ensuring that the measurements are valid. The concept of measurement traceability in chemical metrology is of recent origin compared with the concept in physical metrology. In view of the ever increasing importance of analytical measurements in the area of the analysis of food, including dairy and other agricultural products, and the acceptability of the analytical results on a global basis, the task of harmonization of procedures and practices has been undertaken by the CIPM and the BIPM. In the United Kingdom, the Department of Trade and Industry has initiated a program called the Valid Analytical Measurement (VAM) initiative. This is an integral part of the United Kingdom's national measurement system and the European infrastructure which will allow analytical laboratories to demonstrate the validity of their data and facilitate the mutual recognition of analytical measurements. One of the key principles of the VAM initiative is that certified reference materials should be used to ensure traceability of measurement results.

A *reference material* is defined as a material or substance one or more properties of which are sufficiently well established to be used for the calibration of an apparatus, the assessment of a measurement method, or the assigning of values to materials. A *certified reference material* is defined as a reference material one or more of whose property values are certified by a technically valid procedure and accompanied by or traceable to a certificate or other documentation which is issued by a certifying body.

A reference material becomes a certified reference material if it meets the following conditions:

- It is certified by a certifying body using a technically valid procedure
- The material is accompanied either by a certificate or associated documentation or by a traceability statement giving its linkage to the certificate or documentation

The process of establishing traceability in chemical metrology is illustrated in Figure 6.6.

The traceability chain starts at the apex with the mole, that is, the SI unit of the amount of a substance. The mole is the amount of a substance which contains as many elementary entities as there are atoms in 0.012 kilogram of carbon 12. Whenever the mole is used as a unit, the elementary entities must be specified. These entities may be atoms, molecules, ions, electrons, other particles, or specified groups of such particles.

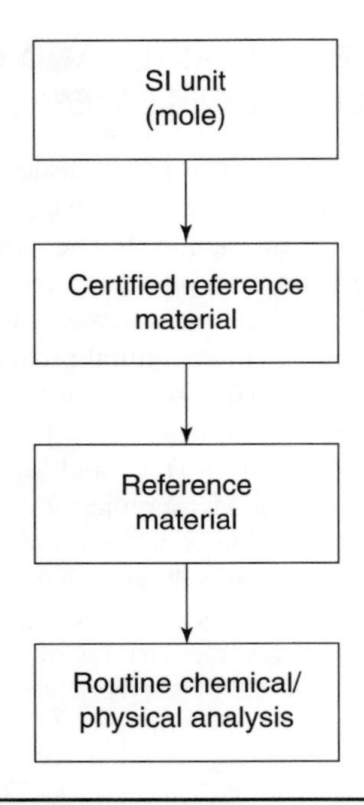

Figure 6.6 Chemical metrology: Traceability chain.

Certified reference materials have to be linked to the mole. This is similar to the realization of primary reference standards based on the definition of the SI unit, as is done in physical metrology. The linking of certified reference materials is done through a definitive method. As the mole is an amount of substance, the definitive methods are techniques which count the particles or atoms present that possess specific chemical composition or characteristics. Examples of the techniques and instrumentation used as definitive methods are isotope detection, mass spectrometry, and spectrophotometry.

One frequently used definitive method is resonance ionization spectroscopy (RIS). This is a highly sensitive analytical measurement method. It employs lasers to eject electrons from selected types of atoms or molecules. The ions and electrons are then detected and counted using various types of instrumentation. The process thus identifies elements or compounds and determines their concentration in a sample. The RIS technique was invented in the 1970s and is widely used in practical measurement systems. The advantage of this technique is that it is highly sensitive and can discriminate between different types of atoms. Its sensitivity is at the one-atom level, as it can count single electrons by atomic counting.

Once the link between the certified reference material and the mole is established, the next step is characterizing the reference material with respect to a certified reference material. This is done through comparison and reference methods. Reference methods are again technically valid methods, and they can be either analytical methods or instrumentation used for spectral analysis. Using reference methods, the characteristics of the secondary reference material are compared with those of the certified reference material. Analogously, the characteristics are transferred to a working reference material by comparison with the secondary reference material, again using the reference method. Ultimately, the routine analysis is validated using the reference materials. Thus, the traceability chain is completed from the SI unit to the analysis done by the actual end user. It can be noted that certified reference materials, secondary reference materials, and working reference materials form a traceability chain in chemical metrology just as primary reference, secondary, and working standards form a traceability chain in physical metrology. Certified reference materials are used for the following purposes in analytical laboratories in addition to establishing a traceability chain:

- Method validation and measurement uncertainty evaluation
- Verification that the analytical method is being used correctly
- Quality control and quality assurance

Valid analytical measurements depend critically on the use of certified reference materials. They ensure the compatibility of analytical measurements as they ensure a link back to the SI unit of measurement. The usefulness of certified reference materials for obtaining valid and traceable analytical measurements is coming to be understood by chemical metrologists.

Types of Reference Materials

There are a variety of applications in which reference materials find use. They are used to support measurements concerned with chemical composition and with biological, clinical, physical, and engineering properties. The NIST in the United States is one of the major manufacturers and suppliers of standard reference materials. The reference materials supplied by the NIST fall into the following broad categories:

- **Standard reference materials for chemical composition:** These reference materials include the subcategories ferrous metals, nonferrous metals, high-purity materials, health and industrial hygiene, inorganics, analyzed gases, fossil fuel, organics, food and agriculture, geological materials and ores, ceramics and glass, cement, and engine wear materials.

- **Standard reference materials for physical properties:** These include ion activity, polymeric properties, thermodynamic properties, optical properties, radioactivity, electrical properties, metrology, ceramics and glasses, and miscellaneous properties.
- **Standard reference materials for engineering purposes:** These include sizing, surface finish, nondestructive evaluation, automatic data processing, fire research, miscellaneous performance, and engineering materials.

The above categories are exhaustive to the extent that they cover all the types of reference materials that have applications in analytical and physical measurements.

The Certification Process for Reference Materials

The reference value assigned to a reference material is obtained by one of the following routes of measurement:

- A previously validated measurement method. As the measurement method is already validated, the certified values have high reliability of accuracy.
- Two or more independent reliable measurement methods. If the values obtained by two independent measurement methods are in agreement, the confidence level of the assigned value becomes high.
- Measurements carried out by a network of cooperating laboratories that are technically competent and thoroughly knowledgeable about the material being tested. The values assigned by various laboratories should be in agreement.

There are a number of organizations in the world which characterize reference materials. In addition to the NIST, some of these organizations are the National Research Council of Canada, the National Research Center for Certified Reference Materials of China, and the National Institute of Environmental Studies of Japan.

The Uncertainty of Reference Materials and Certified Reference Materials

In analytical laboratories, reference materials and certified reference materials play the same role that measurement standards play in physical measurement. These materials should be accompanied by a statement of the uncertainty of the characteristics they represent. Reference materials are generally prepared in batches for which property values are determined with a stated uncertainty.

The uncertainty aspect of reference materials and certified reference materials should be considered when they are being used for the verification of analytical methods, quality control, and quality assurance in analytical laboratories.

ILAC GUIDANCE ON CALIBRATION AND TRACEABILITY

The International Laboratory Accreditation Cooperation is an apex international organization that coordinates the activities of national laboratory accreditation bodies. ILAC has published a guidance document, ILAC-G2:1994, on the traceability of measurements. This document contains guidance and assistance for organizations on how to comply with the traceability requirements of various quality management standards. The document defines the following essential elements of traceability:

1. **An unbroken chain of comparison:** This chain must go back to a standard acceptable to all parties concerned, usually a national or international standard.
2. **Measurement uncertainty:** The measurement uncertainty for each step in the traceability chain must be calculated according to a defined method and must be stated so that an overall uncertainty for the whole chain may be calculated.
3. **Documentation:** Each step in the chain must be performed according to documented and generally acknowledged procedures; the results must equally be documented.
4. **Competence:** Laboratories or other bodies performing one or more steps in the chain must supply evidence for their technical competence (for example, by demonstrating that they are accredited).
5. **Reference to SI units:** The "appropriate" standards must be primary standards for the realization of SI units.
6. **Recalibration:** Calibration must be repeated at appropriate intervals; the length of these intervals depends upon a number of variables (for example, the uncertainty required, frequency of use, manner of use, and stability of the equipment).

TRACEABILITY REQUIREMENTS AS STIPULATED IN THE ISO STANDARDS ON QUALITY MANAGEMENT SYSTEMS

The International Organization for Standardization (ISO) has published the following two quality management system standards. One standard is applicable to manufacturing and service organizations, and the other is applicable to testing and calibration laboratories.

- ISO 9001:1994, *Quality systems—Model for quality assurance in design, development, production, installation, and servicing,* which has now been replaced by ISO 9001:2000, *Quality management system requirements*
- ISO/IEC Guide 25:1990, which has now been replaced by ISO/IEC 17025:1999, *General requirements for the competence of calibration and testing laboratories*

ISO 9001 stipulates that measuring equipment shall be calibrated or verified at specified intervals or prior to use against measurement standards traceable to international or national measurement standards; where no such standard exists, the basis used for calibration or verification shall be recorded. The traceability requirement is similar in the 1994 version and the 2000 version of the ISO 9001 standard.

ISO/IEC Guide 25 as well as ISO/IEC 17025 have identified "measurement traceability" as one of the factors affecting the quality of measurement results. Guide 25 stipulates that the measurements made by the laboratory should be traceable to national standards of measurement. The new standard ISO/IEC 17025 has further elaborated the requirements for calibration laboratories. It states that calibration and measurements should be traceable to an International System of units (SI) through national measurement standards. This elaboration has emphasized the following two aspects:

- Primacy of SI units over national measurement standards, as is also obvious from Figure 6.2
- As all national measurement standards are physical representations of SI units, the requirement also implies international traceability of calibration results

REFERENCES

Aggarwal, N. K. 1996. Uncertainty in dimension measurement. In *Quality assurance through laboratory-accreditation,* 13–25. New Delhi: Commonwealth-India Metrology Centre NPL.

Bandyopadhya, A. K. 2001. NPL New Delhi status report of APMP sponsored regional pneumatic pressure comparison. In *Proceedings of the Third International Conference on Metrology in the New Millennium and Global Trade,* 321–27. New Delhi: Metrology Society of India, NPL.

Brookman, Brian. 1995. The Office of Reference Materials, Laboratory of the Government Chemist: A focal point for the supply of certified reference materials. *VAM Bulletin,* no. 12:5–6.

Chaudhary, S. K. 1995. Usage of reference materials. In *Proceedings of workshop on the role of metrology in quality management and quality improvement,* 191–97. New Delhi: Commonwealth-India Metrology Centre NPL.

Dudley, R. A., N. M. Ridler, and J. M. Williams. 2000. Internet-based calibrations of electrical quantities at the United Kingdom National Physical Laboratory. *Cal Lab: The International Journal of Metrology* 7 (November–December), no. 6:22–25.

Fleming, John. 1995. DTI renews support for the VAM initiative. *VAM Bulletin,* no. 12:3–5.

———. 1995. Reference material: Bringing order out of chaos. *VAM Bulletin,* no. 12:8–9.

Fluke Corporation. 1994. *Calibration philosophy in practice.* 2nd ed. Evertt WA: Fluke Corporation.

Garner, Ernest L., and Stanley D. Rasberry. 1993. What's new in traceability. *Journal of Testing and Evaluation,* April, 1–19.

Harris, Georgia L. 1996. Answers to commonly asked questions about mass standards. *Cal Lab: The International Journal of Metrology* 3 (November–December), no. 6:35–37.

International Laboratory Accreditation Cooperation (ILAC). 1994. ILAC-G2:1994, *Traceability of measurements:* ILAC.

International Organization for Standardization (ISO). 1994. ISO 9001:1994, *Quality systems—Model for quality assurance in design, development, production, installation, and servicing.* Geneva, Switzerland: ISO.

———. 2000. ISO 9001:2000, *Quality management systems–Requirements International Organization for Standardization,* Geneva, Switzerland: ISO.

International Organization for Standardization and International Electrotechnical Commission. 1990. ISO/IEC Guide 25:1990, *General requirements for the competence of calibration and testing laboratories.* Geneva, Switzerland: ISO.

———. 1999. ISO/IEC 17025:1999, *General requirements for the competence of testing and calibration laboratories.* Geneva, Switzerland: ISO.

King, Bernard. 2000. The selection and use of reference materials. *NABL News,* April 2000: 3–7 (published by the National Accreditation Board for Testing and Calibration Laboratories, Department of Science and Technology, Government of India).

Lombard, Michael A. 1999. Traceability in time and frequency metrology. *Cal Lab: The International Journal of Metrology* 6 (September–October), no 5:33–40.

McHenry, Robert, ed. 1993. *The New Encyclopaedia Britannica.* Vol. 24, 361–64. Chicago: Encyclopaedia Britannica.

Mittal, M. K., J. C. Biswas, and A. S. Yadav. 1999. Traceability and testing/calibration in the field of AC power and energy at NPL India. In *Proceedings of the National Conference on Test Engineering and Metrology 1999,* 61–65. New Delhi: STQC.

National Institute of Standards and Technology (NIST). *NIST standard reference materials catalogue 1992–93,* 260. Gaithersburg, MD: NIST.

Pan Xiu Rong. 1995. A view on the traceability of certified values of chemical composition RMs. *VAM Bulletin,* no. 12:18–19.

Ram Krishna. 1996. Temperature standard and calibration facilities at NPL New Delhi. In *Quality assurance through laboratory accreditation,* 48–56. New Delhi: Commonwealth-India Metrology Centre NPL.

Schunke, Adolf Z. 2001. National standards for chemical measurements in Germany. In *Proceedings of the Third International Conference on Metrology in the New Millennium and Global Trade,* 305–10. New Delhi: Metrology Society of India, NPL.

Sharma, D. C. 1996. Mass measurement and metrological requirement. In *Quality assurance through laboratory accreditation,* 29–35. New Delhi: Commonwealth-India Metrology Centre NPL.

Sharp, De Wayne B. 1999. Measurement standards. In *The measurement, instrumentation and sensors handbook,* ed. John G. Webster, 5-1 to 5-12. CRC Press, Springer-Verlag GMBH, IEEE Press.

Sydenham, P. H. 1982. *Handbook of measurement science.* Vol. 1, *Standardization of measurement fundamentals and practices,* 49–94. New York: John Wiley and Sons.

Verma, Ajit Ram. 1987. *National measurements and standards.* New Delhi: National Physical Laboratory.

Walker, Ron. 1995. Reference materials: The gateway for accuracy. *VAM Bulletin,* no. 12, 7–8.

Wason, V. P. 1995. Temperature measurement. In *Proceedings of workshop on the role of metrology in quality management and quality improvement,* 89–97. New Delhi: Commonwealth-India Metrology Centre NPL.

Yang Xiaoren. 1995. Regional collaboration in metrology. In *Proceedings of workshop on the role of metrology in quality management and quality improvement,* 80–82. New Delhi: Commonwealth-India Metrology Centre NPL.

7

Measurement Uncertainty and Its Evaluation: Evolving Concepts

MEASUREMENT UNCERTAINTY: LIMITS TO ERROR

A measurement result whose accuracy is entirely unknown is worth nothing. Accuracy depends upon the error of measurement. It is also accepted that the actual error of a measurement result is unknown and unknowable. This is a contradictory situation. Without knowing its error, one cannot utilize a measurement result confidently for making a decision, but at the same time its error is unknown and unknowable. Metrologists have found a solution to this contradiction in the application of statistical tools. If the absolute value of error cannot be determined, the "limits to error" can always be "inferred" by using rudimentary statistical techniques. Here the term *limits to error* means the maximum value of expected error. This is a worst-case situation in the sense that the error will not exceed the limit. The actual error will be less than the limits-to-error value. Another term used in the context of limits to error is *inferred.* When we infer something, we are not sure about its correctness, so any inference about limits to error always has a risk of being incorrect. If the risk of being incorrect is reasonably low, the knowledge about the limits to error is better than not knowing anything about the quantity of error at all.

The "limits to error" of a measurement result is in fact the uncertainty of measurement. The uncertainty is generated by two factors:

- One factor is the precision of the measurement process by which the result has been derived. This is evaluated through repeated application of the measurement process to obtain repeat measurements of the same parameter. Precision is the characteristic of the measurement process linked to the closeness of repeat measurements among themselves.

Thus, a numerical index of precision should come from the variability of the repeat measurement observations. The population standard deviation σ is the most appropriate numerical index of the precision of a measurement process.

- The other factor which contributes to uncertainty is the unknown component of systematic error. A numerical indication of its contribution to uncertainty is obtained from the analysis of the systematic error components of the measurement process. The bias of the measuring equipment is determined by the calibration process, and this known systematic error is always corrected for. However, there is always an unknown component of systematic error which contributes to uncertainty.

For example, during the calibration of a DC reference voltage source of 250 millivolts, it is observed that the value is 250.05 millivolts when measured with a highly accurate reference meter. Thus, the voltage source has a bias of 0.05 millivolts, which is a known component of systematic error and can be corrected for. However, 250.05 millivolts is not an absolute value. There is an unknown component Δv to this value such that the measured value is 250.05 $\pm\Delta v$ millivolts and not 250.05 millivolts. The objective of the metrologist is to find the reasonable bounds, or limits, to the value of Δv caused by imprecision and unknown systematic error. This value can be ascertained based on the metrologist's experience with the measurement process, knowledge of the effect of other influence factors, such as the environment, on the measurement process, and the dispersion of measured values when the process is repeated.

UNCERTAINTY REPORTING: THE NEED FOR HARMONIZED PRACTICES

The calibration of testing and measuring equipment ensures that measurements made by them can be relied upon. The uncertainty of calibration results at each echelon of a national measurement system should therefore be known. The reporting of calibration uncertainty has long been a usual practice of calibration laboratories.

It has also long been common knowledge among metrologists that the imprecision of a measurement process and the unknown component of systematic error cause uncertainty in measurement results. However, until the 1950s there was no clear understanding of how measurement uncertainty was to be evaluated and expressed. Calibration laboratories reported uncertainty according to their own understanding. The users of the calibration results as well as the laboratories thus faced problems due to the following:

- Inadequate information on the uncertainty value
- Nonuniformity in the procedures used to evaluate uncertainty by various calibration laboratories

The need for calibration laboratories to give adequate information on the calibration certificate provided to clients has been expressed by Harry H. Ku (1968) in the following words:

> In making a statement of uncertainty of the result of calibration, the calibration laboratory transmits information to its clients on the particular item calibrated. It is logical, then, to require the transmitted information to be meaningful and unambiguous and to contain all the relevant information in the possession of the laboratory. The information content of the statement of uncertainty determines to a large extent the worth of a calibrated value. A common deficiency in many statements of uncertainty is that they do not convey all the information a calibration laboratory has to offer, information acquired through ingenuity and hard work. This deficiency usually originates in two ways.
>
> 1. Loss of information through oversimplification.
> 2. Loss of information through the inability of the laboratory to take into account information accumulated from its past experience.
>
> With the increasingly stringent demands for improved precision and accuracy of calibration work, calibration laboratories as a whole just cannot afford such luxury.

The oversimplification of uncertainty statements is a common practice in many calibration laboratories. A calibration result is generally stated as $\bar{x} \pm U$ where $\bar{x}$ is the assigned value of the parameter being calibrated and $\pm U$ is an indication of the uncertainty associated with the assigned value $\bar{x}$, which is taken as the average of a number of repeat observations. A statement such as $\bar{x} \pm U$ may lead to different interpretations by the users of the calibration result in the absence of a clear and concise statement of what is intended to be conveyed. Some clients may interpret the statement as the bounds to the inaccuracy, inferring that the measured parameter lies between $\bar{x} - U$ and $\bar{x} + U$. Others may take it as the standard error of the mean, thus inferring that the uncertainty of $\bar{x}$ is $\pm 3U$. Some may take it as the standard deviation of the observations. Thus, the expression of a measurement result in the form of $\bar{x} \pm U$ without a statement of what is intended to be conveyed creates confusion in the minds of users. Such oversimplification is not desirable. Moreover, the users of calibration results are not always professional metrologists. They may be shop floor personnel in industry in the majority of cases. Thus, confusion in the minds of users due to oversimplification of calibration results could be the rule rather than the exception. In addition to this confusion, there is another problem with quoting a figure for overall uncertainty. The various uncertainty components based on which the overall figure has been derived are not identified individually. For a sophisticated customer, this information may be of great value and may even be employed by the client while making use of the measurement result. Failing to quote information that is already available for the sake of simplicity may result in substantial waste of effort and resources.

For example, the single uncertainty figure $\pm U$ does not indicate its components. These could be the cumulative effect of random variation in repeat observations, an unknown component of systematic error, uncertainty due to variation in environmental conditions compared with the reference conditions, uncertainty due to the interposition of reference values, and so on. If the individual components of uncertainty are specified as part of a calibration report, the information would be of great use for a sophisticated client. The client would know which components are dominant and which are not and could use the information in making decisions. This information is known to the calibration laboratory but is not made available to the client. Thus, oversimplification in the uncertainty reporting of calibration results leads to confusion and loss of information.

In view of these practical difficulties, metrologists agreed that there was a need for the harmonization of concepts and practices pertaining to the evaluation and reporting of measurement uncertainty. Leroy W. Tilton of the U.S. National Bureau of Standards (NBS), now the National Institute of Standards and Technology (NIST), made early attempts in this direction. In the early 1950s, Tilton developed guideline recommendations for the internal use of the Optics and Metrology Division of the NBS on how accuracy should be reported. The concept evolved within the division was further extended to other divisions of the NBS. In 1955, an NBS task group consisting of Tilton, Malcolm W. Jensen of the Office of Weights and Measures, and Churchill Eisenhart of the Applied Mathematics Division made certain recommendations on the accuracy statements of NBS testing and calibrations. These were applicable to all measurement results reported by the NBS. The recommendations, with some additions and illustrations, were incorporated as chapter 23, "Expression of Uncertainties of Final Results," of NBS Handbook 91, *experimental statistics,* published in 1963. This handbook brought together in a single volume all the material on experimental statistics prepared at the NBS.

Handbook 91 gave guidelines for the expression of uncertainty of final results which were applicable for all types of measurement results in testing as well as calibration. However, the evaluation and reporting of the uncertainty of measurement is of particular importance for calibration laboratories, as during the process of calibration not only are the values of parameters propagated from the reference equipment to the equipment under calibration but uncertainties are also propagated. To bring uniformity to the reporting of uncertainty by calibration laboratories, Eisenhart prepared guidelines based on the NBS handbook but made more appropriate to calibration work. These guidelines were published as "Expression of Uncertainties of Final Results" in *Science* in 1968. Subsequently, Harry H. Ku summarized the guidelines in his paper "Expression of Imprecision, Systematic Error and Uncertainty Associated with a Reported Value," published in *Measurement and Data* in 1968.

EISENHART'S GUIDELINES ON THE EVALUATION AND REPORTING OF MEASUREMENT UNCERTAINTY IN CALIBRATION LABORATORIES

Eisenhart's guidelines were based on the then prevailing concept that measurement uncertainty was the result of two distinct characteristics of the measurement process. One was the imprecision and the other was the accuracy of the measurement process. Until the 1960s, the prevailing concept about the precision and accuracy of a measurement process was slightly different from how it is conceived today. A measurement process was seen as characterized by two independent attributes: precision and accuracy. *Precision* referred to the closeness of successive independent measurements of a measurand by repeated application of a measurement process under specified conditions, whereas *accuracy* was linked to the bias or systematic error of the measurement process, implying its capability to measure what it was intended to measure. The greater the bias, the less was the accuracy of a measurement process. Now the term *accuracy* is not used in the above context but has a wider meaning. Part 1 of the international standard ISO 5725-1 on the accuracy of measurement methods and results defines accuracy as the closeness of agreement between a test result and the accepted reference value. This definition is supplemented by a note which states that the term *accuracy* when applied to a set of test results involves a combination of random components and a common systematic error or bias component. Accuracy is thus viewed as a characteristic of a measurement process consisting of precision as well as bias components. A process is considered to be accurate only if it is precise as well as unbiased. A measurement process that generates repeat observations of a characteristic which are not close enough to one another cannot generate an accurate result. In fact, a new word has been coined to replace the word *bias*. Bias is now characterized by the attribute "trueness." *Trueness* indicates the positive aspect of a measurement process, in contrast to bias, which indicated a negative aspect. The old concept of accuracy has been changed to mean trueness. The greater the bias of a measurement process, the less true it is. The term *accuracy* is now used in a broader context which includes both precision and trueness. Many times a measurement result is derived as an average of repeat measurement observations rather than a single observation. When the result is an average, it is likely that it is close to the conventional true value but that the repeat measurements are not close enough to one another. Such a measurement process may be considered to be true but not accurate, as the process is not precise.

The key aspects of Eisenhart's guidelines follow:

- Imprecision and systematic errors are to be treated separately. Imprecision and systematic errors are two distinctly different components of inaccuracy and need different treatment and interpretation in usage. If both contribute significantly to the inaccuracy of a result, separate numerical values should be quoted to represent the imprecision and the bounds of systematic error of the reported result.
- The terms *standard deviation* and *standard error* should be used only to denote a canonical value for the measurement process which is derived based on considerable experience gained with the measurement process in the past. If, on the contrary, the standard deviation and standard error are calculated based on limited experience, the terms *computed standard deviation* and *computed standard error* should be used.
- The uncertainties of fundamental constants and those of national primary standards are encountered in calibration laboratories operating at the apex level and in those deriving their values directly from national metrology institutions. These uncertainties creep into the process by their inherent nature. Since these uncertainty components affect all results obtained by the specific method, they need not be taken into consideration in the quotation of the uncertainty components. The inclusion of these components would make the quoted uncertainty figure unduly more inaccurate. However, a statement should be made to the effect that these uncertainty components have been excluded from the quoted figure.

Indices of Imprecision and Systematic Error

According to Eisenhart's guidelines, uncertainty statements have two aspects: the imprecision statement and the systematic error or bias statement. Ku has identified certain indices that represent the measure of error due to imprecision and the measure of systematic error. The details of these indices for various characteristics of the measurement process and measurement results follow.

Indices Pertaining to the Imprecision of the Measurement Process

Imprecision is linked with the dispersion of repeat observations obtained from the measurement process. As discussed in chapter 3, the standard deviation of a set of data is a good indicator of its dispersion. The standard deviation obtained from a large number of repeat observations is represented by σ, which can be obtained only if the measurement process has been in operation for quite some time. Generally, these data are available to calibration laboratories but they do not make use of the wealth of information they hold in uncertainty reporting. This is what Ku has referred to as the inability of a laboratory to take into account information accumulated from its experience.

In a calibration laboratory, a calibration procedure is repeated to calibrate similar types of testing or measuring equipment. The same reference standard may be used for the calibration of a large number of similar pieces of equipment. For example, a reference DC voltage source may be used to calibrate the voltage parameter of a number of high-resolution digital multimeters of the same make and model but with different serial numbers. During the performance of repeat calibrations, valuable information is generated about the measurement process. If the precision of the measurement process remains unchanged, the standard deviation of the measurement process can be computed based on the set of the data observed on similar pieces of equipment. The variance of various sets of data can be calculated. The weighted mean of these variances weighted by their respective degrees of freedom can then be calculated as the variance of the measurement process. From the calculated value of the variance, an accepted or canonical value of the standard deviation σ of the measurement process can be obtained. The value of the standard deviation established in this way can serve as an index of the precision of the measurement process, which would be treasured information in any calibration laboratory. This standard deviation value that is derived based on experience is an essential element of the statement of uncertainty, but the approach is not practiced by many calibration laboratories.

Thus, σ is an indicator of imprecision. If a large number of data are not available, the value of σ is estimated after calculating the standard deviation s of a small number of observations. There are many ways in which the imprecision of a measurement process can be expressed. This also depends upon how the measurement result is derived. The measurement result can be based on any of the following:

- A single observation made using measuring equipment
- The arithmetic mean of a number of repeat observations
- The grand average of a number of averages
- A calculation based on the functional relationship of several measured variables

A number of indices of imprecision suggested by Ku are given in Table 7.1.

Table 7.1 Indices of imprecision.

Form of result	Index of imprecision
Single observation	The standard deviation σ of a single observation is the index of imprecision. If σ is not known, then the standard deviation(s) of the sample with the associated degree of freedom is the index of imprecision.
Arithmetic mean $\bar{x}$ of n observations	The following measures of error can be used in defining the imprecision of the measurement result: • If σ is known, one index of imprecision is the standard error, i.e., $\sigma / (\sqrt{n})$. • If σ is known and the number of observations is large, the ordinarily used bounds of imprecision are the two-sigma or three-sigma limits. • If the data points are normally distributed, the two-sided confidence interval at a specified confidence level is an index of imprecision. • If σ is known for normally distributed data points, the probable error of the reported value, i.e., $0.6745\sigma / (\sqrt{n})$, is an index of imprecision. However, probable error should not be used if σ is not known. Probable error is a confidence interval corresponding to a 50% confidence level. • The mean deviation or average deviation of measurements from the mean calculated from the sample can serve as an index of imprecision.
Grand average of m averages, each computed from n measurements	If σ is known, then the indices specified above can be used as measures of imprecision. However, if σ is not known, the pooled variance s_p^2 is calculated based on m variances s_i^2 of all data in the set, each containing n observations as per the equation $s_p^2 = \Sigma s_i^2 / m$. In this case, instead of σ, s_p should be used as an estimate of the standard deviation with $m(n-1)$ degrees of freedom.
Calculation based on the functional relationship of several measured variables	In this case, in calculating the result, the means of these variables obtained based on repeat observations should be used. The standard error is calculated by use of the propagation of error formula. This index is appropriate when the errors of measurement are small compared with the values of the variables measured. The standard error of the means of the variables is used in the formula. The number of measurements from which these standard errors are computed should also be reported.

Indices Pertaining to the Systematic Error of the Measurement Process

If the reported value is the numerical value resulting from a measurement process, then the measure of the error due to systematic error is expressed as the reasonable bounds ascribed to the value. The expression of reasonable bounds on the systematic error implies that the systematic error will remain within the bounds most of the time. This can also be expressed as the limits to systematic error. The Eisenhart's guidelines recommend that systematic error be ascertained in one of the following ways:

- Previous reliable establishment of systematic error.
- Estimation of systematic error based on experience or judgment.
- Combination of a number of elemental systematic errors. In this case, the method of combination, such as the simple sum of bounds or the square root of the sum of squares, should be explicitly stated.
- Establishment based on uncertainty in calibrated values. In this case, the meaning of the systematic and random components of uncertainty should be ascertained from the calibration laboratory and a statement made accordingly. Stating the correct interpretation facilitates the making of decisions on the use of these components.

The indices of imprecision of a measurement process are evaluated based on repeat observations from the measurement process. Thus, the indices are obtained from within the measurement process itself. However, to obtain the indices of systematic error we may have to look to external sources beyond the measurement process. One example of a source of information on reliably established systematic error is the technical specification of measuring equipment supplied by the manufacturer. The technical specification contains information on various sources of systematic error that could affect the accuracy of measurement results. How these components should be combined is again a matter on which metrologists do not agree. Some metrologists feel that the quoted uncertainty figure should be safe, implying that the measurement result should at no time go beyond the uncertainty limits. They therefore feel that the addition of all the components of systematic error is the best value of uncertainty due to systematic error. Another group of metrologists believes that this is against the laws of the probabilistic nature of the measurement process and hence that all these error components should be combined as a square root of the sum of squares. It is, however, important that, if the systematic error is a combination of a number of elemental systematic errors, no matter how they are combined, the guidelines stipulate that the method of combination—the simple sum of the bounds or the square root of the sum of squares—should be stated in the calibration certificate.

Another source of information on systematic uncertainty is the uncertainty of the calibrated value. This information is again obtained from a source external to the measurement process, which may be a calibration process within the same calibration laboratory or from an external calibration laboratory.

If information on systematic error is not available from these sources, the calibration laboratory has to estimate the systematic error based on experience or judgment. Estimating systematic error is a real challenge for metrologists and requires a very clear understanding of the physical principles involved in measurement, the technological operations of the measurement setup, and the statistical aspects of the measurement process.

REPORTING UNCERTAINTY STATEMENTS: EISENHART'S GUIDELINES

Eisenhart's guidelines were basically intended for use by calibration laboratories. The objective was to ensure that the uncertainty statement which formed part of a calibration result was unambiguous and contained adequate information so that the user of the calibration result could use it optimally. The guidelines stipulated that as an index of uncertainty, the reported result should be quantified by a quasi-absolute type of statement rather than just a numerical value of the $\pm b$ type.

If the reported value is a numerical value resulting from a measurement process, then the uncertainty statement should consist of two indices as measures of error, one indicating the imprecision of the measurement process and the other the systematic error. The guidelines formulated by Churchill Eisenhart on how uncertainty statements for various situations were to be reported are given below. A few of the examples given by Eisenhart in his paper are also presented. In the guidelines, Eisenhart envisaged four situations and recommended four distinct forms of uncertainty statements corresponding to those situations:

- Systematic error and imprecision both negligible
- Systematic error not negligible, imprecision negligible
- Neither systematic error nor imprecision negligible
- Systematic error negligible, imprecision not negligible

Systematic Error and Imprecision Both Negligible

In this case, the statement should contain an explicit expression of correctness to the last significant figure. The reported result should be given correctly to the number of significant figures consistent with the accuracy requirement of the situation.

Systematic Error Not Negligible, Imprecision Negligible

In this case, the uncertainty statement should be based on either reliably established systematic error or systematic error estimated from experience or judgment:

- A single quasi-absolute type of statement that places bounds on the result's inaccuracy should qualify the reported result.
- These bounds should be stated to no more than two significant figures.
- The reported result itself should be given to the last place affected by the stated bounds, unless it is desired to indicate and preserve such relative accuracy or precision of a higher order that the result may possess for a certain particular use.
- The accuracy statement should be given in sentence form except when a number of results of different accuracies are presented in tabular form. In such a case, the result should be given in the form $a \pm b$ with an appropriate explanatory remark.
- The fact that the imprecision is negligible should be stated explicitly.

The following are examples of appropriate uncertainty statements in this situation:

- . . . is accurate within $\pm x$ units.
- The uncertainty in this value does not exceed . . .

Neither Systematic Error nor Imprecision Negligible

In this case, the uncertainty statement should give indices of both the systematic error and imprecision, as follows:

- The reported result should be qualified by a quasi-absolute type of statement that places bounds on its systematic error and a separate statement of its standard error or probable error or of an upper bound on these, if reliable determination of such a value is available; otherwise, a computed value of standard error or probable error should be given together with a statement of the number of degrees of freedom on which it is based.
- The bounds to the result's systematic error and the measure of its imprecision should be stated to no more than two significant figures.

- The reported result itself should be stated at most to the last place affected by the finer of the two qualifying statements, unless it is desired to indicate and preserve such relative accuracy or precision of a higher order that the result may possess for a certain particular use.
- The qualification of the reported result with respect to its imprecision and systematic error should be given in sentence form except when the results of different precisions or results with different bounds to their systematic errors are presented in tabular arrangement. If it is necessary or desirable to indicate the results' respective imprecisions or the bounds to their respective systematic errors, such information may be given in a parallel column or columns with appropriate identification.

The following are examples of appropriate uncertainty statements in this situation:

- The standard error of these values does not exceed 0.000004 inch, and their systematic error is not in excess of 0.00002 inch.
- . . . with a standard error of x units and a systematic error of not more than $\pm y$ units.

Systematic Error Negligible, Imprecision Not Negligible

In this case, the uncertainty statement should be presented as follows:

- Qualification of the reported value should be limited to a statement of its standard error or of an upper bound thereto, whenever a reliable determination of such value or bound is available. Otherwise, a computed value of the standard error should be given together with a statement of the number of degrees of freedom on which it is based.
- The standard error or upper bound thereto should be stated to no more than two significant figures.
- The reported result itself should be stated to the last place affected by the stated value or bounds to its imprecision, unless it is desired to indicate and preserve such relative precision of a higher order that the result may possess for a certain particular use.
- The qualification of a reported result with respect to its imprecision should be given in sentence form except when results of different precisions are presented in tabular arrangement and it is necessary or desirable to indicate their respective imprecisions, in which event such information may be given in a parallel column or columns with appropriate identification.
- The fact that the systematic error is negligible should be stated explicitly.

The following are examples of appropriate uncertainty statements in this situation:

- . . . with a computed standard error of x units based on v degrees of freedom.
- . . . with an overall uncertainty of ±6.5 km/sec derived from a computed standard error of 1.5 km/sec based on 9 degrees of freedom (the number 6.5 is equal to 4.3×1.5 where 4.3 is the value of the Student t factor for 9 degrees of freedom with a 99.8% confidence level).

LIMITATIONS OF EISENHART'S GUIDELINES

Eisenhart's guidelines were the first attempt to harmonize the concepts and practices of measurement uncertainty evaluation and reporting. The guidelines gave detailed recommendations on the expression of uncertainties for use by calibration laboratories. The recommendations were very specific, with detailed information on various situations, and helped in the harmonization of concepts and practices, thereby removing confusion. However, certain concepts presented in the guidelines are contrary to current concepts:

- The guidelines envisaged uncertainty as consisting of the two independent components of imprecision and accuracy, reflected in the systematic error. Now accuracy is conceived in a broader sense, as explained earlier. The average user of measurement results sometimes needs a single uncertainty figure. The guidelines are silent on how to combine the two uncertainty components pertaining to imprecision and systematic error if needed.
- The guidelines did not define measurement uncertainty as a metrological characteristic, and the term was used interchangeably with the term *error.*
- The terms *systematic error* and *bias* were used in identical senses in the guidelines. In the present context, systematic error consists of two parts, one known and the other unknown. The known systematic error is the bias of the measurement, which can be determined through the calibration process and corrected for. The unknown systematic error is a matter of concern to the metrologist in evaluating uncertainty, also called systematic uncertainty.
- The guidelines did not define the criteria for the negligibility of the indices of imprecision and systematic error.

RANDOM AND SYSTEMATIC UNCERTAINTY: GUIDELINES FOR ESTIMATION AND REPORTING

Churchill Eisenhart contributed a great deal by bringing the issue of measurement uncertainty into focus in calibration laboratories. By the 1960s, the study of uncertainty had developed into a discipline of knowledge within the science of measurement. The subject of the estimation and reporting of the uncertainty of measurement results became more focused in the 1970s and 1980s in most of the developed and developing countries, primarily because of the following factors:

- There was an increased need for the comparison of measurement results from various quarters in order to determine compliance with the requirements of a specification or standard. This became important in view of increasing international trade.
- Ensuring uniformity in the inter comparison of the national standards maintained by various national metrology institutions became important in order to facilitate international traceability.
- Technological advances led to a need for measurement in new areas such as materials science and manufacturing technologies.
- Increased awareness arose on the part of calibration laboratories of the importance of uncertainty statements as part of calibration results, and in addition the clients of calibration laboratories started demanding uncertainty statements with calibration results.

Generally, metrologists working in apex-level calibration laboratories are well versed in the application of statistical techniques to the evaluation of measurement uncertainty. This covers both the "why" and "how" aspects of the statistical techniques used in measurement uncertainty evaluation. However, a need was felt for guidelines on the evaluation of measurement uncertainty and its reporting that did not involve too much statistics so that the guidelines would be used by calibration laboratories at all levels uniformly. A number of guideline documents were prepared by various agencies to meet these objectives, including the following:

- British Calibration Service Document 3003, *The Expression of Uncertainty in Electrical Measurements*
- Ministry of Defence, United Kingdom, Defence Standard 00-26, no. 1, dated 15 February 1982, *Guide to the Evaluation and Expression of Uncertainties Associated with Results of Electrical Measurements*
- Commonwealth Science Council Documents NOCSC(80) MS-8 and CSC(80) MS-9, *Guidelines for Estimation and Statement of Overall Uncertainty in Measurement Results*

Most of these documents are identical in content and approach. They are primarily based on the NBS document; *Experimental statistics,* however, some changes in terminology, classification of uncertainty components, and so forth have been introduced.

Random and Systematic Uncertainties

Errors have been historically classified as random or systematic. Random errors are caused by random variations in the outcome of the measurement process due to various influencing factors which affect the process individually and collectively. This is an inherent characteristic of the measurement process. Variations due to random effects can be minimized at a cost but cannot be eliminated altogether. Systematic error, on the other hand, is generally constant and is due to limitations of the measurement setup and the measurement method, among other factors. Since errors and uncertainty have a cause-and-effect relationship, uncertainty is also classified as random or systematic.

Random Uncertainty

This is uncertainty in the measurement result caused by random variation in the outcome of the measurement process due to various sources of random errors. It is analogous to the imprecision of the measurement process as defined in Eisenhart's guidelines. Random uncertainty is evaluated by statistical analysis of repeat observations.

Systematic Uncertainty

This is uncertainty caused by various sources of systematic errors present in the measurement process. A known systematic error or bias of the measurement setup can always be corrected, but there are other sources such as the inherent inaccuracy of measuring equipment which cannot be corrected. Systematic uncertainty components cannot be analyzed based on statistical analysis of repeat observations. This type of uncertainty is analogous to the uncertainty of systematic error as defined in Eisenhart's guidelines. Some representative sources of systematic uncertainty are as follows:

- Uncertainty reported in the calibration certificate of a measuring instrument, measurement standards, or reference materials
- Uncertainty due to the interpolation of reference values obtained from calibration charts
- Uncertainty due to the operation of a piece of equipment or a standard under different climatic conditions than those at which it was calibrated
- Inherent uncertainty of a measurement method

- Hysterisis of measuring equipment
- Resolution of digital measuring equipment
- Different impedance conditions in radio frequency measurements

ESTIMATION OF RANDOM UNCERTAINTY

Case 1: Known Standard Deviation σ of the Calibration Process

Calibration laboratories use various setups for calibrating specific types of measuring equipment. For example, a multifunction calibrator is used for calibrating digital multimeters of a specific make and model. The data generated during the calibration of various pieces of equipment of that specific make and model can be used to estimate the standard deviation σ of the calibration process. Since pieces of equipment of the same make and model have identical technical specifications and behave the same way technically, the value of σ obtained in this way can be used for all equipment of the same make and model.

In this process, large number of data is gathered during repeat observations obtained on the various pieces of measuring equipment using the same calibration process. If we intend to estimate σ for a specific value, say 50.0 volts DC at a range of 100 volts DC, we have data on repeat observations of 50.0 volts DC as obtained on m pieces of equipment. If s_1^2, s_2^2 . . . and so on, are variances of the data for the same number of observations obtained on various pieces of equipment while calibrating 50.0 volts DC at a 100-volt range, the standard deviation σ of the calibration process can be estimated by the following equation:

$$\sigma^2 = (s_1^2 + s_2^2 \ldots + s_m^2) / m$$

If m is large, the estimated σ will approach the true σ of the calibration process. Once the value of σ has been estimated, it can be used to estimate the random uncertainty component rather than recomputing it every time the equipment is calibrated by that calibration process. If the number of observations is not the same for the calculation of the variances, the weighted mean of the variances weighted by their respective degrees of freedom can be calculated. From this value of variance, σ can be calculated.

The calibration result is quoted as a mean of n repeat observations x_1, x_2, . . . x_n such that $\bar{x}$ = is equal to $(x_1 + x_2 + \ldots + x_n) / n$. If σ is the standard deviation of the calibration process, the standard deviation of the mean is equal to

$\sigma / \sqrt{n}$. We know from the theory of normal distribution that the true value of the measurand lies in the following interval:

$$\bar{x} + k\sigma / \sqrt{n} > \mu > \bar{x} - k\sigma / \sqrt{n}$$

$$\text{or} \quad \mu + k\sigma / \sqrt{n} > \bar{x} > \mu - k\sigma / \sqrt{n}$$

where μ is the so-called true value of the parameter.

Thus, the random uncertainty is given by $k\sigma / \sqrt{n}$ where n is the number of repeat observations based on which the calibration result is estimated. σ is the known standard deviation of the calibration process, and k is a multiplier constant whose value depends upon the confidence level at which the random uncertainty is being estimated. The value of k is 1.96 for a 95% confidence level, 2.00 for a 95.5% confidence level, 2.58 for a 99% confidence level, and 3.00 for a 99.7% confidence level.

Case 2: Unknown Standard Deviation σ of the Calibration Process

In this situation, the standard deviation s of a small number of repeat observations is calculated to provide an estimate of the calibration process standard deviation σ. The number of repeat observations is small, generally on the order of five. As the normal distribution is applicable to large samples, the multiplying factors obtained from the normal distribution cannot be used with a small sample size. To find the random uncertainty, the standard deviation is multiplied by the t factor of the Student t distribution.

If $x_1, x_2, \ldots x_n$ are repeat observations, n being on the order of five, the average of these repeat observations $\bar{x}$ is the calibration result and the standard deviation s is obtained from $s^2 = \Sigma(x_i - \bar{x})^2 / (n - 1)$. From the theory of Student t distribution, the true value of the parameter is expected to lie within the following interval:

$$\bar{x} + ts / \sqrt{n} > \mu > \bar{x} - ts / \sqrt{n}$$

$$\text{or } \mu + ts / \sqrt{n} > \bar{x} > \mu - ts / \sqrt{n}$$

The random uncertainty is therefore estimated as $ts / \sqrt{n}$, and the multiplier factor t is the Student t factor, whose values are available in statistical tables. One such table is given in Appendix B of this book. The value of t depends upon the following factors:

- The number of degrees of freedom of the repeat observations, which is one less than the number of repeat observations. As the degrees of freedom increase, the value of t decreases rapidly in the beginning, but the rate of decrease also decreases with the increase in the degrees of freedom. This implies that with a greater number of observations, not only does the t factor decrease but the $s / \sqrt{n}$ factor also decreases. Thus, the random uncertainty also decreases with an increase in the number of repeat observations. For large values of degrees of freedom, on the order of 30 and above, the t values tend toward the k values of normal distribution. For infinite degrees of freedom, the t values and k values are the same.

- The confidence level at which the uncertainty is desired to be estimated. The value of the t factor increases with a higher confidence level. Thus, for the same set of repeat observations, the random uncertainty will be greater at a higher confidence level than at a lower confidence level.

Steps in the Estimation of the Random Uncertainty U_r

The following steps are used to estimate the random uncertainty U_r:

1. Take n repeat observations of the specific parameter under the same conditions. The value of n should be between 3 and 10. Let the repeat observations be $x_1, x_2, \ldots x_n$
2. Determine $\bar{x}$ the arithmetic mean of the n observations

$$\bar{x} = (x_1 + x_2 + \ldots + x_n) / n$$

   $\bar{x}$ is the estimate of the true value of the parameter under measurement
3. Determine the sample variance and standard deviation s of the repeat observations with the equation

$$s^2 = \Sigma(x_i - \bar{x})^2 / (n - 1)$$

4. The sample standard deviation s is the estimate of the standard deviation of the population parameter
5. Determine the standard error of the mean as $s / \sqrt{n}$
6. Decide at what confidence level the uncertainty is to be estimated and determine the number of degrees of freedom, which is equal to $n - 1$
7. Determine the Student t factor for a given confidence level and the number of degrees of freedom
8. The random component of uncertainty U_r is given by

$$U_r = ts / \sqrt{n}$$

ESTIMATION OF SYSTEMATIC UNCERTAINTY

Various factors contribute to the systematic component of uncertainty. The most important of these are the measuring apparatus, the operating procedure, and the inherent characteristics of the device under calibration. As measurement uncertainty is a probabilistic concept, the evaluation of systematic uncertainty also requires knowledge about the probability distribution of the sources of systematic error. In random uncertainty, the standard deviation is calculated based on repeat observations. To estimate systematic uncertainty, information on the standard deviation of the probability distribution is needed. This information is evaluated based on the statistical behavior of these uncertainty components and on experience of their behavior as specified by equipment manufacturers, calibration laboratories, and users.

The following steps are used to estimate the systematic uncertainty U_s:

1. Identify all the sources of systematic error that are contributing to the systematic uncertainty.
2. Determine their probability distributions and determine their standard deviations separately. If σ_{s1}, σ_{s2}, etc . . . are the standard deviations of various uncertainty components, the standard deviation of the combined uncertainty σ_s is given by

$$\sigma_s = \sqrt{(\sigma_{s1}^2 + \sigma_{s2}^2 + \ldots)}$$

3. The systematic uncertainty is evaluated by the formula $U_s = k\,\sigma_{s.}$ Where k is a multiplier obtained from a normal table, the limiting value of t for infinite degrees of freedom is also equal to k and can be obtained from the t-table. The value of k should be used for the same confidence level as that used for estimating $U_{r.}$
4. If all the systematic uncertainty components follow a rectangular probability distribution with semi-ranges, a_1, a_2, . . . and so on, then from the theory of rectangular distribution, the variance is given by

$$\sigma_s^2 = (a_1^2 + a_2^2 + \ldots) / 3$$

Standard deviation of combined uncertainty = σ_s

$$U_s = k\sigma_s$$

5. If the systematic uncertainty component U_s is found to be greater than the arithmetic sum of all the semi-ranges, then a dominant component

of uncertainty is believed to be present. In this case, the largest semi-range is taken out as the dominant component *ad,* and the systematic uncertainty is estimated as follows:

$$U_s = ad + U'_s.$$

U'_s is computed in the same way as U_s except that the dominant component *ad* is taken out and the combined variance is calculated using the remaining variance components.

OVERALL UNCERTAINTY

Normally the random and systematic components of uncertainty should be quoted separately. Quoting them separately helps in determining the relative dominance of the random and systematic uncertainty components and may be helpful in subsequent decision making. This practice is in line with Eisenhart's guidelines on quoting the uncertainty due to imprecision and the uncertainty due to systematic error separately. If it is desired to quote a single value of uncertainty, the random and systematic uncertainties should be combined as follows.

- If there is no dominant component of systematic uncertainty:

$$U = \sqrt{(U_r^2 + U_s^2)}$$

- If a dominant component of systematic uncertainty is present:

$$U = ad + \sqrt{(U_r^2 + U'^2_s)}$$

The result is expressed as $\bar{x} \pm U$ at a specified confidence level.

SELECTION OF THE CONFIDENCE LEVEL

The confidence level indicates the probability that the value of the measurand lies within the stated interval around the assigned result. Since the confidence level is a probability, it can vary between 0% and 100%. A 100% confidence level means that the true value is certain to lie within the stated interval. If we intend to state the uncertainty at a higher confidence level, the *t* value increases and the uncertainty figure also increases. At too high a confidence level, the uncertainty figure becomes too large and loses its information content, particularly if the number of observations is small.

For example, suppose that the uncertainty is estimated based on three repeat observations. From the table in Appendix B, it is noted that the *t* value for a 95% confidence level is 4.30, whereas for a confidence level of 99.73% it is 19.21; thus, the ratio of the *t* values for the 99.73% and 95% confidence levels is 4.47. The uncertainty figure is therefore multiplied by a factor of 4.47. An uncertainty figure of 1% of the measurement result at the 95% confidence level becomes 4.47% at the 99.73% confidence level. Based on this, we can conclude that the measurement result lies within 1% of the measurement result 19 out of 20 times on average, and it lies within 4.47% 27 out of 10,000 times on average. Too low a confidence level can create a false feeling about the quality of a measurement result, because the estimated uncertainty will be low. Normally, if the measurement result is within the stated value 95% of the time, or 19 out of 20 times, one is satisfied with it. In many situations, if the confidence level is not stated with the uncertainty figure, there is every chance that the measurement result will be interpreted wrongly. Generally, a confidence level of 95% is adequate for normal measurement applications. A high confidence level such as 99.73% is usually required only in very specific applications.

SELECTION OF THE NUMBER OF OBSERVATIONS

It can be observed from the *t* values given in Appendix B that the *t* factor for a single confidence level decreases with an increasing number of observations. The decrease is rapid in the beginning and slowly stabilizes as the number of observations increases. Thus, with an increased number of repeat measurements, the measurement uncertainty decreases. The number of measurements to be made should be decided based on the cost involved in taking the measurements and on the uncertainty requirements for a particular measurement application. Generally, the number of observations varies between 3 and 10.

EXAMPLES OF UNCERTAINTY EVALUATIONS

The inclusion of an uncertainty statement is mandatory in a calibration result. The following examples illustrate how the uncertainty of a calibration result is estimated based on the methods described in this chapter.

Example 1

A 10.000-volt DC reference voltage source is used to calibrate a digital multimeter. The uncertainty of the source according to its technical specifications is

±(120 ppm + 150 microvolts). The repeat observations of the reference voltage source as measured by the digital multimeter are 9.992, 9.990, 9.996, 9.994, and 9.991. It is desired to evaluate the various uncertainty components and the overall uncertainty at the 95% and 99% confidence levels.

The mean value of the above observations is given by

$\bar{x} = (9.992 + 9.990 + 9.996 + 9.994 + 9.991) / 5 = 9.9926.$

The calibration result is 9.9926, implying that the digital multimeter reads a 10.000-volt DC source as 9.9926 volts.

Evaluation of random uncertainty U_r:

Estimated standard deviation $s = 2.4083 \times 10^{-3}$

$U_r = ts / \sqrt{n}, s / \sqrt{5} = 1.077 \times 10^{-3}$

$t = 2.78$ at 95% confidence level and 4 degrees of freedom

$t = 4.60$ at 99% confidence level and 4 degrees of freedom

$U_r = 2.78 \times 1.077 \times 10^{-3} = 0.003$ volt at the 95% confidence level

$U_r = 4.60 \times 1.077 \times 10^{-3} = 0.005$ volt at the 99% confidence level

Evaluation of systematic uncertainty U_s:

120 ppm of 10.000 volts = 0.0012 volt

Uncertainty according to technical specifications = 0.0012 + 0.000150 = 0.00135 volt

Assuming that this uncertainty follows a normal distribution at the 95% confidence level, the multiplying factor $k = 1.96$.

Standard deviation of the systematic uncertainty component $\sigma_s = 0.00135 / 1.96 = 0.0007$ volt

$U_s = 0.0007 \times 1.96 = 0.00135$ volt at the 95% confidence level

$U_s = 0.0007 \times 2.586 = 0.00180$ volt at the 99% confidence level

Evaluation of overall uncertainty U:

$U = 3.29 \times 10^{-3} = 0.0033$ volt at the 95% confidence level

$U = 5.31 \times 10^{-3} = 0.0053$ volt at the 99% confidence level

Example 2

A 1-milliohm standard four-terminal resistor is used to calibrate a double Kelvin bridge. The following systematic uncertainty components are known for the value of the 1-milliohm standard resistor.

- The component of systematic uncertainty quoted in its calibration certificate: $a_1 = \pm 1 \times 10^{-6}$ ohm with a rectangular probability distribution
- The component of systematic uncertainty due to instability between calibration intervals: $a_2 = \pm 2 \times 10^{-6}$ ohm with a rectangular probability distribution

The repeat observations of the standard resistor value as measured by the Kelvin bridge in ohms are 0.0010014, 0.0010018, 0.0010014, 0.0010016, and 0.0010012.

Mean of the observations $\bar{x} = 0.00100148$ ohm

Evaluation of U_r:

Estimated standard deviation $s = 2.28 \times 10^{-7}$

$U_r = (2.78 \times 2.28 \times 10^{-7}) / \sqrt{5} = 2.83 \times 10^{-7}$ ohm at the 95% confidence level

$U_r = (4.60 \times 2.28 \times 10^{-7}) / \sqrt{5} = 4.69 \times 10^{-7}$ ohm at the 99% confidence level

Evaluation of U_s:

Standard deviation $\sigma_s = (\sqrt{2^2 + 1^2}) / \sqrt{3} = 1.29 \times 10^{-6}$

$U_s = 1.96 \times 1.29 \times 10^{-6} = 2.53 \times 10^{-6}$ ohm at the 95% confidence level

$U_s = 2.58 \times 1.29 \times 10^{-6} = 3.33 \times 10^{-6}$ ohm at the 99% confidence level

The sum of semi-ranges $a_1 + a_2 = 3 \times 10^{-6}$

Since U_s (at the 99% confidence level) > $a_1 + a_2$, a dominant systematic component of uncertainty is present which is 2×10^{-6}.

Hence, $U_s = 2.53 \times 10^{-6}$ ohm at the 95% confidence level

$U_s = 2 \times 10^{-6} + 2.58 \times / \sqrt{1/3} \times 10^{-6} = 3.49 \times 10^{-6}$ ohm at the 99% confidence level

Evaluation of total uncertainty U:

$U = \sqrt{(2.53^2 + 0.283^2)} \times 10^{-6} = 2.55 \times 10^{-6}$ ohm at the 95% confidence level

$U = 2 + \sqrt{(1.49^2 + 0.469^2)} \times 10^{-6} = 3.56 \times 10^{-6}$ ohm at the 99% confidence level

Example 3

A low-power-factor wattmeter with a range of 2.5 watts is calibrated by a power calibrator at a power factor of 0.2. The power from the calibrator is adjusted so as to get a value of 2.5 watts in the wattmeter under calibration. The repeat observations of the power setting in the power calibrator are 2.5977, 2.5984, 2.5982, 2.5977, and 2.5971.

The uncertainty of the power setting in the power calibration at 2.5 watts is 0.0013 watts.

Average of the observations $\bar{x} = 2.5978$ watts

Estimate of standard deviation $s = 5.07 \times 10^{-4} = 0.0005$ watts

Evaluation of U_r:

$U_r = (2.78 \times 0.0005) / \sqrt{5} = 0.0006$ watt at the 95% confidence level

$U_r = (4.60 \times 0.0005) / \sqrt{5} = 0.0010$ watt at the 99% confidence level

Evaluation of U_s:

Standard deviation $\sigma_s = 0.00075$ watt

$U_s = 1.96 \times 0.00075 = 0.0015$ watt at the 95% confidence level

$U_s = 2.58 \times 0.00075 = 0.0019$ watt at the 99% confidence level

Evaluation of total uncertainty U:

$U = \sqrt{(0.0006^2 + 0.0015^2)} = 0.0016$ watt at the 95% confidence level

$U = \sqrt{(0.001^2 + 0.0019^2)} = 0.0021$ watt at the 99% confidence level

THE NEED FOR A HARMONIZED APPROACH TO MEASUREMENT UNCERTAINTY EVALUATION AND REPORTING

The need to report measurement uncertainty as part of a measurement result was first felt at the U.S. National Bureau of Standards, and the early guidelines for its evaluation were formulated by Churchill Eisenhart, Harry H. Ku, and others. They conceived measurement uncertainty as having two distinct components, originating from imprecision and systematic error. Both the components were required to be reported separately.

Subsequently, efforts were made to establish a clear-cut distinction between error and uncertainty as two different attributes of a measurement result. Uncertainty was attributed to two types of error: random and systematic. The resulting uncertainties were called random and systematic uncertainties. Random uncertainty was analogous to imprecision, and systematic uncertainty was analogous to the systematic error of Eisenhart's earlier approach. However, in the newer approach, it was accepted that measurement uncertainty could be quoted as a single figure, called overall uncertainty. Rules were also evolved for combining various uncertainty components into a single figure.

This approach to evaluating random, systematic, and overall uncertainty was practiced by metrologists globally and remained in use until the mid-1990s. The approach, however, had certain limitations, such as a lack of specific criteria for classifying random and systematic uncertainties and nonuniformity in combining them. In addition, there were no specific guidelines on how to report an uncertainty statement and how to select a confidence level. Thus, the interpretation of an uncertainty statement depended on the concerned individual's level of understanding of the concept. This resulted in difficulty in comparing measurement results and in mismatches in metrological communication.

This difficulty has been overcome through the efforts of international metrology organizations and a number of national metrology institutions. With international consensus, a new guidance document for the evaluation and expression of uncertainty in measurement was published in 1995. This document, which covers the harmonized approach to this subject, is dealt with in chapter 8.

REFERENCES

Churchill, Eisenhart. 1968. Expression of uncertainties of final results. *Science* 10:69-1201 to 72-1204.

International Organization for Standardization (ISO)and International Electrotechnical Commission (IEC). 1999. ISO/IEC 17025:1999-12-15, *General requirements for the competence of testing and calibration laboratories.* Geneva, Switzerland: ISO.

Ku, Harry H. 1968. Expression of imprecision, systematic error and uncertainty associated with a reported value. *Measurement and Data,* July–August, 73-72 to 77-76.

National Physical Laboratory of India. 1989. *Guidelines for estimation and statement of overall uncertainty in measurement results.* New Delhi: Department of Science and Technology, Government of India.

National Physical Laboratory of the United Kingdom. 1995. *The expression of uncertainty and confidence in measurement for calibration.* NIS 3003, ed. 8. Middlesex, England: NAMAS NPL.

8

Guide to the Expression of Uncertainty in Measurement (GUM): An International Consensus

LACK OF CONSENSUS ON UNCERTAINTY EVALUATION: ISSUES OF CONCERN

The importance of reporting uncertainty as part of a measurement result is accepted by all concerned. Metrologists have made efforts to harmonize the methods for uncertainty evaluation and its reporting, and various approaches and methods for this are described in the previous chapter. However, the method of evaluating and reporting the uncertainty of measurement results has always been a subject on which consensus could not be reached among metrologists. The disputes pertain to the following aspects of uncertainty evaluation.

Safe Versus Realistic Evaluation

The evaluation of uncertainty of measurement is an involved task. One needs not only a clear understanding of the technological aspects of the measurement process but also a good understanding of statistical analysis of measurement data. Most metrologists consider this a problematic task. The orthodox view has been that the reported value of uncertainty should be safe. This implies that the measurement result should always be expected to lie in the interval covered by the reported uncertainty even if the uncertainty quoted is on the higher side. It means that one must not err on the side of the uncertainty being too small; however, erring on the other side is considered acceptable. According to the orthodox view, all uncertainty components should be added, and the quoted uncertainty should be the sum of all individual uncertainty components. Other schools of metrologists have believed that the addition of all uncertainty components to arrive at an overall figure of uncertainty goes against the laws of probabilistic behavior. They therefore have advocated that evaluation of overall uncertainty should be based on the principle of the root of the sum of the squares of the various uncertainty components, as is done while combining variances in statistical analysis.

To arrive at a judicious decision based on measurement result and associated uncertainty, the uncertainty figure needs to be realistic rather than safe. A safe uncertainty figure may amount to overstatement of uncertainties. This may result in the unnecessary rejection of a costly product. In the case of a calibration laboratory, the overstatement of uncertainties may lead to unacceptability of the services of that laboratory by its clients. Sometimes one may be compelled to purchase costly equipment unnecessarily due to overstatement of uncertainty. Similarly an understatement of the uncertainty of measurement may lead to undeserved overconfidence in the measurement result, which may result in serious problems. Thus, both understatement as well as overstatement of the uncertainty figure is undesirable; uncertainty of the measurement result has to be realistic.

Random Versus Systematic Uncertainty

Measurement uncertainties have been classified as random and systematic, a classification primarily derived from the classical categorization of measurement errors as random and systematic. Since the origin and nature of random and systematic errors are different, uncertainties associated with them were also considered to be of a different nature. The rules for their combinations were also different: for example, random uncertainties were combined using the root of the sum of the squares, but the same was not considered applicable to systematic uncertainties, which were combined as simple addition. Some metrologists were of the opinion that systematic uncertainties cannot be analyzed by statistical methods. The guidelines provided by Churchill Eisenhart required that standard error as a measure of precision and systematic error should be reported separately.

In fact, the classification of uncertainties as random and systematic is artificial. Systematic uncertainties can be evaluated by statistical means, just as random uncertainty can. For example, the uncertainty quoted in the calibration certificate of a standard was considered to be a systematic uncertainty. While calibrating the standard the higher-echelon calibration laboratory arrives at an uncertainty figure based on repeat measurements. Thus, in a hierarchical calibration system the random uncertainty reported by a higher-echelon laboratory forms part of the total uncertainty. This uncertainty of the reference standard is taken as systematic uncertainty by a lower-echelon laboratory. Thus, the random and systematic uncertainties are not uniquely identifiable, and they change position depending upon where one is in the echelon chain. This concept is confusing, and it has not lent clarity to the uncertainty concepts.

Moreover, there has been no consensus on how the so-called random and systematic uncertainty components are to be reported. Consider the following two cases:

- **Coverage factor and confidence level:** The overall uncertainty figure is evaluated by multiplying the standard deviation of the mean by a multiplying factor t for random uncertainty and k for systematic uncertainty. Random uncertainty depends on the confidence level and the degree of freedom, but there have been no guidelines on at what confidence level random uncertainty is to be reported and how many repeat measurements should be taken to evaluate it.
- **Uncertainty reporting:** No consensus existed regarding how an uncertainty statement is to be reported. The practice was just to quote a figure $\pm U$ along with the measurement result. In the absence of detailed information as to how this figure was arrived at, it was difficult for the user of the measurement result to extract information on the quality of the measurement result and make a decision judiciously. Information such as the value of individual uncertainty components, degrees of freedom, confidence level, and so forth were generally not provided. In metrology many times the uncertainty figure is required to be transferred to another measurement result, particularly in calibration laboratories. In the absence of detailed information, it was not easy to transfer the uncertainty figures.

Problems in Intercomparison

The lack of consensus among metrologists on methods for evaluating and reporting uncertainties makes comparing measurement results difficult. Meaningful conclusions regarding comparison of measurement results can be drawn only if the measurement uncertainties for all the results being compared have been evaluated by the same method and reported in the same way by all concerned. Sometimes measurement results must be compared among themselves, and other times against the requirements of a standard or specification.

Comparison Among Themselves

The measurement results obtained on national standards maintained by national metrology laboratories are compared among themselves to ensure international traceability. This task is coordinated by the International Bureau of Weights and Measures (BIPM). The lack of consensus on the evaluation and reporting of uncertainty of measurement has been affecting the international comparison of measurement standards.

Measurement results are also compared among themselves with a view toward ascertaining the technical competence of accredited testing and calibration laboratories. This comparison is done under the aegis of laboratory accreditation bodies. During the process, measurement results from various laboratories are compared and any laboratory reporting results outside the statistical limits of tolerance is considered to be an outlier and is asked to investigate the cause for this. This exercise when used by an accreditation body to assess the competence of testing and calibration laboratories is called proficiency testing. This is one of the effective measurement assurance techniques in use.

Measurement results are also intercompared during the process of calibration. Measurement results obtained by equipment under calibration are compared against the values attributed to the reference standard used for calibration to make a decision on compliance of the equipment with its accuracy specifications.

Comparison Against Specification

In manufacturing and service organizations, measurement results are compared against the requirement of a standard specification. The objective of that comparison is to make a decision regarding the product's compliance or noncompliance with the requirements of the specification.

The intercomparison of measurement results either among themselves or against the requirements of a specification is done with a view toward making judicious decisions. In the absence of a uniform approach to estimating and reporting measurement uncertainty, obtaining a consensus decision can be difficult as we can compare only similar entities.

THE NEED FOR A HARMONIZED APPROACH TO UNCERTAINTY EVALUATION AND REPORTING

The acceptance of the International System of Units (SI) harmonized measurement systems globally. For intercomparison of measurement results and with the need for their common understanding, it was considered essential to evolve a harmonized method of evaluating and reporting the uncertainty of measurement results that would be accepted worldwide. To achieve global acceptability, it was considered essential that the harmonized method be *universal* (that is, applicable to all types of measurements in physical and chemical metrology and all types of input data used in measurement) and *straightforward* (that is, easy to understand and implement). The uncertainty figure obtained by this method should be *transferable* and *internally consistent. Transferable* means that if a measurement result is used in another measurement, it should be possible to use the uncertainty figure of the first result in evaluating the uncertainty of the second measurement result. *Internally consistent* means that the measurement uncertainty should be derivable from the components that contribute to it as well as independent of how these components are grouped and the decomposition of the components into subcomponents.

In view of the importance of international traceability of national standards of measurements, the International Committee for Weights and Measures (CIPM) in 1977 requested that the BIPM address this problem in association with national metrology laboratories and make recommendations on a harmonized approach to uncertainty evaluation. Based on the coordinated efforts of the BIPM and deliberations among a number of national metrology laboratories, it was agreed that it is important to arrive at an internationally accepted procedure for evaluating and expressing measurement uncertainty and combining individual uncertainty components into a single total uncertainty. However, no consensus could be reached regarding what methods should be used for this purpose. To evolve a consensual approach on this issue, the BIPM appointed a working group with the participation of experts from 11 national metrology laboratories. That group, called the Working Group on the Statement of Uncertainties, developed a set of recommendations titled Recommendation INC-1 (1980), *Expression of Experimental Uncertainties.* These recommendations were approved by the CIPM in 1981 and reaffirmed in 1986. They are as follows:

1. The uncertainty in the result of measurement generally consists of several components that may be grouped into two categories according to the way in which their numerical value is estimated.
   A. Those which are evaluated by statistical method
   B. Those which are evaluated by other means
   There is not always a simple correspondence between the classification into categories A or B and the previously used classification into "random" and "systematic" uncertainties. The term "systematic uncertainty" can be misleading and should be avoided. Any detailed report of the uncertainty should consist of a complete list of the components, specifying for each method used to obtain its numerical value.
2. The components in category A are characterized by the estimated variance s_i^2 (or the estimated "standard deviation" s_i), and the number of degrees of freedom; where appropriate the covariances should be given.

3. The components in category B should be characterized by quantities u_j^2, which may be considered as approximations to the corresponding variances, the existence of which is assumed. The quantities u_j^2 may be treated like variances and the quantities u_j like standard deviations. Where appropriate, the covariances should be treated in a similar way.
4. The combined uncertainty should be characterized by the numerical value obtained by applying the usual method for the combination of variances. The combined uncertainty and its components should be expressed in the form of "standard deviations."
5. If for a particular application it is necessary to multiply the combined uncertainty by a factor to obtain an overall uncertainty, the multiplying factor must always be stated.

The CIPM reviewed the recommendations of the BIPM Working Group on the Statement of Uncertainties and adopted the following recommendations in 1981:

- The BIPM should attempt to apply its working group's recommended approach to uncertainty estimation while carrying out international comparison under its auspices
- The proposals are to be diffused widely
- Other interested organizations are to be encouraged to examine and test these proposals and send their comments to the BIPM
- The BIPM is to report back on the application of these proposals after two or three years

In considering the recommendations of the working group, the CIPM noted the following and recognized that the working group's proposal might form the basis of an eventual agreement on the expression of uncertainties:

- There is a need to find an agreed-upon way of expressing measurement uncertainties in metrology
- Many organizations have devoted a lot of effort to this issue over the years
- The discussion of the working group has resulted in good progress in finding an acceptable solution

GUIDE TO THE EXPRESSION OF UNCERTAINTY IN MEASUREMENT (*GUM*)

The working group recommendations had only outlined the approach and concepts; there was still a necessity for a detailed guidance document on this subject which specified the methods and procedures for evaluating and reporting measurement uncertainty. The document was expected to be used globally and was expected to harmonize practices so that there would be no ambiguity associated with the evaluation and interpretation of quoted uncertainty values. The CIPM referred the task of preparing this document to the International Organization for Standardization (ISO).

The ISO is an international body primarily engaged in the preparation of standards and specifications in various technological areas related to products, systems, and methods of measurement. The task was thus assigned to the ISO Technical Advisory Group on Metrology (TAG-4). This group includes representatives from other international organizations having an interest in metrology, viz., the International Electrotechnical Commission (IEC), the International Organization of Legal Metrology (OIML), the International Union of Pure and Applied Chemistry (IUPAC), the International Union of Pure and Applied Physics (IUPAP), and the International Federation of Clinical Chemistry (IFCC).

TAG-4 established a working group (WG-3) to prepare a guidance document on the statement of uncertainties based on the BIPM working group's Recommendation INC-1 (1980). The guide was expected to provide rules on the expression of measurement uncertainty for use with standardization, calibration, laboratory accreditation, and metrology services. The guide was also expected to do the following:

- Promote full information on how uncertainty statements are derived
- Provide a basis for international comparison of measurement results

The working group ISO/TAG-4/WG-3 consisted of experts nominated by the BIPM, the IEC, the ISO, and the OIML. The working group prepared the guidance document called *Guide to the Expression of Uncertainty in Measurement* (*GUM,* for short) in 1993. That document has been jointly published by a number of international organizations, viz., the BIPM, the IEC, the IFCC, the ISO, the IUPAC, the IUPAP, and the OIML. The document is based on international consensus; a corrected version of the document was reprinted in 1995.

Applicability of *GUM*

The guide is applicable to a broad spectrum of measurements at various levels of accuracy and in many fields from the shop floor of an industry to fundamental research. Here are some areas where the guide finds application:

- Quality control and quality assurance in production
- Complying with and enforcing laws and regulations
- Conducting basic research and applied research and development in science and engineering

- Calibrating standards and instruments and performing tests throughout a national measurement system in order to achieve traceability to national standards
- Developing, maintaining, and comparing international and national physical reference standards including reference materials

A measurement result and its uncertainty may be used for different purposes (for example, to draw a conclusion about the compatibility of a result with other similar results, to establish tolerance limits in a manufacturing process, or to decide if a certain course of action may be safely undertaken). However, the guide's authors have made it very clear that the guide addresses only how measurement uncertainty is to be evaluated and reported. It does not address how the uncertainty of a measurement result, once evaluated, may be used.

GUM and Recommendation INC-1 (1980)

GUM was prepared with a view toward implementing the recommendations of Recommendation INC-1 (1980), which the CIPM approved in 1981. The first recommendation states that the various uncertainty components should be classified into two categories according to the way in which their numerical value is estimated, rather than as random and systematic uncertainties as was the accepted classification until publication of the guide. Thus, the type-A and type-B classification as recommended by the BIPM working group is based purely on the method of evaluation rather than on an uncertainty's source of origin. *GUM*[1] has defined the two terms as follows:

- **Type-A evaluation (of uncertainty):** Method of evaluation of uncertainty by the statistical analysis of a series of observations.
- **Type-B evaluation (of uncertainty):** Method of evaluation of uncertainty by means other than the statistical analysis of a series of observations.

Another recommendation states that uncertainty components in category A should be characterized by estimated variance or standard deviation and degrees of freedom. Similarly, uncertainty components in category B should also be characterized by variance and standard deviation, the existence of which is assumed. To meet these recommendations *GUM* has brought in the concept of standard uncertainty as the basic unit of uncertainty which should be used for the evaluation of combined standard uncertainty and overall uncertainty, which has been named *expanded uncertainty. GUM* thus envisages that there will be a standard uncertainty component corresponding to each influence factor that contributes to uncertainty. The definitions of various uncertainty terms used in *GUM*[1] are as follows:

- **Standard uncertainty:** Uncertainty of the result of a measurement expressed as a standard deviation.
- **Combined standard uncertainty:** Standard uncertainty of the result of a measurement when that result is obtained from the values of a number of other quantities, equal to the positive square root of the sum of the terms, the terms being the variances or covariance of these other quantities weighed according to how the measurement result varies with change in these quantities.
- **Expanded uncertainty:** Quantity defining an interval about the result of a measurement that may be expected to encompass a large fraction of the distribution of values that could reasonably be attributed to the measurand. This was termed *overall uncertainty* in Recommendation INC-1 (1980).
- **Coverage factor:** Numerical factor used as a multiplier of the combined standard uncertainty in order to obtain an expanded uncertainty. The coverage factor k is typically in the range 2 to 3.

PHILOSOPHY OF *GUM* AND NEW DEFINITION OF UNCERTAINTY AND ASSOCIATED TERMS

The uncertainty of measurement reflects the lack of exact knowledge about the value of a measurand. Before the publication of *GUM* the uncertainty of measurement was defined as "an estimate characterizing the range of values within which the true value of a measurand lies" in the *International Vocabulary of Basic and General Terms in Metrology,* 1st edition (1984, abbreviated as *VIM*). This definition of uncertainty is linked to the term *true value.* The "range of values within which the true value lies" also gives an indication of the possible error of measurement.

The terms *measurand, error,* and *uncertainty* are frequently misunderstood. The measurand is a quantity subject to measurement; it needs to be defined and then realized. If the realized measurand is not perfectly realized, as per the definition of a measurand, the measurement result needs to be corrected for this deficiency. The measurement result also needs to be corrected for all other recognized systematic effects. The final corrected result has been viewed as a best estimate of the "true" value. *GUM*, however, recommends not to use the term *true value,* because *true* implies something illusory which we are searching for without finding. In fact, the word *true* is redundant, as the final result is the best estimate of the value of the measurand. Thus, what we are intending to find out is the value of a measurand and not its "true" value.

The preceding concepts can be summarized as follows:

- *True value* is a misnomer. What we are interested in measuring is the value of a measurand, and the measurement result is the best estimate of the value of the measurand.
- Since a measurement result is an estimate, error is always associated with it. This is because of the imperfect measurement process due to (a) random variability of the measurement process and (b) inadequate correction for systematic effects. The possibility also exists of an unrecognized systematic effect due to incomplete knowledge of the physical world. The error is unknown and unknowable.
- Though an estimated measurement result is assigned as a single value, there are an infinite number of values dispersed about the estimated result which could be attributed to the measurand with varying degrees of credibility. This dispersion is the uncertainty of the measurement result. The uncertainty due to random effect and known systematic effect can be evaluated.
- Since *GUM* is also applicable to fundamental research, the presence of an unknown systematic effect that might be influencing the measurement process but not known to us due to incomplete knowledge of the physical world has been accepted. In most of the practical measurement situations where *GUM* has applicability, the presence of unknown systematic effects is not envisaged. This includes measurements taken in testing and calibration laboratories and the intercomparison of national metrology standards.

A New Definition of the Uncertainty of Measurement

The definition of the uncertainty of measurement was therefore changed in the 1993 edition of *VIM*. It is defined there as the "parameter associated with the result of a measurement that characterizes the dispersion of the values that could reasonably be attributed to a measurand."

The 1984 edition of *VIM* linked uncertainty with the "range of values within which true value lies." The present definition defines it as a "parameter of the result of a measurement" characterizing "the dispersion of values"; the definition is qualified using the concept of *reasonability*. *GUM* has conceived the uncertainty of a measurement result differently than did the concepts that prevailed before the publication of the guide.

The 1993 definition is supplemented by explanatory notes as follows:

- The parameter may be, for example, a standard deviation (or a given multiple of it) or the half-width of an interval having a stated level of confidence.

- Uncertainty of measurement comprises, in general, many components, some of these components may be evaluated from the statistical distribution of the results of series of measurements and can be characterized by experimental standard deviation. The other components, which can also be characterized by standard deviations, are evaluated from assumed probability distributions based on experience or other information.
- It is understood that the result of measurement is the best estimate of the value of the measurand and that all components of uncertainty including those arising from systematic effects, such as components associated with corrections and reference standards, contribute to the dispersion.

MATHEMATICAL MODELING OF THE MEASUREMENT PROCESS

A measurement process can be represented by a mathematical relationship of the following form:

$$y = f(x_1, x_2 \ldots\ldots\ldots\ldots\ldots x_n)$$

where y represents the measurement result which is the best estimate of the value of the measurand and uncertainty associated with it. It has been arrived at based on measurements taken on the realized measurand after necessary corrections for known systematic effects have been applied; y is not measured directly but determined from other quantities x_1, x_2,..... and so on. Since the value of y depends upon the values of x_1, x_2,.and so on, y is called a dependent variable whereas x_1, x_2, and so on are called independent variables. Function f shows a functional relationship between the independent and dependent variables. The variables x_1, x_2, and so on are random variables with either a known or assumed probability distribution; x_1, x_2, and so on are also called input variables, whereas y is called an output variable. Our objective of measurement is to find the best estimate of the measurand represented by y_0 and the uncertainty associated with it represented by Δy. Thus,

$$y = \bar{y} \pm \Delta y$$

How measurement processes are represented mathematically is shown in the following examples.

Example 1

Let us take the simplest case of measuring the mass of an object using a nonautomatic weighing instrument of accuracy class III of document OIML R76-1. The measurement process can be expressed mathematically as

$$y = x_1 + x_2$$

where x_1 indicates the observation on the weighing instrument and x_2 is the uncertainty of the measurement of the weighing instrument expressed as maximum permissible error. If $x_1 = 400$ grams and $x_2 = 0.5$ grams, then $y = 400 \pm 0.5$ grams. The result is based on a single observation (that is, a single value of x_1), and x_2 is due to inadequate knowledge about correction due to systematic effect and has been obtained based on the requirement stipulated in document OIML R76-2. The result has been obtained based on type-B evaluation of uncertainty. The variable x_2 is a random variable with zero expectation and a known probability distribution and standard deviation.

If we intend to measure the mass more accurately we will also take into consideration the random variations in the repeat observations of x_1, take the average of these values as the best estimate of the mass, and also take the random variability into consideration while evaluating the uncertainty of the measurement.

Example 2

Let us take the measurement of the output of a high-precision DC voltage source using a six-digit digital multimeter (DMM). The measurement process can be described mathematically as

$$y = x_1 + x_2 + x_3 + x_4$$

where x_1 represents a random variable associated with repeat observations of measurement of voltage by the DMM. It has an expectation $\bar{x}_1$, a standard deviation s_{x1}, and variance s_{x1}^2. The variable x_2 is a random variable representing the inherent inaccuracy of the DMM. Its expectation is zero, and its probability distribution is either known or assumed. Its standard deviation and variance can be evaluated using the technical specifications of the DMM as supplied by the manufacturer. The variable x_3 may represent another random variable representing inaccuracy due to temperature difference. Accuracy specification is claimed by the manufacturer for operation at a specified temperature. For measurements at a different temperature, a source of uncertainty represented by variable x_3 is introduced. Another random variable, x_4, represents inaccuracy due to measurements made beyond a certain period of time after calibration. Generally, accuracy specifications stipulate that they are valid for a certain time after calibration. If equipment is used beyond that time, inaccuracy increases.

Generally the random variables x_2, x_3, x_4, and so forth will have zero expectation but known probability distributions, and hence known standard deviations and variances.

Example 3

Now let's look at the measurement of electric current by using a standard resistor and then measuring voltage across the standard resistance when current is passed through it. The functional relationship is given as

$$I = V / R$$

Here the input variables are voltage V and resistance R, which have their own expectations and standard deviations. These standard deviations characterize the uncertainty associated with the variables V and R. I is the output variable.

EVALUATION OF STANDARD UNCERTAINTY

Standard uncertainty is the basic elementary unit, which is expressed as a standard deviation. For each influence factor that contributes to uncertainty, standard uncertainty is calculated. Combined and expanded uncertainties are obtained from the value of various standard uncertainty elements. Standard uncertainty is evaluated either by the type-A evaluation method or the type-B evaluation method.

Type-A Evaluation of Standard Uncertainty

As explained earlier, type-A evaluation of uncertainty is based on statistical analysis of repeat observations of the measurand obtained under the same conditions of measurement. Standard uncertainty is expressed as standard deviation. Let x_1 be an input variable to the measurement process. If x_{11}, $x_{12}, \ldots \ldots \ldots \ldots \ldots x_{1i} \ldots \ldots \ldots \ldots x_{1n}$ are n repeat observations of x_1, then the best possible estimate of the random variable is given by its expectation or the arithmetic mean or average $\bar{x}_1$ of the observations:

$$\bar{x}_1 = \Sigma x_{1i} / n$$

The estimated variance s_{x1}^2 of the probability distribution of x_1 is given by

$$s_{x1}^2 = \Sigma(x_{1i} - \bar{x}_1)^2 / (n - 1)$$

The positive square root ofs_{x1}^2 is called the experimental standard deviation and characterizes the dispersion of observed values about their mean $\bar{x}_1$. The best estimate of the variance of mean (that is, $\bar{x}_1$) is given by

$$\sigma^2_{(\bar{x}1)} = \sigma^2_{(x1)} / n$$

The experimental standard deviation of mean $\sigma_{(\bar{x}1)}$ is the positive square root of $\sigma^2_{(\bar{x}1)}$. The expression $\sigma_{(\bar{x}1)}$ quantifies how well $\bar{x}_1$ estimates the measurand result, and it is used as a measure of the uncertainty of $\bar{x}_1$. The terms $\sigma_{(\bar{x}1)}$ and $\sigma^2_{(\bar{x}1)}$ are the type-A standard uncertainty and type-A variance respectively. The type-A standard uncertainty has $n - 1$ degrees of freedom.

Type-B Evaluation of Standard Uncertainty

Type-B evaluation is uncertainty evaluation by means other than statistical analysis of a series of observations. Type-B uncertainty is evaluated by scientific judgment based on all the available information about the variability of the variables contributing to the uncertainty. Type-B evaluation is based on probability distribution, which must be adopted based on knowledge of the behavior of the factor contributing to the uncertainty. If the distribution is known or assumed to be known based on past experience, the standard uncertainty can be evaluated by the knowledge of the distribution. Various probability distributions that were discussed in chapter 3 find wide applicability in evaluation of type-B uncertainty. Some of the distributions that are frequently used in type-B evaluation are uniform distribution, trapezoidal/triangular distribution, and normal distribution.

Type-B evaluation of uncertainty is as reliable as type-A evaluation, particularly if type-A evaluation is based on a small number of independent observations. Moreover, type-B evaluation saves a lot of scientific effort, as in this we rely on past experience and scientific judgments of professionally expert persons and organizations such as calibration laboratories, manufacturers, and so on. If a metrology laboratory has the time and resources at its command, it can evaluate all sources of uncertainty by statistical analysis of a series of independent observations. This could be done by conducting a series of experiments using different kinds of instruments, different methods of measurement, different applications of methods, and different approximations in theoretical models of measurement. Still, the information thus generated may not be better than what we get from the vast treasure of already-available information. The already-available knowledge based on which type-B evaluation can be carried out may consist of the following:

- Previous measurement data
- Manufacturer's specification
- Data provided in calibration certificates
- Uncertainties assigned to reference data, and so on

Examples of Type-B Evaluation of Standard Uncertainty

Standard uncertainty measure is characterized by standard deviation, so our objective is to find out the standard deviation for the random variable contributing to uncertainty based on the already-available information.

Case 1

The available information on the manufacturer's specification or calibration certificate is in the form of "quoted uncertainty as a specific multiple of standard deviation." In this case, to obtain the standard deviation the quoted uncertainty is divided by the stated specific multiple. The value of standard deviation thus obtained is the estimated value of standard uncertainty obtained by type-B evaluation.

Case 2

In this case, the quoted uncertainty is defined as an interval at a specific percentage of confidence level, say 90%, 95%, or 99%. In this case, it is appropriate to assume that normal distribution was used to calculate the quoted uncertainty unless it is otherwise specifically stated that some other distribution has been used. We know from the theory of normal distribution that if σ is the standard deviation of the data, 90% of the observations remain within $\pm 1.64\sigma$ of the mean, 95% of the observations remain within $\pm 1.96\sigma$ of the mean, and 99% of the observations remain within $\pm 2.58\sigma$ of the mean. Thus, to obtain the value of the standard deviation the quoted uncertainty is divided by 1.64 for a confidence level of 90%, by 1.96 for a confidence level of 95%, and by 2.58 for a confidence level of 99%. For other values of confidence level, to obtain the standard deviation from the quoted uncertainty the dividing factors can be obtained from standard normal table. If the result is quoted at a 50% confidence level, the dividing factor to obtain the standard deviation from the quoted uncertainty is 0.675. For a 68% confidence level the dividing factor is unity. The standard deviation thus obtained is the value of the standard uncertainty obtained by type-B evaluation.

Case 3

In this case, the quoted uncertainty is $\pm a$, such that the probability of the random variable lying within $\pm a$ from the mean value is equally likely and is zero beyond $\pm a$. This is the rectangular probability distribution with the half-width a. In such a case the variance of the random variable is given by $a^2 / 3$, and its standard deviation is $a / \sqrt{3}$, which is the estimate of type-B evaluation of standard uncertainty.

Case 4

In this case, the quoted uncertainty follows a trapezoidal distribution with a base of width $2a$ and a top of width $2a\beta$, where $0 \leq \beta \leq 1$. The variance of the distribution is $a^2(1 + \beta^2) / 6$. For $\beta = o$, the distribution becomes triangular and its variance is equal to $a^2 / 6$. The standard uncertainty in these cases is equal to the positive square root of the previously cited variance values.

Random and Systematic Versus Type-A and Type-B Evaluations of Uncertainties

Uncertainty has traditionally been conceived of as primarily due to random variations in repeat observations and inadequate determination of correction for systematic effects, called random and systematic uncertainties. The evaluation of random uncertainty was based on the assumption that the repeat observations are independent random variables, but in practical measurement situations there are many influencing factors which have a strong nonrandom element in their variations. If the common services of the laboratory such as electric supply voltage and frequency, water pressure, temperature, relative humidity, and so forth are influencing the measurement process, their variation is not random in nature. Similarly, in selecting the least significant figure of a digital indication, it is sometimes difficult not to select unknowingly a personally preferred value of that digit. Thus, what we presume to be a purely random event is not so. Similarly, the uncertainty figure for inadequacy in the correction for systematic effect would also have been determined based on statistical analysis of repeat observations, which also have some randomness associated with them. The systematic uncertainty figure is based on a study of some random variable. Thus, it is incorrect to classify the uncertainty components as random and systematic.

However, there is a difference in their methods of evaluation. In one case, we evaluate the uncertainty from a frequency-based distribution of a repeat number of observations. In the other case, we evaluate the uncertainty based on knowledge of a prior probability distribution. It is sometimes erroneously believed that the concept of probability is applicable only to events that can be repeated a large number of times under essentially the same conditions with a probability that indicates the relative frequency with which the event will occur. The concept of probability is equally valid in the case of an a priori probability distribution, which is based on past experience and hence has a degree of belief that an event will happen in a particular way.

Hence, the BIPM working group, in its recommendation, suggested classifying uncertainties the way they are evaluated—that is, as type A and type B—rather than as systematic and random, as had been the practice till then, thereby removing the ambiguity.

DETERMINING COMBINED STANDARD UNCERTAINTY: THE LAW OF PROPAGATION OF UNCERTAINTY

GUM has harmonized the methods of evaluating the uncertainty of measurements using type-A and type-B evaluation methods. Thus, all the standard uncertainty components arising from various variables which contribute to the uncertainty can be evaluated by either the type-A or type-B method of uncertainty evaluation. The evaluated value of standard uncertainty is characterized by the variance or standard deviation of the random variable.

To obtain the value of combined standard uncertainty, *GUM* has recommended the use of the law of propagation of uncertainty.

Law of Propagation of Uncertainty: Multivariate Taylor Series Expansion

The law of propagation of uncertainty is derived based on the Taylor series expansion of a functional relationship frequently used in differential calculus. A measurement process can be modeled mathematically as follows:

$$y = f(x_1, x_2 \ldots\ldots x_n)$$

where x_1, x_2, and so on are the number of measurable input quantities and y is the quantity which is finally being measured. Each input variable x_i has its own probability distribution, which could be frequency based or an a priori probability distribution. As a probability distribution is characterized by its mean value and its variance, each of the variables x_i has a mean value $\overline{x}_1$ and a variance $u^2_{(xi)}$. Similarly, the output variable y has a mean value $\overline{y}$ and a variance $u^2_c(y)$. The output variable y also has its own probability distribution, which depends on the probability distribution of various input variables and the nature of their functional relationships.

Having known the mean values $\bar{x}_i$ of various input variables, the best estimate of the output variable $\bar{y}$ is obtained substituting the values of $\bar{x}_i$ instead of x_i in the equation representing the functional relationship, and thus a value of $\bar{y}$ is obtained. The value $\bar{y}$ is thus the best estimate of the quantity under measurement:

$$\bar{y} = f(x_1, x_2, \ldots x_n)$$

evaluated at $x_1 = \bar{x}_1$, $x_2 = \bar{x}_2$, and so on.

If the function $f(x_i)$ is continuous and has derivatives around the mean value of input variables, multivariate Taylor series expansion can be used to find out the variance of the output variable y.

The Taylor series expansion is a common method of expressing nonlinear functional equations as linear functions and gives excellent results if higher-order terms are extended to infinity. However, in most cases relating to physical measurements, the Taylor series is truncated either at the first order or, rarely, at the second order. The truncation error in most cases is negligible. Thus, the functional relationship $y = f(x_i)$ can be written as follows using Taylor series:

$$y - \bar{y} = (\partial f / \partial x_1)(x_1 - \bar{x}_1) + (\partial f / \partial x_2)(x_2 - \bar{x}_2) + \ldots\ldots\ldots\ldots\ldots + (\partial f / \partial x_n)(x_n - \bar{x}_n)$$

$$= \Sigma(\partial f / \partial x_i)(x_i - \bar{x}_i)$$

The terms $\partial f / \partial x_i$ are the partial derivatives of the function with reference to various variables, and the value of $\partial f / \partial x_i$ is evaluated at the mean value of various variables (that is, $x_i = \bar{x}_i$).

The expression $y - \bar{y}$ represents the error term and is a random variable with the same probability distribution as that of y except that its location has been shifted by $\bar{y}$. So $y - \bar{y}$ will have the same variance as that of y. The variance of y gives an indication of the combined standard uncertainty.

The variance of y is obtained by the solution of the following equation:

$$\sigma_y^2 = \int (y - \bar{y}) f_y dy$$

where σ_y^2 is the variance of y that is the combined variance and f_y is the probability distribution function of the dependent variable y. The probability density function of y depends upon the individual density functions of the independent variables x_i, and y is a function of these independent variables.

Without going into the details of the statistical theory, the variance of y for two situations follows.

Case 1: Uncorrelated Input Quantities

If two or more input quantities are independent and are not correlated, the combined standard uncertainty $u_c(y)$ is the positive square root of the combined variance $u_c^2(y)$, which is obtained from the following equation:

$$u^2_c(y) = \Sigma(\partial f / \partial x_i)^2 \, . \, u^2(x_i)$$

The term $\partial f / \partial x_i$ is the partial derivative of the function f with respect to various input variables x_i and evaluated at the mean values of the input variables $x_i = \overline{x}_i$.

The term $u(x_i)$ is the standard uncertainty for the component x_i and can be obtained either by the type-A or type-B method of evaluation of standard uncertainty. The summation extends over all the input variables that are contributing to the uncertainty of y. The term $u_c(y)$ is the estimate of the standard deviation of probability distribution of y.

The previously cited equation is called the *law of propagation of uncertainty.*

If the nonlinearity of f is significant, higher-order terms in the Taylor series expansion must be included in the expression for $u^2_c(y)$. If the distribution of x_i is symmetrical about the mean, then the following second-order terms must be included in the calculation of $u^2_c(y)$:

$$\Sigma\Sigma[\tfrac{1}{2}(\partial^2 f / \partial x_i \partial x_j)^2 \cdot u^2(x_i) \cdot u^2(x_j)]$$

where $\partial^2 f / \partial x_i \partial x_j$ is the second-order partial derivative of function f first with respect to x_i and subsequently with respect to x_j.

Sensitivity Coefficients

The term $\partial f / \partial x_i$ is the partial derivative of $y = f(x_1, x_2 ...... x_n)$. This is obtained by differentiating the function with respect to the input variable x_i and evaluating the partial derivative at mean values of input variables. A small change Δx_i in input variable x_i, with other variables remaining constant, produces a change $(\Delta y)_i$ in the output variable:

$$(\Delta y)_i = (\partial f / \partial x_i) \cdot \Delta x_i$$

Thus, if $u(x_i)$ is the standard uncertainty due to the input variable x_i, and $u_i(y)$ is the corresponding uncertainty in the output variable y, $u_i(y)$ and $u(x_i)$ will have the following relationship:

$$u_i(y) = (\partial f / \partial x_i) \cdot u(x_i)$$

The combined variance $u^2_c(y)$ can be viewed as the sum of the terms representing the variance of the output estimate y generated by the variance of each input estimate x_i:

$$u^2_c(y) = \Sigma u^2_i(y) = \Sigma(\partial f / \partial x_i)^2 \cdot u^2(x_i)$$

The terms $\partial f / \partial x_i$ are defined as sensitivity coefficients and describe how the output estimate y varies with change in the values of input estimates.

Case 2: Correlated Input Quantities

The relationship between two random variables is called *correlation.* If change in the value of one random variable also causes change either in the same direction or opposite direction in the value of another random variable, a correlation is said to exist between those two variables. For example, a measurement process may be influenced by temperature as well as relative humidity of the environment. The temperature of the environment and its relative humidity are correlated as the relative humidity changes with the change of temperature.

The mutual dependence of two random variables which are correlated is characterized by a parameter called *covariance.* The term *covariance* is analogous to variance of a random variable, and its value is estimated from n repeated simultaneous observations.

If (x_1, y_1), (x_2, y_2), and so on are simultaneous observations of two correlated random variables, their covariance $s(x, y)$ is estimated from the following equation:

$$s(x_i, y_i) = \Sigma(x_i - \bar{x}) \cdot (y_i - \bar{y}) / (n - 1)$$

Correlation Coefficient

The term *correlation coefficient* is more frequently used as a measure of the mutual dependence of two variables rather than covariance. It is a measure of relative mutual dependence of two variables and is equal to the ratio of their covariance to the positive square root of their variances. The correlation coefficient is represented by r, and it is estimated by the following equation:

$$r(x_i, y_i) = r(y_i, x_i) = s(x_i, y_i) / s(x_i)s(y_i)$$

The correlation coefficient is a pure number and its value lies between –1 and +1.

The law of propagation of uncertainty for the correlated input variables is given as follows:

$$u^2_c(y) = \Sigma(\partial f / \partial x_i)^2 \cdot u^2(x_i) + 2\Sigma\Sigma(\partial f / \partial x_i)(\partial f / \partial x_j) \cdot u(x_i, x_j)$$

Where $u(x_i, x_j)$ is the estimated covariance of x_i and x_j, the value of $u(x_i, x_j)$ is estimated from the preceding equation. In terms of the correlation coefficient, the law of propagation of uncertainty can also be written as follows:

$$u^2_c(y) = \Sigma(\partial f / \partial x_i)^2 \cdot u^2(x_i) + 2\Sigma\Sigma(\partial f / \partial x_i)(\partial f / \partial x_j) \cdot u(x_i)u(x_j)r(x_i, x_j)$$

Advantages of Adopting the Law of Propagation of Uncertainty

The advantages of adopting the law of propagation of uncertainty are as follows:

All components of uncertainty have been treated in the same way as variance or standard deviation.

The combined standard uncertainty figure $u_c(y)$ of a result of a measurement can be readily incorporated into the uncertainty of another result if the first result was used to derive the second result, again by the application of the law of uncertainty propagation.

The combined standard uncertainty can be used as a basis for evaluating overall uncertainty at a specified confidence level.

DETERMINING EXPANDED UNCERTAINTY

The combined standard uncertainty is a good measure of the uncertainty of a measurement result. However, in certain regulatory situations, such as health and safety, the requirements need to stipulate that the measured quantity be within a specified interval. Any value beyond this interval is considered to be in noncompliance with regulatory requirements. Such requirements are also stipulated in commercial transactions and in industries where acceptance or rejection of a product is based on whether the measured quantity lies within this interval. Such a situation has been envisaged by the BIPM working group in its recommendation as *overall uncertainty.*

GUM has termed it *expanded uncertainty,* and it is denoted by U. The expanded uncertainty is a confidence interval and is obtained by multiplying the combined standard uncertainty $u_c(y)$ by a coverage factor k such that $U = k \cdot u_c(y)$.

If $\bar{y}$ is the best estimate of the measurement result, it is implied that the value of the quantity under measurement lies between $\bar{y} - U$ and $\bar{y} + U$ with a high level of confidence. The coverage factor k is chosen based on the desired level of confidence. In general the value of k will range from 2 to 3.

REPORTING THE UNCERTAINTY OF MEASUREMENT: GUIDELINES

Because a measurement result is used by its user, *GUM* recommends that adequate information on the measurement result and its uncertainty be given so that there is no ambiguity in the mind of the user of the measurement result. It is therefore recommended to err on the side of providing too much information rather than too little.

General Guidelines for Reporting Uncertainty

GUM sets forth the following general recommendations:

- It is recommended that the measurement result and its uncertainty contain all the information necessary for reevaluation of the measurement (thus, that information is available to users of the measurement result). This recommendation applies to all levels of the measurement hierarchy, viz., the BIPM; national standards laboratories; various echelons of calibration laboratories; measurements made in trade, industry, and other laboratories; and measurements made in regulatory activities. As we move down the measurement hierarchy, less information is required.
- The calibration result is reported on a certificate, which also contains information on the uncertainty of measurement. The uncertainty is evaluated by reference to some existing document generally published by a national metrology laboratory or a national laboratory accreditation body. In such a case, it is necessary that the measurement procedure is compatible with that document.
- A great many measurements are made in trade and regulatory activities without an explicit report of measurement uncertainty. Many of those measurements are covered under legal metrology and are carried out with instruments which are subjected to legal inspection. Those instruments are known to be in conformance with certain norms and regulations—for example, norms evolved by the International Organization of Legal Metrology. The uncertainties of such measurements may be inferred from these regulations and norms.

Specific Guidelines for Reporting Uncertainty

The uncertainty of measurement results can be reported either as a combined standard uncertainty or expanded uncertainty. If the combined standard uncertainty is known, the user of the measurement result can always calculate the expanded uncertainty by multiplying the combined standard uncertainty with the coverage factor k. Similarly, if the expanded uncertainty and its associated coverage factor are known, one can calculate the combined standard uncertainty by dividing the expanded uncertainty with coverage factor k. Either way, it should be unambiguously communicated that the quoted measurement uncertainty is a combined standard uncertainty or expanded uncertainty. However, it is essential that the measurement result and its uncertainty contain all the necessary information so that the user can reevaluate the measurement result. For this, the user will need information on definition of measurand, its estimation as a measurement result, and uncertainties contributed by various factors. *GUM* therefore recommended that certain information be included when reporting uncertainty.

If the reported uncertainty is the combined standard uncertainty, the following should be reported:

- Definition of the measurand and its estimation as measurement result
- Unambiguous statement that the reported uncertainty is combined standard uncertainty and its value
- Units of measurement results and combined standard uncertainty
- If appropriate, the relative standard uncertainty that is the ratio of combined standard uncertainty and the measurement results
- If deemed useful, the effective degree of freedom
- If deemed useful, the type A and type B combined standard uncertainties and their effective degrees of freedom
- If applicable, correlation coefficients of input quantities

If the reported uncertainty is expanded uncertainty, the following should be reported:

- Definition of measurand and its estimation as measurement result
- The combined standard uncertainty and coverage factor k used to obtain expanded uncertainty
- The measurement result to be reported as "estimated measurement result ± expanded uncertainty U"
- Appropriate level of confidence associated with the interval

The *GUM* has also recommended that the combined standard uncertainty and expanded uncertainty should be estimated to two significant digits. In certain cases, it may be necessary to retain additional digits to avoid errors in subsequent calculations.

PROCEDURE FOR EVALUATION OF UNCERTAINTY IN ACCORDANCE WITH *GUM*

Here is the step-by-step procedure for evaluating the uncertainty of a measurement according to *GUM*:

- Define the measurand.
- Identify all input quantities that influence the measurement process and contribute to uncertainty. A measurand can be directly measured using measuring equipment—for example, measuring pH value with a pH meter or measuring electric voltage using a digital multimeter. However, sometimes a measurand is calculated from intermediate values, which are either measured or obtained from external sources such as a calibration certificate or table of constants. The calculation of a measurand is done based on a mathematical expression showing the relationship between the measurand and intermediate values. For example, the volume V of a cylinder as a measurand is calculated based on the measurement of its length l and diameter d. The relationship expression in this case will be $V = \pi(d / 2)^2 l$. The uncertainties of intermediate values are important sources that contribute to overall uncertainty. In addition, there are other factors which do not appear in the relationship expression but contribute to uncertainty. For example, while measuring the length of a gage block, the temperature at which the measurement is being made may contribute to uncertainty.
- Establish a mathematical relationship between the input quantities and the quantity desired to be measured. This should cover the intermediate parameters as well as other influence factors contributing to uncertainty.
- Determine the estimated value of the input quantity x_i. Each input quantity x_i is to include corrections for all known systematic effects that significantly influence the estimate y of the measurand.
- Evaluate the standard uncertainty $u(x_i)$ for each input estimate either by type-A evaluation of standard uncertainty or type-B evaluation of standard uncertainty.
- Evaluate sensitivity coefficients.
- If the values of input quantities are correlated, evaluate their correlation coefficient.
- Calculate the estimate $\overline{y}$ of the measurand from the functional relationship f and the estimate $\overline{x}_i$ of input quantities.
- Evaluate the combined standard uncertainty of the measurement result y from the standard uncertainties and sensitivity coefficients of input quantities using the law of propagation of uncertainties. If input

parameters are correlated, find their correlation coefficients. Find the standard combined uncertainty using the law of propagation of uncertainties for correlated input quantities.

- If reporting expanded uncertainty is required, multiply the combined standard uncertainty by a coverage factor k which is in the range of 2 to 3. Select k on the basis of the desired level of confidence associated with the interval.
- Report the result of measurement $\bar{y}$ together with its standard uncertainty $u_c(y)$ or expanded uncertainty U, as necessary.

The uncertainty evaluation process is shown in Figure 8.1.

EXPANDED UNCERTAINTY AND CONFIDENCE INTERVAL

The combined standard uncertainty $u_c(y)$ multiplied by a coverage factor k gives the value of expanded uncertainty U. The value of U is used to find the confidence interval of the measurement result, which is the dispersion between which the measurement result is expected to lie with a certain confidence level. Thus, in $U = k\ .u_c(y)$, k is the coverage factor, and if $\bar{y}$ is the best estimate of the value of the measurand, the measurement result is expected to lie between $\bar{y} - U$ and $\bar{y} + U$.

Choice of Confidence Level

All uncertainty evaluations are an estimation, and hence they should be viewed as approximates only. During the process of type-A evaluation of uncertainty we estimate the standard deviation of the mean of the repeat measurement observations. With a new set of observations we will not only get a different mean but also a different estimate of the standard deviation of the mean. Thus the estimate of type-A uncertainty has its own variability, resulting in uncertainty of uncertainty. This relative standard deviation is 30% for 30 repeat observations and 10% for 50 observations. Therefore, one need not be too conservative in specifying the confidence levels. Generally a 95% confidence level is adequate for most applications. Moreover, according to the reporting guidelines given in *GUM*, even if expanded uncertainty is quoted, one can always find the combined standard uncertainty from the knowledge of the probability distribution and degrees of freedom. However, one has to be very cautious when evaluating confidence intervals at a confidence level of 99% and higher. With the increasing confidence level the confidence interval increases considerably with a lessening of the number of degrees of freedom. At a confidence level of 99.99%, sometimes the measurement result loses its relevance.

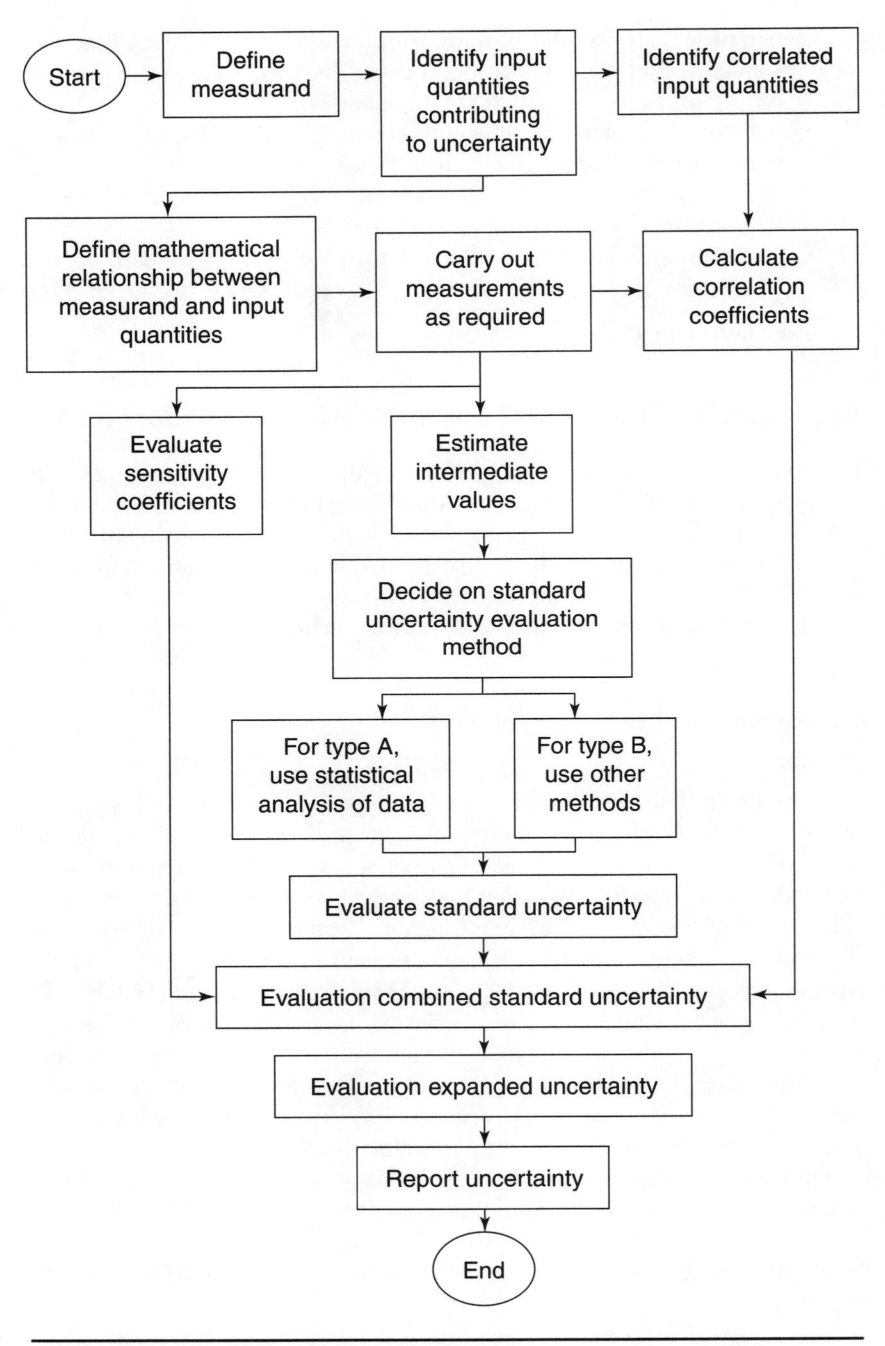

Figure 8.1 Uncertainty evaluation process.

Choice of Coverage Factor for Normally Distributed Data

If the measurements follow normal distribution, which is usually the case, the following coverage factors should be used for various levels of confidence.

Confidence level	68.27	90	95	95.5	99	99.73
Coverage factor	1.0	1.645	1.960	2.00	2.526	3.0

Use of *t* Distribution to Obtain Coverage Factors

The Student *t* distribution is a very useful sampling distribution in obtaining the confidence interval, particularly if the sample size is small. In most of the measurement applications the number of repeat observations are small, and hence *t* distribution can be used to obtain the coverage factor for evaluating the expanded uncertainty. The value of *t* factor $t_p(\nu)$ depends upon the chosen confidence level $p\%$ and the number of degrees of freedom ν. The values of $t_p(\nu)$ for various degrees of freedom ν and confidence levels of 68.27%, 90%, 95%, and 99% are given in the table in Appendix B, which has been reproduced from *GUM*.

If this result is to be quoted at a confidence level of $p\%$ and for ν degrees of freedom, the value of $t_p(\nu)$ is obtained from the table. Thus, expanded uncertainty $U = t_p(\nu) \cdot u_c(y)$. This implies that the measurement result lies between $\bar{y} - t_p(\nu) \cdot u_c(y)$ and $\bar{y} + t_p(\nu) \cdot u_c(y)$ with a confidence level of $p\%$, where $\bar{y}$ is the best estimate of the measurement result. The term $u_c(y)$ is the combined standard uncertainty, and $t_p(\nu)$ is the Student *t* factor for ν degrees of freedom and $p\%$ of confidence level. For a large number of observations, the coverage factor $t_p(\nu)$ assumes the value of coverage factor *k*, which is applicable to normal distribution.

Degrees of Freedom and Effective Degrees of Freedom

For a single quantity which is estimated by the arithmetic mean of *n* observations, the degree of freedom is $n - 1$. However, if *y* is the output variable which is a function of input variables and the variance of *y* is estimated based on the propagation of uncertainty formula, then the distribution of *y* is approximated by the *t* distribution. In such a case the output variable *y* has effective degrees of freedom given by the Welch-Satterthwaite formula as shown here:

$$\nu_{\text{eff}} = u_c^4(y) \,/\, \Sigma\{u_i^4(y) \,/\, \nu_i\} \leq \Sigma\nu_i$$

where $u_i(y) = c_i u(x_i)$ and $u_c^2(y) = \Sigma u_i^2(y)$

The variable ν_i is the degree of freedom of x_i. The value ν_{eff} may not be an integer, so to obtain $t_p(\nu_{eff})$ the value is found by interpolation or for a value of ν next below ν_{eff}.

STRUCTURE OF *GUM*

GUM consists of nine chapters and 10 annexures. They are as follows:

Chapter 0: Introduction introduces the document and explains the objective for its preparation. This chapter also explains the requisites for the ideal method of evaluating and expressing uncertainty and covers INC-1 (1980) recommendations.

Chapter 1: Scope defines the scope of the document.

Chapter 2: Definitions consists of basic definitions pertaining to various uncertainty terminology used in *GUM*.

Chapter 3: Basic Concepts contains basic concepts of measurement uncertainty and elaborates on the BIPM working group's Recommendation INC-1 (1980).

Chapter 4: Evaluating Standard Uncertainty discusses the mathematical modeling of the measurement process and how type-A and type-B standard uncertainties have to be evaluated.

Chapter 5: Determining Combined Standard Uncertainty covers the methods used to obtain combined standard uncertainties in accordance with the law of propagation of uncertainty in various situations where the input quantities are correlated or uncorrelated.

Chapter 6: Determining Expanded Uncertainty covers the method whereby the measurement result is expressed as an interval, called expanded uncertainty.

Chapter 7: Reporting Uncertainty contains guidelines on the contents of the uncertainty statement, so that the user of the measurement result can use the information judiciously.

Chapter 8: Summary of Procedure for Evaluating and Expressing Uncertainty summarizes the various steps that are to be followed in the evaluation and reporting of measurement uncertainty.

Annexure A: Recommendations of Working Group and CIPM covers the details of the BIPM working group's Recommendation INC-1 (1980) and the CIPM's recommendations on Recommendation INC-1 (1980).

Annexure B: General Metrological Terms contains definitions of various metrological terms.

Annexure C: Basic Statistical Terms and Concepts covers various statistical concepts and terms used in evaluating measurement uncertainty.

Annexure D: True Value, Error, and Uncertainty describes the metrological aspects of the aforementioned terms and graphically illustrates them.

Annexure E: Motivation and Basis for Recommendation INC-1 (1980). Since the guide is based on the BIPM working group's Recommendation INC-1 (1980), this annexure sets forth the logic and concepts that led to the recommendations.

Annexure F: Practical Guidance on Evaluating Uncertainty Components gives guidelines on type-A evaluation of standard uncertainty and type-B evaluation of standard uncertainty in various situations.

Annexure G: Degree of Freedom and Levels of Confidence deals with the methods of estimating expanded uncertainty at a given confidence level given the combined standard uncertainty. It also deals with determination of the coverage factor and effective degrees of freedom.

Annexure H: Examples consists of worked-out examples of uncertainty evaluation and reporting in various situations.

Annexure I: Glossary of Principal Symbols.

Annexure J: Bibliography.

GUM: AN IDEAL METHOD

An ideal method of evaluating and expressing the uncertainty of a measurement was expected to have the following attributes. Let us see how *GUM* meets those attributes.

- **Universal:** *GUM* is applicable to a broad spectrum of measurements in physical as well as chemical metrology. These include measurements made at various levels of accuracy from the shop floor of industries to fundamental research. Measurements made in various technological areas are covered in the guide. Thus, *GUM* methods are universal in application.
- **Coherent:** Just as the SI brought coherence to all scientific and technological measurements, *GUM* brought coherence to methods for uncertainty evaluation and reporting, as is seen in its acceptance by international organizations having an interest in metrology.

It was also desired that the uncertainty figure arrived at by the ideal method should have the following attributes.

- **Internally consistent:** All uncertainty figures are first evaluated as standard uncertainties, and the combined uncertainty is then derived from these components. This is also independent of how standard uncertainty components are grouped. Thus, the method generates an uncertainty figure which is internally consistent.
- **Transferable:** It is possible to use directly uncertainty evaluated for one result in evaluating the uncertainty of another result which uses the first result in its derivation. Thus, the method generates an uncertainty figure which is transferable.

UNCERTAINTY BUDGET

GUM stipulates that the amount of information to be supplied in reporting a measurement result and its uncertainty should be sufficient for the user of the measurement result to be able to evaluate it should that be necessary. The amount of information also depends upon the intended use of the measurement result. *GUM* also recommends that while quoting a measurement result and its uncertainty, all the uncertainty components should be listed and documented fully as to how they were evaluated.

The uncertainty evaluation of a measurement result starts with the identification of all sources of error that are contributing to uncertainty, and the next step is the evaluation of standard uncertainty for all of the uncertainty components either by type-A evaluation or type-B evaluation. The information about various uncertainty components and their quantitative values are generally presented in tabular form. That information is called the uncertainty budget. The uncertainty budget gives us a quick glance at the totality of information on various uncertainty components and their values. It may also contain information on type-A or type-B evaluation, associated sensitivity coefficients, and their degrees of freedom.

EXAMPLES OF UNCERTAINTY EVALUATION USING *GUM*

Example 1: Evaluating and Reporting the Uncertainty of Calibration of a 1-Ohm Standard Resistor

The Measurement Problem

A 1-ohm standard resistor is calibrated using another standard resistor as a reference standard. Both of the resistors are connected in a series with a stable DC source, and the same current is passed through them both. A digital voltmeter is used to measure voltage drop across both resistors. Through a change-over switch, voltages across the resistors are measured immediately one after the other.

Mathematical Model

The value R_x is the value of the standard resistor under calibration, and R_s is the value of the reference standard resistor. The voltages across them are V_x and V_s respectively. Since the same current is flowing in both the resistors,

$$V_x / R_x = V_s / R$$

or

$$R_x = V_x \cdot R_s / V$$

The intermediate parameters V_x and V_s are obtained based on the repeat observations of the voltmeter readings.

Factors Contributing to Uncertainty

- Uncertainty in repeat measurements of V_x
- Uncertainty in repeat measurements of V_s
- Uncertainty due to resolution of digital voltmeter
- Uncertainty of assigned value of reference standard
- Uncertainty due to stability of reference standard
- Uncertainty due to difference in temperature of the laboratory and reference temperature of the calibration of reference standard

The value of the reference standard and its uncertainty are obtained from its calibration certificate, which is given as $R_s = 1.000004 \pm 2.5 \times 10^{-6}$ ohms at 20 degrees Celsius (°C). The repeat observations of V_x and V_s, the mean, and the standard deviation of the data are as follows:

V_x (millivolts)	V_s (millivolts)
99.91963	99.92478
99.91970	99.92474
99.91968	99.92470
99.91977	99.92492
99.91986	99.92488
99.92058	99.92561
99.92055	99.92558
99.92008	99.92506
99.91987	99.92479
99.91968	99.92468

$$\text{Mean of } V_x = 99.919940$$

$$\text{Mean of } V_s = 99.924973$$

$$\text{Standard deviation of } V_x = 3.448 \times 10^{-4}$$

$$\text{Standard deviation of } V_s = 3.333 \times 10^{-4}$$

$$\text{Standard deviation of mean of } V_x = 3.448 \times 10^{-4} / \sqrt{10} = 1.0904 \times 10^{-4}$$

$$\text{Standard deviation of mean of } V_s = 3.333 \times 10^{-4} / \sqrt{10} = 1.054 \times 10^{-4}$$

$$\text{Degrees of freedom} = 9$$

Calibration Result

$$R_x = V_x \cdot R_s / V_s = 99.919940 \times 1.000004 / 99.924973 = 0.999954 \text{ ohms}$$

Calculation of Sensitivity Coefficients

$$\partial R_x / \partial V_x = R_s / V_s = 1.000004 / 99.924974 = 0.01$$

$$\partial R_x / \partial V_s = V_x \cdot R_s / V_s^2 = 99.91994 \times 1.000004 / 99.924974^2 = 0.01$$

$$\partial R_x / \partial R_s = V_x / V_s = 99.91994 / 99.924974 = 0.9999496 = 1.0$$

Type-A Evaluation of Standard Uncertainty

Standard uncertainty due to variability of mean of $V_x = u(V_x) = 1.0904 \times 10^{-4}$

Standard uncertainty due to variability of mean of $V_s = u(V_s) = 1.054 \times 10^{-4}$

Type-B Evaluation of Standard Uncertainty

1. Uncertainty due to resolution of digital voltmeter = 0.1 parts per million (ppm)
   Distribution: rectangular
   Standard uncertainty = $u(r) = 0.1 / \sqrt{3} = 0.0577 \times 10^{-6}$
2. Calibration uncertainty of the reference standard = 2.5 ppm
   Distribution: normal confidence level is 95%; coverage factor is 2
   Standard uncertainty = $u(R_s) = 2.5 / 2 = 1.25 \times 10^{-6}$

3. Uncertainty due to stability of reference standard = 5 ppm
   Distribution: rectangular
   Standard uncertainty = $u(s) = 5 / \sqrt{3} = 2.8867 \times 10^{-6}$
4. Uncertainty due to relative change in standard resistance value due to change in temperature compared with calibration temperature of reference standard = $u(t) = 0$ as measurements were made at 20°C

Estimation of Uncertainty of Measurement

All of the preceding standard uncertainty terms and sensitivity coefficients are given in Table 8.1.

The combined uncertainty is obtained from the following equation

$$\Sigma(c_i u_i)^2 = 12.1957 \times 10^{-2}$$

Table 8.1 Uncertainty budget: Summary of sensitivity coefficients and standard uncertainty terms.

Source of uncertainty	Sensitivity coefficient c_i	Standard uncertainty u_i	$c_i u_i$	$(c_i u_i)^2$
Measurement of V_x				
1. Variation in mean of observations $u(V_x)$	0.01	1.0904×10^{-4}	1.0904×10	1.1890×10^{-12}
2. Resolution of digital voltmeter $u(r)$		0.0577×10^{-6}	0.000577×10^{-6}	3.3293×10^{-19}
Measurement of V_s				
1. Variation in mean of observations $u(V_s)$	0.01	1.054×10^{-4}	1.054×10^{-6}	1.1109×10^{-12}
2. Resolution of digital voltmeter $u(r)0$		0.0577×10^{-6}	0.000577×10^{-6}	3.3293×10^{-19}
Reference standard				
1. Uncertainty of calibration $u(R_s)$	1.0	1.25×10^{-6}	1.25×10^{-6}	1.5625×10^{-12}
2. Stability of reference standard $u(s)$		2.8867×10^{-6}	2.8867×10^{-6}	8.3333×10^{-12}
3. Relative change in standard resistance $u(t)$		0	0	0

The combined standard uncertainty $u_c(R_x) = \sqrt{(12.1957 \times 10^{-12})} = 3.4922 \times 10^{-6}$ ohm.

Degrees of Freedom

The effective degrees of freedom as per the Welch-Satterthwaite equation are given by

$$\nu_{eff} = u_c^4(y) / \Sigma\{u_i^4(y) / \nu_i\}$$

In the present case,

$$\nu_{eff} = (3.4922 \times 10^{-6})^4 / [\{(1.0904 \times 10^{-6})^4 / 9\} + \{(0.0577 \times 10^{-6})^4 / \infty\} + \{(1.0154 \times 10^{-4})^4 / 9\} + \{(1.25 \times 10^{-6})^4 / \infty\} + \{(0.0577 \times 10^{-6})^4 / \infty\} + \{(2.8867 \times 10^{-6})^4 / \infty\}]$$

The effective degrees of freedom = 148.73 / 0.2751 = 541.

Expanded Uncertainty

Since the degrees of freedom are 541, which can be considered large, the coverage factor can be chosen from the normal table. For a 95.5% confidence level, $k = 2$.

$$\text{Expanded uncertainty } U = 2 \times 3.4922 \times 10^{-6} = 6.9844 \times 10^{-6} = 0.0000069 = 0.000007$$

$$\text{Calibration result} = 0.999954 \pm 0.000007 \text{ ohm}$$

Reporting the Uncertainty

When reporting uncertainty, one must give enough information to ensure that there is no ambiguity in the mind of the user of the calibration result as to how the uncertainty figure has been arrived at. The aforementioned result could be expressed as

$$R_x = 0.999948 \pm 0.000007 \text{ ohm}$$

where ±0.000007 ohm is the expanded uncertainty (the expanded uncertainty has been derived based on a combined standard uncertainty of 0.0000035 and a coverage factor $k = 2$) corresponding to about a 95% confidence level.

Effect of Correlated Input Variables

In the present case the voltages V_x and V_s (that is, the voltages across the unknown standard resistor and the known reference standard) are measured at about the same time. As the measurements are taken at the same time using the same digital voltmeter, the input variables V_x and V_s may be correlated. There is a chance that while measuring, the digital voltmeter might read both the voltages either on the higher side or the lower side simultaneously.

In such a situation the formula for combined standard uncertainty should also take the correlation factor into consideration. Since there are only two variables V_x and V_s which are correlated, the following expression has to be evaluated and taken into account while calculating the value of combined standard uncertainty.

The additional term due to correlation of measured variables is $2(\partial R_x / \partial V_x) \cdot (\partial R_x / \partial V_s) \cdot u(x_i, x_j)$. The term $u(x_i, x_j)$ is the covariance of the mean $= \Sigma(x_i - \bar{x}_i)(x_j - \bar{x}_j) / n(n - 1)$. In the present case $x_i = V_x$ and $x_j = V_s$.

The calculation of $(x_i - \bar{x}_i)(x_j - \bar{x}_j)$ is shown in the following table.

$(x_i - \bar{x}_i)$	$(x_j - \bar{x}_j)$	$(x_i - \bar{x}_i)(x_j - \bar{x}_j)$
(99.91963 – 99.91994)	(99.92478 – 99.924974)	6.014×10^{-8}
(99.91970 – 99.91994)	(99.92474 – 99.924974)	5.166×10^{-8}
(99.91968 – 99.91994)	(99.92470 – 99.924974)	7.124×10^{-8}
(99.91977 – 99.91994)	(99.92492 – 99.924974)	0.918×10^{-8}
(99.91986 – 99.91994)	(99.92488 – 99.924974)	0.752×10^{-8}
(99.92058 – 99.91994)	(99.92561 – 99.924974)	40.704×10^{-8}
(99.92055 – 99.91994)	(99.92538 – 99.924974)	24.766×10^{-8}
(99.92008 – 99.91994)	(99.92506 – 99.924974)	1.204×10^{-8}
(99.91987 – 99.91994)	(99.92479 – 99.924974)	1.288×10^{-8}
(99.91968 – 99.91994)	(99.92468 – 99.924974)	7.644×10^{-8}

The calculations for various terms are as follows:

$$\Sigma(x_i - \bar{x}_i)(x_j - \bar{x}_j) = 95.58 \times 10^{-8}$$

$$u(x_i, x_j) = 95.58 \times 10^{-8} / 10(10 - 1) = 1.062 \times 10^{-8}$$

$$2(\partial R_x / \partial V_x) \cdot (\partial R_x / \partial V_s) \cdot u(x_i, x_j) = 2 \times 0.01 \times 0.01 \times 1.062 \times 10^{-8}$$

$$= 2.124 \times 10^{-2}$$

$$\text{Total } u_c^2 = 12.1957 \times 10^{-12} + 2.124 \times 10^{-12} = 14.3197 \times 10^{-12}$$

$$\text{Combined standard uncertainty } u_c = 3.7841 \times 10^{-6}$$

By taking the correlation of input variables, the combined standard uncertainty has increased to 3.7841×10^{-6} from 3.4922×10^{-6}.

$$\text{Expanded uncertainty } U = 2 \times 3.7841 \times 10^{-6}$$

$$= 7.5682 \times 10^{-6}$$

$$= 0.0000076$$

Example 2: Uncertainty Evaluation for Calibration of a Micrometer Using a Gage Block, Grade 0

The Measurement Problem

A micrometer is calibrated using a gage block of grade 0 as a reference standard. The length of the gage block is measured using the micrometer to assign it a value corresponding to the length of the gage block.

The calibrated value of the gage block is reported as 12.9001 ±0.0002 millimeters (mm) at 20°C ±1 °C.

Temperature at which the calibration was carried out = 24 °C.

The gage block was measured using the micrometer, and the following repeat observations in millimeters were obtained at different times: 12.900, 12.901, 12.899, 12.901, 12.900, 12.899, and 12.901.

Mathematical Model

The following mathematical equation represents the measurement process:

$$l = l_s \pm x_1 \pm x_2 \pm x_3 \pm x_4$$

where l is the intended calibration result (that is, the value of the length of the gage block of a length of 12.9 mm as measured by the micrometer); l_s represents the standard reference length of the gage block, which is reported as 12.9001 ±0.0002 mm at 20°C ±1°C; and x_1, x_2, x_3, and x_4 are various error factors that contribute to the uncertainty of the measurement result.

Factors Contributing to Uncertainty

1. Uncertainty in repeat measurements of l.
2. Uncertainty of calibration of the gage block (that is, ±0.0002 mm). Since the reference of the gage block could be anywhere within this interval, a rectangular distribution for this uncertainty component is assumed.
3. The variable x_1 represents the uncertainty component due to uncertainty of the temperature measurement as reported in the calibration certificate. The coefficient of thermal expansion of the gage block is reported to be $11.5 \times 10^{-6}/°C$. The ±1°C temperature uncertainty on the reported value at 20°C will create an uncertainty component equal to $12.9 \times 1 \times 11.5 \times 10^{-6} = 148.35 \times 10^{-6}$ mm.
   This can also be assumed to have rectangular distribution.
4. The variable x_2 represents the uncertainty component due to the difference between the temperature at which the calibration was carried out and the temperature at which the reference value is reported. In this case the difference in temperature is $24 - 20 = 4$°C. The coefficient of thermal expansion of the micrometer material is reported to be $10 \times 10^{-6}/°C$. This difference in the coefficient of thermal expansion for gage block and micrometer material will create an uncertainty component equal to $12.9 \times 4 \times (11.5 \times 10^{-6} - 10.0 \times 10^{-6}) = 77.4 \times 10^{-6}$ mm. This uncertainty component can also be assumed to have rectangular distribution.

   Two additional components of uncertainty are errors introduced by the flatness of the micrometer faces and parallelness of the micrometer faces. This is determined by the number of fringes appearing on the anvils when different recommended thicknesses of optical flats are gripped between the anvils of the micrometer. Each fringe amounts to an error of 0.3 microns (μm). During this part of the calibration process, two fringes have been observed.
5. The variable x_3 is an uncertainty component due to the flatness of the micrometer faces, which is equal to 2×0.3 μm = 0.6 μm = 0.0006 mm.
6. The variable x_4 is an uncertainty component due to parallelness of the micrometer faces, which is equal to 2×0.3 μm = 0.6 μm = 0.0006 mm.

Having identified all of the uncertainty components, the standard uncertainty for various components can be evaluated as follows:

Mean of repeat observation $\bar{x} = 12.9001$, the assigned calibration value.

Type-A Evaluation of Standard Uncertainty

$$\text{Standard deviation of observations } s_x = 8.9981 \times 10^{-4}$$

$$\text{Standard deviation of mean } s_{\bar{x}} = 3.4009 \times 10^{-4}$$

$$\text{Standard uncertainty due to repeatability } u(l) = 3.400 \times 10^{-4} \text{ mm} = 0.340 \ \mu\text{m}$$

Type-B Evaluation of Standard Uncertainty

Standard uncertainty due to calibration of reference standard

$$u(l_s) = 0.0002 / \sqrt{3} = 1.1547 \times 10^{-4} \text{ mm}$$

$$\text{Standard uncertainty due to } x_1 = u(x_1) = 148.35 \times 10^{-6} / \sqrt{3}$$
$$= 0.856 \times 10^{-4} \text{ mm}$$

$$\text{Standard uncertainty due to } x_2 = u(x_2) = 77.4 \times 10^{-6} / \sqrt{3} = 0.447 \times 10^{-4} \text{ mm}$$

$$\text{Standard uncertainty due to } x_3 = u(x_3) = 0.0006 / \sqrt{3} = 3 \times 10^{-4} \text{ mm}$$

$$\text{Standard uncertainty due to } x_4 = u(x_4) = 0.0006 / \sqrt{3} = 3 \times 10^{-4} \text{ mm}$$

Sensitivity Coefficients

Since the functional relationship between various independent variables is a simple additive relationship, all the sensitivity coefficients c_i will be equal to 1.

The uncertainty budget for the preceding data is given in Table 8.2.

Table 8.2 Uncertainty budget.

Source of uncertainty	Standard uncertainty component	Sensitivity coefficient c_i	Standard uncertainty $u_i(x)$	$c_i \cdot u_i(x)$	$\{c_i \cdot u_i(x)\}^2$
Repeatability	$u(l)$	1	3.4009×10^{-4}	3.4009×10^{-4}	11.566×10^{-8}
Reference standard	$u(l_s)$	1	1.1547×10^{-4}	1.1547×10^{-4}	1.332×10^{-8}
x_1	$u(x_1)$	1	0.856×10^{-4}	0.856×10^{-4}	0.748×10^{-8}
x_2	$u(x_2)$	1	0.447×10^{-4}	0.447×10^{-4}	0.200×10^{-8}
x_3	$u(x_3)$	1	3.000×10^{-4}	3.000×10^{-4}	9.000×10^{-8}
x_4	$u(x_4)$	1	3.000×10^{-4}	3.000×10^{-4}	9.000×10^{-8}

Combined Standard Uncertainty

$$u_c^2 = \Sigma\{c_i \cdot u_i(x)\}^2 = 31.845 \times 10^{-8}$$

$$u_c = 5.643 \times 10^{-4} \text{ mm} = 0.0006 \text{ mm}$$

The calculation of the degrees of freedom as per the Welch-Satterthwaite equation is as follows:

$$\nu_{\text{eff}} = u_c^4(y) / \Sigma\{u_i^4(y) / \nu_i\}$$

$$= (0.0006^4) / [\{(0.0034)^4 / 6\} + \{(0.00012)^4 / \infty\} + \{(0.000086)^4 / \infty\} + \{(0.000045)^4 / \infty\} + \{(0.0003)^4 / \infty\} \times 2]$$

$$= (1.296 \times 10^{-13}) / (2.227 \times 10^{-15})$$

$$= 58$$

Expanded Uncertainty

In view of the high value of the degree of freedom, a coverage factor $k = 2$ corresponding to a confidence level of 95.5% can be used:

$$\text{Expanded uncertainty } U = 2 \times 0.0006 = 0.0012 \text{ mm}$$

Reporting the Uncertainty

The measured length of the gage block is equal to 12.9001 mm ±0.0012 mm, where 0.0012 mm is the expanded uncertainty at a 95.5% confidence level. The expanded uncertainty has been arrived at based on the combined standard uncertainty figure of 0.0006 μm and a coverage factor of $k = 2$.

Example 3: Uncertainty of the Current Setting of a 10-Ampere DC Current Source

The Measurement Problem

A current of a nominal value of 10 amperes is passed through a standard resistor. The voltage drop across the standard resistance is measured using a 6.5-digit digital voltmeter. The value of the current I is calculated based on the voltage measurements on the digital voltmeter and the known value of the standard resistor.

Mathematical Model

The mathematical equation representing the measurement process is as follows:

$$I = V / R$$

where R = a standard resistor of a nominal value of 1 ohm, and V = the voltage drop across resistor R.

The value of the standard resistor is 1.000010 $\pm \overline{0}.000024$ohms.

Factors Contributing to Uncertainty

- Reproducibility of voltage measurements
- Inaccuracy of the voltmeter
- Uncertainty of the value of the reference standard
- Uncertainty due to drift of the standard resistor

The DC voltage is measured at different times, and the repeat observations are 9.99983, 9.99977, 9.99985, 9.99990, 9.99992, 9.99987, 9.99988, 9.99992, 9.99993, and 9.99994.

Mean of the preceding observations } = 9.99988

Standard deviation of the preceding data = 5.31246×10^{-5}

Value of standard resistance as per the calibration certificate

= 1.000010 $\pm \overline{0}.000024$ohms

So,

$$I = 9.99988 / 1.000010 = 9.99978 \text{ amperes}$$

Type-A Evaluation of Standard Uncertainty

Standard uncertainty due to repeatability of voltage measurement

$$u(V) = (5.31246 \times 10^{-5}) / 10^{0.5} = 1.67995 \times 10^{-5}$$

Type-B Evaluation of Standard Uncertainty

1. Uncertainty due to accuracy of digital voltmeter = 8 ppm. This has a normal distribution and is specified at a 95.5% confidence level. Standard uncertainty $u(Ac) = 8 / 2$ ppm = 4×10^{-6}.

2. Uncertainty due to calibration of reference standard = 24 ppm. This is assumed to have a rectangular distribution.
   Standard uncertainty $u(R) = 24 / \sqrt{3} = 13.856 \times 10^{-6}$.
3. Uncertainty due to drift in the value of the standard resistor = ±50 ppm. This is also assumed to have a rectangular distribution.
   Standard uncertainty $u(d) = 50 / \sqrt{3} = 28.868 \times 10^{-6}$.

Sensitivity Coefficients

Calculation of sensitivity coefficients is as follows:

$$I = V / R$$

$$\partial I / \partial V = 1 / R = 1.0$$

$$\partial I / \partial R = -V / R^2 = -9.99958$$

The various standard uncertainty terms and sensitivity coefficients are given in Table 8.3.

The combined uncertainty can be calculated as follows:

$$u_c^2 = \Sigma(c_i u_i)^2 = 102824.56 \times 10^{-12}$$

Combined standard uncertainty $u_c = 320.66 \times 10^{-6}$ amperes = 0.000321

Table 8.3 Uncertainty budget.

Source of uncertainty	Sensitivity coefficient c_i	Standard uncertainty u_i	$c_i \cdot u_i$	$(c_i \cdot u_i)^2$
Voltage measurement Repeatability of voltage observations $u(V)$	1.0	1.67995×10^{-5}	1.67995×10^{-5}	282.22×10^{-12}
Accuracy of digital voltmeter $u(Ac)$	1.0	4×10^{-6}	4×10^{-6}	16×10^{-12}
Calibration of reference standard $u(R)$	9.99958	13.856×10^{-6}	138.554×10^{-6}	19197.2×10^{-12}
Drift in standard resistance value $u(d)$	9.99958	28.868×10^{-6}	288.67×10^{-6}	83329.14×10^{-12}

$$\text{Effective degrees of freedom} = (320.66 \times 10^{-6})^4 / [\{(282.22 \times 10^{-6})^4 / 9\} + \{(4 \times 10^{-6})^4 / \infty\} + \{(138.534 \times 10^{-6})^4 / \infty\} + \{(288.67 \times 10^{-6})^4 / \infty\}]$$

$$= 1.05725 \times 10^{10} / 704869373 = 15$$

Expanded Uncertainty

Since the effective degree of freedom is only 15, expanded uncertainty can be obtained based on a coverage factor obtained from the Student t distribution corresponding to a 95% confidence level.

From the table in Appendix B, $t = 2.13$.

$$U = 2.13 \times 0.000321 = 0.000684$$

Reporting the Uncertainty

The measured current is equal to 9.99978 ±0.000684 amperes. The uncertainty figure of 0.000684 is the expanded uncertainty and has been derived from a combined standard uncertainty of 0.000321 and a coverage factor of $t = 2.13$ corresponding to a 95% confidence level.

Example 4: Evaluating and Reporting the Uncertainty of Measurement Results for Calibration of a Torque Sensor with Indicator

The Measurement Problem

An instrument called a torque sensor with indicator is calibrated by the in-house calibration laboratory of an industry. The specification of the indicator states its range as 0–100 nanometers (Nm) and its accuracy as 1% of full-scale deflection (fsd).

Thus, the measuring uncertainty of this equipment in the entire range of measurement will not exceed 1 Nm. This uncertainty is due to the inherent design limitation of the torque sensor and indicator. The reference standard used for calibration is a torque calibration system. The specification of the torque calibration system states its accuracy as ±0.1% of the reading.

Factors Contributing to Uncertainty

- Repeatability of indicator readings
- Uncertainty of the torque calibration system

Mathematical Model

$$T = T_r + x$$

where T = the measured value by the torque indicator, T_r = the reference torque value set in the torque calibrator, and x = the uncertainty contribution due to repeatability.

For calibration, a reference torque value of 100 Nm is set in the calibrator, and the same is read on the indicator. Six repeat measurements are taken over a period of time, and the same are given as follows: 105.9, 105.7, 105.9, 105.6, 105.7, and 105.6.

The calibration result for the torque indicator will be reported as the average of the above observations:

$$\bar{x} = 105.73 \text{ Nm}$$

Type-A Evaluation of Standard Uncertainty

$$\text{Standard deviation of repeat observations} = 0.1366$$

$$\text{Standard uncertainty } u(x) = 0.01366 \,/\, \sqrt{3} = 0.06 \text{ Nm}$$

Type-B Evaluation of Uncertainty

One uncertainty component is the uncertainty of the reference standard, which is equal to 0.1 Nm, with normal distribution at about a 95% confidence level.

$$u(T_r) = 0.1 \,/\, 2 = 0.05 \text{ Nm}$$

Sensitivity Coefficients

In view of the additive relationship, both of the sensitivity coefficients are unity.

Combined Standard Uncertainty

$$(0.06^2 + 0.05^2)^{0.5} = 0.078 \text{ Nm}$$

In this case, the combined uncertainty is due to two factors only, viz., the repeatability of the measurement observations and the uncertainty of the reference standard.

Effective Degrees of Freedom

From the Welch-Satterthwaite equation:

$$\nu_{eff} = 7.8 \times 10^{-2} \times 4 \,/\, \{(6 \times 10^{-2})^4 \,/\, 5 + (5 \times 10^{-2})^4 \,/\, \infty\} = 14$$

Expanded Uncertainty

To obtain expanded uncertainty, it would be appropriate to obtain the coverage factor from the Student t distribution. For a 95% confidence level and 14 degrees of freedom, from the table in Appendix B:

$$t = 2.14$$

$$\text{Expanded uncertainty } U = 2.14 \times 0.078 = 0.17 \text{ Nm}$$

As per the technical specification of the torque sensor, the accuracy is stated as ±1% of the full-scale deflection. Thus, the maximum permissible error while reading 100 Nm is ±1.0 Nm. The uncertainty due to the calibration process is 0.17 Nm, which is much less than the allowable figure of ±1%. However, there is a permanent bias in the indicator. The bias is equal to 105.73 – 100 = 5.73 Nm. The torque sensor with indicator measures precisely but not accurately due to its inherent bias. However, because the bias is known during the process of calibration, it can always be corrected. If the industry has to make a decision about the use of this torque sensor with indicator for a specific measurement application, it will depend upon the criticality of the torque measurements, the cost of new equipment, and the company's own faith in its calibration process

Reporting the Uncertainty

The value of torque of 100 Nm as measured on the torque sensor with indicator is 105.73 ±0.17 Nm. The uncertainty of 0.17 Nm is an expanded uncertainty evaluated based on a combined standard uncertainty of 0.078 Nm evaluated and a coverage factor of the Student t factor = 2.14 corresponding to 14 degrees of freedom and a 95% confidence level of the Student t distribution.

Example 5: Evaluation of Measurement Uncertainty for Calibration of a Wattmeter at 500 Watts

The Measurement Problem

A wattmeter is calibrated using a power calibrator. The power is adjusted in the power calibrator to read 500 watts in the wattmeter. The power meter is set for a voltage range of 250 volts, current range of 10 amperes, and a power factor of 0.2

Factors Contributing to Uncertainty

- Repeatability of reference values that read 500 watts in the wattmeter
- Uncertainty of the power setting in the power calibrator

Mathematical Model

$$W = W_r + x$$

where W = value measured by wattmeter, W_r = reference power set in power calibrator, and x = uncertainty due to repeatability.

Repeat observations of the power calibrator settings as obtained at different times are as follows: 498.6, 498.5, 498.6, 498.6, 498.8, 498.6, 498.8, 498.6, 498.7, and 498.6.

The arithmetic mean of the preceding observations $\bar{x} = 498.64$ watts

Standard deviation of the preceding data = 0.0966

Type-A Evaluation of Standard Uncertainty

$$\text{Standard uncertainty } u(x) = 0.0966 / \sqrt{10} = 0.0305$$

$$\text{Type-A standard uncertainty } u_1 = 0.0305$$

Type-B Evaluation of Standard Uncertainty

The uncertainty of the power setting of the power calibrator is 608 ppm, and its distribution is assumed to be rectangular.

$$\text{Standard uncertainty } u(W_r) = (608 / \sqrt{3}) \times 10^{-6} = 0.000351$$

Sensitivity Coefficients

Since the relationship is linear among independent and dependent variables, the sensitivity coefficients are equal to 1.

Combined Standard Uncertainty

$$\text{Combined standard uncertainty } [(0.0305)^2 + (0.000351)^2]^{1/2} = 0.0305$$

It is to be noted that the combined standard uncertainty is the same as the type-A standard uncertainty due to repeatability, as the other component of uncertainty due to the power calibrator setting is insignificant compared with the uncertainty of repeatability. The effective degrees of freedom will be 9 (that is, one less than the number of repeat observations).

Expanded Uncertainty u

To obtain the expanded uncertainty, the coverage factor has to be taken from the Student t table given in Appendix B. For a 95% confidence level and 9 degrees of freedom, $t = 2.26$.

$$\text{Expanded uncertainty } U = 2.26 \times 0.0305 = 0.07$$

Reporting the Uncertainty

The value of power as measured by the wattmeter is 498.64 ±0.07 watts. The uncertainty of ±0.07 watt is the expanded uncertainty derived based on the combined standard uncertainty of 0.0305 and a coverage factor equal to 2.26 corresponding to a 95% confidence level and 9 degrees of freedom of the Student t distribution.

Example 6: Uncertainty Evaluation with Three Variables

The Measurement Problem

Electric power consumed by a load is calculated with intermediate measurements of voltage, current, and power factor separately, which are measured using three different pieces of measuring equipment. The following observations have been made:

$$\text{Voltage} = 110 \text{ volts AC}$$

$$\text{Current} = 5.5 \text{ amperes AC}$$

$$\text{Phase angle} = 35^\circ$$

Calculate the power consumed by the load and its uncertainty of measurement.

Mathematical Model

The power consumed by the load is calculated from the following relation:

$$P = V \cdot I \cdot \cos \phi$$

where V is the voltage across the load, I is the current passing through the load, and $\cos \phi$ is the power factor of the load.

Factors Contributing to Uncertainty

- Inaccuracy of voltage measurement
- Inaccuracy of current measurement
- Inaccuracy of power factor measurement

The measurement accuracy according to the technical specifications of the measuring equipment is as follows:

- AC voltage measurement: ±0.5% of reading
- AC current measurement: ±1% of reading
- Phase angle: ±1° at 35°

Power consumed P by the load is calculated from the formula given earlier:

$$P = 110 \times 5.5 \times \cos 35$$

$$P = 110 \times 5.5 \times 0.819 = 495.495 \text{ watts}$$

Now the author intends to evaluate the uncertainty of the preceding result using the uncertainty propagation formula with the preceding uncertainty figures, which have been obtained from technical specifications and will need type-B evaluation.

Type-B Evaluation of Standard Uncertainties

Standard Uncertainty for V

Uncertainty in the measurement of voltage $V = 0.5 \times 110 / 100 = 0.55$ volt.

Assuming that the specification has been quoted at a 99.73% confidence level, corresponding to a coverage factor of $k = 3$, the standard uncertainty of voltage measurement $u(V) = 0.55 / 3 = (0.1833)$.

Standard Uncertainty for Current I

Uncertainty in the measurement of current I is as follows:

$$1 \times 5.5 / 100 = 0.055$$

Standard uncertainty of the current measurement is as follows:

$$u(I) = 0.55 / 3 = (0.0183)$$

Standard Uncertainty for Power Factor Cos ϕ

$\phi = 35° \pm 1°$

ϕ varies between 34° and 36°

cos 34° = 0.829, cos 35° = 0.819, cos 36° = 0.809

Let x be another variable equal to cos ϕ:

$$x = \cos \phi$$

The variable x varies between 0.819 ±0.010 when ϕ varies between 34° and 36°.

Assuming that the phase angle lies between 34° and 36° with equal probability, then x can be assumed to have a rectangular distribution with the semi-range ±0.010.

$$\text{Uncertainty of } x = \pm 0.010$$

$$\text{Standard uncertainty of } x = u(x) = 0.010 / \sqrt{3} = (0.0058)$$

Evaluation of sensitivity coefficients is as follows:

$$P = V \cdot I \cdot \cos \phi = V \cdot I \cdot x$$

$$\partial P / \partial V = I \cdot x = 5.5 \times 0.819 = 4.5045$$

$$\partial P / \partial I = V \cdot x = 110 \times 0.819 = 90.09$$

$$\partial P / \partial x = \partial P / \partial \phi \cdot \partial \phi / \partial x = V \cdot I \cdot \sin \phi (1 / \sin \phi) = V \cdot I = 605$$

Uncertainty Budget

The uncertainty budget is shown in Table 8.4.

Combined Standard Uncertainty

$$u_c^2 = \Sigma(c_i u_i)^2 = 15.7127$$

$$u_c = 3.9639$$

Combined standard uncertainty of power measurement = 3.9639 watts

Expanded Uncertainty

Expanded uncertainty for a coverage factor k = 3 corresponding to about a 99.73% confidence level is shown as follows:

$$U = 3 \times 3.9639 = 11.892 \text{ watts}$$

Measurement Result

Power consumed by the load = 495.495 ±11.892 watts

Relative uncertainty = 11.892 × 100 / 495.495 = 2.4%

Table 8.4 Uncertainty budget.

Source of uncertainty	Standard uncertainty u_i	Sensitivity coefficient c_i	$c_i u_i$	$(c_i u_i)^2$
Inaccuracy of voltage measurement	0.1833	4.5045	0.8257	0.6817
Inaccuracy of current measurement	0.0183	90.09	1.6486	2.7180
Inaccuracy of power factor measurement	0.0058	605	3.509	12.3130

Effect of Nonlinearity in Functional Relationship

To evaluate the effect of higher-order terms in a Taylor series expansion, the effect of the following second-order term on standard uncertainty can be evaluated:

$$\Sigma\Sigma\tfrac{1}{2}[(\partial^2 P / \partial x_i \partial x_j)^2]u^2(x_i)u^2(x_j)$$

There are three variables, viz., v, I, and x. Therefore, the following second-order term has to be considered:

$$\tfrac{1}{2}(\partial^2 P / \partial V \partial I)^2 \cdot u^2(V) \cdot u^2(I)$$

$$+ \tfrac{1}{2}(\partial^2 P / \partial V \partial x)^2 \cdot u^2(V) \cdot u^2(x) + \tfrac{1}{2}(\partial^2 P / \partial x \partial I)^2 \cdot u^2(x) \cdot u^2(\mathrm{I})$$

Second-order sensitivity coefficients are as follows:

$$(\partial^2 P / \partial V \partial I) = x = 0.819$$

$$(\partial^2 P / \partial V \partial x) = I = 5.5$$

$$(\partial^2 P / \partial x \partial I) = V = 110.0$$

Putting these values in the expression for second-order Taylor series terms:

$$\tfrac{1}{2}(0.819)^2.(0.1833)^2.(0.0183)^2 + \tfrac{1}{2}(5.5)^2.(0.1833)^2.(0.0058)^2.$$
$$+ \tfrac{1}{2}(110)^2(0.0183)^2(0.0058)^2$$

$$= 3.7737 \times 10^{-6} + 17.0953 \times 10^{-6} + 68.1575 \times 10^{-6} = 89.0265 \times 10^{-6}$$
$$= 0.000089$$

$$u_c^2 = 15.7127 + 0.000089 = 15.712789$$

$$u_c = 3.9639$$

It has thus been observed that the exclusion of second-order terms in the Taylor series expansion in the present case has not had any effect on the combined uncertainty figure of the measurement result.

EXPRESSING THE UNCERTAINTY OF MEASUREMENT RESULTS CONTAINING UNCORRECTED BIAS

The method for evaluating uncertainty proposed by *GUM* is applicable to all measurements including industrial measurements. The guide also recommends that all known bias should be corrected for before uncertainty is evaluated. However, at the level of the shop floor of an industry it may not be economically possible to correct the bias. For example, when the number of measurements is very large, correcting each measurement is time consuming and prone to error, and operators may need to be specially trained to do so. These situations are not uncommon in industrial measurement where automated measurement systems are used. In such cases, even if the bias is known the operator is unable to modify the behavior of the measurement system. *GUM* has not dealt with such situations, which are not uncommon in industrial measurements. Steven D. Phillips and Keith R. Eberhardt of the National Institute of Standards and Technology (NIST) and Brian Parry of the Boeing Corporation have proposed guidelines for expressing the uncertainty of measurement results containing uncorrected bias. The guidelines are meant to be applied only in a situation where applying a correction for a known measurement bias would be costly but increasing the measurement uncertainty to allow for uncorrected bias would still result in an acceptable uncertainty statement. The recommended method is explained in the following paragraphs.

In a usual situation, if y is a measurement result of a measurand Y with expanded uncertainty U, the uncertainty interval is given by $y - U \leq Y \leq y + U$. In a case where the measurement result y has a known bias δ, the corrected result $y_c = (y - \delta)$.

The uncertainty interval desired (see previous paragraph) will also be applicable for y_c such that $y_c - U \leq Y \leq y_c + U$ gives the uncertainty interval. This is equivalent to an uncertainty interval of $y - (U + \delta) \leq Y \leq y + (U - \delta)$, where y is the measurement result with uncorrected bias.

This way of expressing uncertainty results in the uncertainty interval being unevenly placed with respect to the measurement result with uncorrected bias. This measurement result can thus be stated as follows:

$$y^{+(U-\delta)}_{-(U+\delta)}$$

If $(U - \delta) = U_+$ and $(U + \delta) = U_-$, the uncertainty interval in the presence of uncorrected bias can be stated as $y - U_- \leq Y \leq y + U_+$.

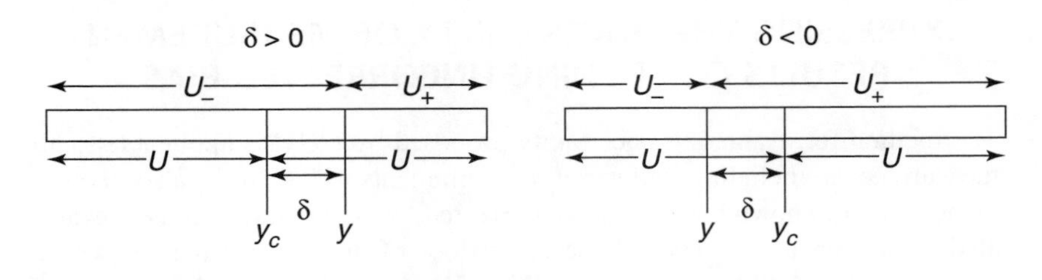

Figure 8.2 Uncertainty intervals with uncorrected bias.

The reference is now taken as y, and whether U_+ or U_- will be larger depends upon the sign of the bias. If the bias is positive, the U_- component is larger compared with U_+, and vice versa, as shown in Figure 8.2.

If the numerical value of the bias δ is more than the uncertainty U than for a positive bias, the value of U_+ becomes negative and the value of U_- becomes greater than $2U$. Similarly, in the event of δ being more than U numerically, for a negative bias the value of U_- becomes negative and the value of U_+ becomes greater than $2U$. Since the uncertainty component can never be negative, dealing with such situations becomes confusing. To deal with such situations, the authors Phillips, Eberhardt, and Parry have proposed an additional requirement to state that uncertainty limits be greater than or equal to zero for all values of δ. Thus, whenever U_+ or U_- is calculated to be negative, it will be taken as zero, and the value of the other component will be taken as the same as has been arrived at based on calculations as given earlier. This additional requirement results in a wider uncertainty interval. Thus, the uncertainty interval in the presence of uncorrected bias will be given as $y - U_- \leq Y \leq y + U_+$.

$$Y = y \begin{array}{l} + U_+ \\ \\ - U_- \end{array}$$

where
$$U_+ = U - \delta \quad \text{if } U - \delta > 0$$
$$= 0 \quad \text{if } U - \delta \leq 0$$

$$U_- = U + \delta \quad \text{if } U + \delta > 0$$
$$= 0 \quad \text{if } U + \delta \leq 0$$

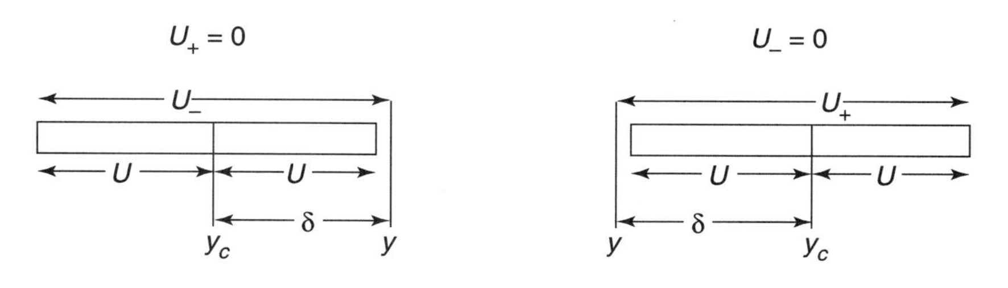

Figure 8.3 Uncertainty interval with uncorrected bias, when δ is numerically greater than U.

The situation is shown in Figure 8.3.

It can be noted from the figure that the large bias results in a one-sided uncertainty interval. The authors have also deliberated on the proposed guidelines and offered examples applicable to various situations. Two of them are given here.

One Type-A Bias

Let the bias be δ_1, which has been obtained based on repeat measurements of a reference standard with a value $R \pm kU_{ref}$, where U_{ref} is the combined uncertainty of the reference standard. If N_1 measurements are made of the reference standard by the measurement setup, δ_1 is the average of the differences between these N_1 measurements and the reference value. If s is the standard deviation of these N_1 measurements, the bias δ_1 will also have a standard uncertainty of $s / \sqrt{N_1}$. If U_1 is the combined standard uncertainty of the measurement result for all other contributing factors other than bias, the repeatability uncertainty of N_1 observations will also be contained in U_1. The combined standard uncertainty U_c, as if corrected for bias of the result, consists of U_1, the standard uncertainty of bias, and the standard uncertainty of the reference standard.

$$U_c = (U_1^2 + s^2 / N_1 + U_{ref}^2)^{1/2}$$

If δ_1 is the bias, $U = kU_c$

Thus, $U_+ = kU_c - \delta_1$ and $U_- = kU_c + \delta_1$

The uncertainty interval is therefore given by the following:

$$y - U_- \leq Y \leq y + U_+$$

One Type-B Bias

Sometimes bias is estimated rather than directly measured. Generally lengths measured on the shop floor of industry are not corrected for temperature variations. The lengths are generally corrected for and reported at 20°C. If the shop floor temperature varies between 20°C and 30°C, it can be assumed to have a uniform distribution with a mean of 25°C and an expanded uncertainty of ±5°C. This distribution will have a standard deviation of $5 / \sqrt{3} = 2.9$°C. It can be stated that the length as measured on the shop floor has been reported at 25°C. Since the result was to be reported at 20°C, the bias δ_2 due to this can be accounted for by the length deviation that is due to the 5°C mean uncorrected thermal expansion. The bias δ_2 will be obtained by multiplying the measured length by the appropriate thermal coefficient of expansion and multiplied further by 5. The standard uncertainty of the length measurement U_t due to temperature variation can be obtained by multiplying the standard uncertainty of the temperature by the appropriate coefficient of thermal expansion. If U_2 is the combined standard uncertainty of measurement, and if the measurement has been corrected back to 20°C, then

$$U_c = (U_2^2 + U_t^2)^{1/2}$$

and

$$\text{bias} = \delta_2$$

The uncertainty interval can be given by the following:

$$y - U_- \leq Y \leq y + U_+$$

The uncertainty as per Phillips, Eberhardt, and Parry's (1997) approach is evaluated by finding the following:

- The combined standard uncertainty U_c, computed as if the measurement result were to be corrected for bias. Thus, U_c also includes the uncertainty of uncorrected bias
- The bias and whether it is positive or negative
- The expanded uncertainty, which includes the effect of bias term

The proposed guidelines also recommend that if there are several sources of uncorrected bias, the biases should be algebraically added together. The resulting net bias is stated together with the combined standard uncertainty. If multiple sources of uncertainty have biases and those biases are not independent, then the degree of overlap of biases is estimated and subtracted from the bias summation. In this case too, the uncertainty of overlap correction is taken into account while calculating the combined standard uncertainty. It is envisaged that the final uncertainty quoted by this method will be greater than or equal to the uncertainty that would be quoted if the bias were corrected. It is also envisaged that the level of confidence for the expanded uncertainty should be at least as great as the level obtained for the case of corrected bias.

GUM AND INTERNATIONAL PERSPECTIVES

GUM received universal acceptance, and based on it many organizations in various countries have published their own guidance documents for use by metrology organizations in those countries. The National Institute of Standards and Technology in the United States has published NIST Technical Note 1297 as a guidance document on the evaluation and expression of uncertainty for NIST measurement results. In addition to Technical Note 1297, a number of other documents have been published on the subject that are based on the concept of the evaluation of measurement uncertainty as proposed in *GUM*. Some of them are listed here:

- NIS 3003 (8 May 1995), *The Expression of Uncertainty and Confidence in Measurement for Calibration,* published by NAMAS (United Kingdom)
- Recommended Practice RP-12 (1 February 1994), *Determining and Reporting Measurement Uncertainties,* published by the National Conference of Standards Laboratories (NCSL) (United States)
- NABL 141, *Guidelines for Estimation and Expression of Uncertainty in Measurement,* published by the National Accreditation Board for Testing of Calibration Laboratories (India)

GUM has already been translated into French, German, Chinese, and Italian, and the translated versions have been published by the ISO. *GUM* methods have been adopted by various regional metrology organizations.

NIST POLICY ON UNCERTAINTY OF MEASUREMENT RESULTS

As noted in the previous section, the NIST has published the document NIST Technical Note 1297 titled *Guidelines for Evaluating and Expressing the Uncertainty of NIST Measurement Results.* It is based on the approach contained in BIPM recommendation (1980) and *GUM*. The guidance document was prepared for use by the NIST staff. It is a crisp document compared with *GUM*. The salient features of Technical Note 1297, which also contains NIST policy on measurement uncertainty, are given in the balance of this section.

The guide is intended to apply to most NIST measurement results including results associated with the following:

- International comparisons of measurement standards
- Basic research
- Applied research and engineering
- Calibrating a client's measurement standards
- Certifying standard reference materials
- Generating standard reference data

The policy applies to all measurement results reported by the NIST. The policy consists of guidelines regarding two aspects: one pertaining to what should be expressed as an index of measurement uncertainty, and the other pertaining to how it is to be reported (that is, what information is to be provided as part of the uncertainty statement).

NIST Policy on Expressing Uncertainty

a. Commonly the combined standard uncertainty u_c should be used for reporting results of determination of fundamental constants, fundamental metrological research, and international comparisons of realization of SI units.
b. In keeping with the practice adopted by other national standards laboratories and several metrological organizations, use expanded uncertainty U to report the result of all NIST measurements other than those for which u_c has traditionally been employed. To be consistent with current international practices, the value of coverage factor k to be used at NIST for calculating U is by convention $k = 2$. This pertains to a confidence level of 95.5%. Values of k other than 2 are only to be used for specific application dictated by established and documented requirements.
c. As an exception, NIST policy also states that any valid method that is technically justified under existing circumstances may be used to determine the equivalent of u_i, u_c, or U. Further, it is recognized that international, national, or contractual agreements to which the NIST is a party may occasionally require deviation from NIST policy. In both cases, the report of uncertainty must document what was done and why.

NIST Policy on Reporting Uncertainty

a. All NIST measurement results are to be accompanied by a quantitative statement of uncertainty. Report U together with the coverage factor k used to obtain it, or report combined standard uncertainty u_c.

b. When reporting a measurement result and its uncertainty, include the following information in the report itself or by referring to a published document:
   (i) A list of all components of standard uncertainty together with their degrees of freedom where appropriate and the resulting value of u_c. The components should be identified according to method used (that is, type-A or type-B evaluation).
   (ii) A detailed description of how each component of standard uncertainty was evaluated.
   (iii) A description of how k was chosen when k is not equal to 2.

c. It is often desirable to provide probability interpretation, such as level of confidence for the interval defined by U or u_c.

d. It is not possible to know the details of the uses a NIST measurement result will be put to; thus, it is usually inappropriate to include in the reported uncertainty any other component that arises from a NIST assessment of how the result might be used.

ENDNOTE

1. © International Organization for Standardization (ISO). This material is reproduced from the *ISO Guide to the Expression of Uncertainty in Measurement* with permission of the American National Standards Institute on behalf of ISO. No part of this material may be copied or reproduced in any form, electronic retrieval system or otherwise or made available on the Internet, a public network, by satellite or otherwise without the prior written consent of the American National Standards Institute, 25 West 43rd Street, New York, NY 10036.

REFERENCES

International Organization for Standardization (ISO). 1995. *Guide to the Expression of Uncertainty in Measurement (GUM).* Geneva, Switzerland: ISO.

National Institute of Standards and Technology (NIST). 1993. Technical Note 1297, *Guidelines for Evaluating and Expressing the Uncertainty of NIST Measurement Results.* Gaithersburg, U.S. Department of Commerce.

Phillips, Steven D., Keith R. Eberhardt, and Brian Parry. 1997. Guidelines for expressing the uncertainty of measurement results containing uncorrected bias. *Journal of Research of National Institute of Standards and Technology* 102, no. 5:577–85.

9

Measurement Assurance Programs (MAPS)

MEASUREMENT SYSTEMS, METROLOGICAL REQUIREMENTS, AND METROLOGICAL CONFIRMATION

A measurement result is obtained as an output of a measurement process. A measurement process is defined as a set of interrelated resources, activities, and influences related to a measurement. The interrelated resources which are part of a measurement process consist of measuring equipment or measurement setup, the measurement procedure, and persons making the measurements and deriving the measurement result. The interrelated activities include preparation, verification, and validation of the measurement procedure and operation and control of the measurement process. The influences that form part of the measurement process are all the influence factors—for example, temperature, pressure, humidity, and so on—that influence the measurement process and ultimately the measurement result. These influence factors cause variability and bias in the measurement process.

The measurement process is also sometimes called the measurement system, which has been defined as the collection of operations, procedures, gages and other equipment, software, and personnel used to assign a number to the characteristics being measured—the complete process used to obtain the measurement result. Measurement systems are used to verify that products including services meet the customers' requirements. This is an important activity used to demonstrate that products are complying with their specifications and that the organization is achieving its quality objectives.

Metrological Requirements

A requirement is a stated need or expectation. A documented requirement is generally an obligatory requirement. A requirement could also be an implied requirement. Implied requirements bear on the level of customer satisfaction. A metrological requirement is the requirement for measurement and measurement result. Two examples of metrological requirements follow:

- Measurement of the insulation resistance of an electric cable at 500 volts DC
- Blood analysis of an individual for cholesterol content

Measurement systems are used to verify product characteristics, so metrological requirements are derived from the specified requirements for the products. The requirements for the products are specified in the products' technical specifications. For example, the technical specifications of a steel rod being manufactured in an automobile industry may include its diameter to be 6.30 ±0.25 millimeters. To measure this, a measurement system is needed. Thus, the metrological requirements which should be met by this measurement system depend upon the technical specifications of the product being tested.

The technical specifications are generally stated in terms of the value and range of the parameter and its tolerance. Therefore, the metrological requirement for a measurement system can be stated in those terms, that is:

- The measurement system should be capable of measuring the specific parameter within the required range
- The uncertainty of the measurement system preferably should be better than one-fourth of the product characteristics' tolerance and in no case worse than one-third of that

The preceding are broad yardsticks for arriving at the metrological requirements from the product specification requirements.

Metrological Confirmation

Having identified the metrological requirements, the next step in the process is the identification of measuring equipment and measurement setup which are capable of measuring the characteristics to the required level and accuracy. The process of ensuring that the measuring equipment is in compliance with the metrological requirement is called metrological confirmation.

The metrological confirmation process consists of the following three stages:

1. Study the product measurement requirements, which depend upon the following:
   a) Product technical specifications
   b) Production process capability if the measurement result is used for statistical process control

After studying the product measurement requirements, the metrological requirements for the measurement system are decided. They are expressed in terms of measurement range, maximum permissible error or uncertainty, class of instrument, and so forth. The following should be ensured:

a) The metrological requirements have been defined unambiguously
b) The risk of bad measurements is within acceptable limits

2. Theoretically study the technical specifications of the measuring equipment to ensure that the equipment's capability matches the metrological requirement.
3. Carry out all activities to ensure that the measuring equipment is in a state fit to measure the desired product characteristics as per the metrological requirements. These activities may include the following:
   a) Repair, maintenance, and adjustment
   b) Periodic calibration and recalibration after repair and adjustment
   c) Sealing for integrity and labeling
   d) Verification of the preceding through objective evidence

   As calibration of measuring equipment is part of the metrological confirmation system, the organization should keep the following points in mind:
   a) The accuracy of a piece of measuring equipment is designed and manufactured into it, but confidence in its continued accuracy comes only from the periodic calibration process.
   b) The reference standard used for calibration should have traceability to national or international standards of measurement. This is a mandatory requirement for the acceptability and validity of a measurement result.
   c) The calibration process should have an adequate test uncertainty ratio (TUR).

QUALITY ASSURANCE OF MEASUREMENT RESULTS

Control of the Measurement System

The metrological confirmation system as explained in the preceding section ensures that the measuring capability of a piece of equipment or the measurement setup is fit for measurement of the characteristics being measured. It also ensures that the equipment is in a state of continued readiness for use in the measurement system. It, however, does not necessarily ensure that the measurement result also complies with the metrological requirement and ultimately with user requirements. The metrological confirmation only ensures one aspect of the measurement system, that is, the measuring equipment. Other factors, viz., the skill of the operator, the procedure of measurement, and the environmental conditions, also affect the quality of the measurement result, and they are not controlled by the metrological confirmation process. The ultimate concern of an organization is the quality of the measurement result and not the quality of the measuring equipment. To ensure that the measurement result is in compliance with metrological requirements and ultimately with the user's requirements, all aspects of the measurement system need to be controlled.

Measurement Assurance

The compliance of the measurement result with metrological requirements is achieved through a process called a measurement assurance process, or MAP. Quality of products and services is a key business issue for most modern organizations. This is achieved through the application of quality management principles. Quality management is part of the overall management function of an organization. The process of quality assurance is a part of the quality management function; it focuses on providing confidence that the quality requirements of a product are fulfilled. The quality of a measurement result is characterized by its uncertainty of measurement. Measurement assurance can thus be defined as a process which focuses on providing confidence that quality requirements for a measurement result are fulfilled. It thus consists of all the activities which are carried out to provide confidence that the measurement system is producing measurement results that comply with metrological requirements. The confidence has to be provided through objective evidence on a continuing basis rather than at an isolated period of time.

Activities for Measurement Assurance

A number of activities have to be performed sequentially to achieve the desired quality of a measurement result. These activities are part of the measurement assurance program. They are performed in various steps, as follows.

Step 1

Understand the product specification requirements, and based on them, decide on the metrological requirements.

Step 2

Design a measurement system that is capable of meeting the metrological requirements. During this design phase, the following issues need to be considered:

- The design is to be undertaken by professionally competent persons
- Measuring equipment should be capable for the purpose
- The uncertainty of the measurement result due to the bias and precision of the measurement system is to be evaluated for various influence factors
- The variability of the measurement system is to be small compared with the tolerance of the product characteristics desired to be measured
- If the measurement system is meant for statistical control of a manufacturing process, the measurement system variability is to be small compared with the manufacturing process variability
- The measurement result must be traceable to national or international standards of measurement
- The cost of making measurements is to be acceptable
- The time required for making measurements must be reasonable

Step 3

Evaluate the measurement system to verify its capability to comply with metrological requirements. During this evaluation phase, the following issues need to be considered:

- All of the measuring equipment which is part of the measurement system must be calibrated with traceability to national or international standards.
- The measurement system should be under statistical control—that is, the variability of the measurement system is due to chance causes only and not attributed to an assignable cause. This can be ensured through the use of $\bar{x}$ and R control charts (discussed later in the chapter).
- The measurement system must have adequate discrimination.
- The bias, linearity, and stability of the measurement system must be consistent over the expected range of measurement and acceptable.
- The repeatability and reproducibility (R&R) of the measurement system must be evaluated and statistically analyzed.
- The ratio of the variability of the measurement system and the product characteristics tolerance or manufacturing process capability must be determined and examined for acceptability.
- The measurement system is to be robust with reference to the environment and other factors influencing the measurement system.
- The evaluation is to be carried out by competent persons knowledgeable about the measurement system.
- The decision regarding acceptability or otherwise of the measurement system for an intended application should be based on all the aforementioned factors.

Step 4

Continually evaluate the measurement process. Having been evaluated and accepted for its capability for its intended application, the measurement system is put to use. However, a onetime evaluation of a measurement system is not enough. The measurement assurance program provides the desired confidence about the measurement system's continuing suitability for its intended application based on objective evidence. Thus, continuous evaluation of the measurement process is an essential requirement of an effective measurement assurance program. That continuous evaluation is done through the following activities:

- A metrological confirmation process through periodic calibration verification.
- Maintenance of measuring equipment.
- Conducting gage R&R studies at regular intervals. If a study indicates doubt about the capability of the measurement system, the reasons for the same have to be investigated, and necessary corrective actions are to be taken to restore the measurement system's capability.
- The use of statistical process control (SPC) charts. SPC charts are a powerful tool to control the capability of a measurement system; they can be used in testing and calibration laboratories as well as in industrial metrology.

The thrust of measurement assurance is toward controlling a measurement system on a continuous basis rather than as an isolated event. Measurement assurance helps to resolve quality problems and provides opportunities for enhancing product quality. It also enhances the quality of calibration and testing processes in testing and calibration laboratories.

MEASUREMENT ASSURANCE PROGRAMS IN TESTING AND CALIBRATION LABORATORIES

To ensure the quality of test and calibration results, laboratories use quality assurance measures in their operations. Such measures can be classified into two categories:

- **Internal quality control measures:** These measures are part of a laboratory's internal operations system and include implementing a documented quality management system, training personnel, validating test and calibration methods, and following other good laboratory practices. In addition to these measures, laboratories need to monitor their operations and results continuously to decide whether results are reliable enough to be reported. Such quality control measures include replicate tests or calibrations using the same or different methods, retesting or recalibration of retained items, and regular use of reference materials. The use of control charts using statistical process control techniques is now becoming popular among metrology organizations.

- **External quality control measures:** These measures need the involvement of other laboratories that provide similar types of testing or calibration services as well as the involvement of external agencies such as laboratory accreditation bodies. These measures include participation of laboratories in interlaboratory comparison and proficiency testing programs.

The details of the statistical control of measurement processes and proficiency testing techniques as measurement assurance programs are given in the following section.

THE MEASUREMENT PROCESS AND STATISTICAL PROCESS CONTROL

Statistical Control of Manufacturing Processes

Dr. Walter Shewhart proposed in 1924 that control charts be used to control manufacturing processes with a view to ensure that the manufactured product complies with the requirements of its specifications. Before the introduction of control charts, inspection was used as a quality control technique to screen out items not meeting specifications. This strategy of detecting defectives is wasteful and uneconomical, whereas by the use of control charts information about the process and product is continuously gathered and analyzed so that action can be taken on the process itself.

Control charts are a graphical means of applying statistical principles to control a manufacturing process. The control chart theory is based on two types of variability affecting a manufacturing process. The first type is random variability, which is due to chance causes, or common causes. Those include a variety of causes that are consistently present but cannot be readily identified. Examples of common causes are variability in raw material, operation of a machine and its operator, the surrounding environment, and so on. Each constitutes a very small component of the total variability. The cumulative effect of all these small variability components is responsible for the random variability of the manufacturing process and is assumed to be inherent in the process. This variability cannot be eliminated altogether, but it can be reduced by improving the process by allocation of additional resources.

There is a second type of variability which can be attributed to some identifiable causes that are not an inherent part of the process and could be eliminated. Those identifiable causes are called special or assignable causes of variation. They could be attributed to competence of a machine operator, nonuniformity in raw material, a broken or degraded tool, and so on. A manufacturing process which is under the influence of common causes of variation only is said to be in a state of statistical control. If an assignable cause of variation is present, the process is said to be out of statistical control. Control charts help in the detection of an unnatural pattern of variation in the data and provide a criterion for detecting an out-of-statistical-control condition.

Statistical Control of Measurement Process

Shewhart, the inventor of the control chart theory, proposed that measurements are like the output of a production process. Churchill Eisenhart also propagated the analogy between a manufacturing process and a measurement process and advocated the use of statistical techniques in the analysis, estimation, and control of measurement data. The concept of statistical control of measurement processes was advanced by a number of statisticians and metrologists. It is accepted that for a measurement process to give meaningful results, it has to be in a state of statistical control.

The common, or chance, causes of variation in a measurement process are various sources of random errors that influence the measurement process. These include the variability in the measuring equipment, how a device being measured is put in the gage, the variability in the functioning of the operator, the environment under which measurements are being made, and so forth. It is difficult to identify a single source of random variation in the measured data; rather, the cumulative effect of small contributions by various sources of random errors contribute to the total variability. Such variability is inherent in the measurement process.

On the other hand, the special, or assignable, causes of variation generate an unnatural pattern of variation in the measurement process, similar to what happens in the manufacturing process, and they need to be detected and eliminated. Examples of assignable causes that can influence a measurement process are use of measuring equipment with bias, uncalibrated measuring equipment, incompetence of the person taking measurements, noncompliance with a measurement method, and so on.

Control charts thus help us detect an unnatural pattern of variation in the measurement process, indicating the presence of an assignable cause influencing that process.

Nature of Shewhart Control Charts

In a manufacturing process control charts are plotted based on the data obtained by sampling the process at approximately regular intervals. An interval is defined in terms of time. The sample is called a subgroup and consists of items from the manufacturing process with identical measurable characteristics. From each subgroup one or more subgroup characteristics are derived—for example, average ($\bar{x}$), range (R), or standard deviation (s) of the subgroup. A control chart is a graph of the values of the derived subgroup characteristics versus the subgroup number. It consists of a central line (CL) located at the reference value. The reference value is usually the long-term value of the characteristics obtained based on past experience with the process and is linked with the target value as per the product specifications.

The control charts also have two control limits on either side of the central line which are called the upper control limit (ULC) and lower control limit (LCL). These control limits are determined statistically and are at a distance of 3σ on each side of the central line. Two of the very frequently used control charts are the $\bar{x}$ and R charts. The symbol $\bar{x}$ represents the average of the subgroup, and R represents the range of the subgroup. R thus represents within-subgroup variability of the characteristics. The value σ is within-subgroup standard deviation when calculated for a large number of subgroups forming the population. This does not include subgroup-to-subgroup variation. The 3σ limits indicate that approximately 99.73% of the subgroup values will be included within the control limits provided the process is within statistical control. There is approximately a 0.27% risk on an average of a plotted point being outside of either the upper or lower control limit when the process is in control. The subgroup size is chosen to ensure that subgroup variability or short-term variability is reflected in the items chosen in the sample. A subgroup size of four or five is considered appropriate. The sampling frequency depends upon the probability of the process average shifting. The sampling frequency is high in the beginning and low once a state of statistical control is reached. Preliminary estimates for the central line and control limits are reached based on 20 to 25 subgroups of size four to five. The $\bar{x}$ and R charts have applications in metrology in controlling the uncertainty of an operating measurement system. The $\bar{x}$ and R control charts are simultaneously drawn. For the process to be in statistical control it is essential that both $\bar{x}$ as well as R should be within the upper and lower control limits. When R is out of control, that implies that process dispersion has increased, whereas $\bar{x}$ being out of control implies that the process has shifted from its target value.

Following are the steps involved in the construction of control charts for the average $\bar{x}$ and range R:

- Pick up the samples in accordance with the sampling frequency and in accordance with the agreed-upon subgroup size, and measure the characteristics desired to be controlled on each item of the subgroup.
- Calculate the subgroup average $\bar{x}$ and the range R.
- Compute the grand average of all the subgroup averages $\bar{x}$ and average range $\bar{R}$ of all the subgroup ranges R.
- Select a suitable scale on graph paper for the $\bar{x}$ and R of the subgroup on the vertical scale and the subgroup number on the horizontal scale. Plot the values of $\bar{x}$ on the average chart and the values of R on the range chart.
- Draw solid horizontal lines at values equal to the grand average for the $\bar{x}$ chart and at $\bar{R}$ for the R chart. These represent the central lines of the control charts.
- Draw control limits as dashed horizontal lines on the chart. For the $\bar{x}$ chart the control lines are drawn at $\bar{x} \pm A_2\bar{R}$, and for the R chart the control lines are drawn at D_3R as the lower control limit and D_4R as the upper control limit. The values of A_2, D_3, and D_4 depend upon the subgroup size n and are given in Appendix C of this book.
- If points on the R and $\bar{x}$ control charts are observed to be out of the upper and lower control limits, it indicates that some assignable causes of variation may be operating. Suitable remedial action is taken to eliminate assignable causes and to prevent their reoccurrence. The data indicating out-of-control conditions are discarded, and the values of the grand average of $\bar{x}$ and $\bar{R}$ are recalculated. The control limits are also recalculated, and a revised control chart is plotted.
- As more and more data become available, the central line and upper and lower control limits are redrawn and periodically reviewed.

Controlling Measurement Uncertainty with SPC

Statistical process control (SPC) techniques are effectively employed in controlling the uncertainty inherent in an operating measurement system. Implementation of SPC to control a measurement process is one of the well-accepted techniques of a measurement assurance program. The application of SPC to control a measurement process needs measurement data such as data on manufactured items from a production process. The measurement data should reflect the short-term variability of the measurement process as well as its long-term stability.

To obtain these data, check standards are used. A check standard is an artifact with known metrological characteristics. Its value and associated uncertainty are known to a high level of accuracy. The value of the check standard is measured by the measurement process from time to time, and the data thus generated are used to plot $\bar{x}$ and R control charts. A check standard is thus an item identical to the one that is being measured by the measurement process except that its value is known beforehand. If it is designed to control a calibration process for a standard resister of a nominal value of 1000 ohms, the check standard could be a standard resistor of a nominal value of 1000 ohms but with its values and uncertainty known. A check standard can be used to control more than one measurement process. If the calibration process is used for a range of magnitude then several check standards should be used and control charts should be plotted for each check standard.

Drawing a Control Chart for Measurement Process Control

The procedure used for plotting a control chart for a measurement process is the same as given earlier except that a check standard is used to generate the data pertaining to variability of the measurement process. Repeat measurements are made on the check standard using the measurement system, similar to the way the measurement system is used for designed measurement.

A subgroup of four or five repeat observations is chosen. The observations in the subgroup represent the short-term variability of the measurement process. Thus, four to five repeat observations on the check standard should be made either in quick succession or within a short interval of time, which depends upon the short-term variability of the measurement process. The subgroup variability gives the indication of the precision and random uncertainty of the measurement process. The next subgroup of observations is taken after a period which depends upon how frequently the measurements are made with the process. It could be every hour, every half-day or day, every week, or every time the measurement system is used for measurement. The subgroup number or the sequence of measurements is plotted on the horizontal axis and $\bar{x}$ and R are plotted on the vertical axis on the average and range charts respectively.

The preliminary charts are plotted based on the repeat measurements of 20 to 25 subgroups, each having four to five repeat observations.

Example

The repeat observations obtained on a check standard by a calibration process used to calibrate standard resistors of a nominal value of 1000 ohms are shown in Table 9.1. The measurement system consisted of a DC resistance bridge in a controlled calibration laboratory environment. The measurements were carried out by a competent professional using an approved calibration procedure.

The central line and the upper and lower control limits for the $\bar{x}$ and R charts are given below.

For a subgroup size of four, $n = 4$, so $A_2 = 0.729$, $D_3 = 0$, and $D_4 = 2.282$.

Average ($\bar{x}$) Chart

Central line (CL) = 1000.009 ohms.

Upper control limit (ULC) = $1000.009 + 0.729 \times 0.018 = 1000.022$.

Lower control line (LCL) = $1000.009 - 0.729 \times 0.018 = 999.995$.

Table 9.1 Data collected on check standard (nominal value = 1000 ohms).

	Measured value				Subgroup mean	Subgroup range
Subgroup	x_1	x_2	x_3	x_4	$\bar{x}$	R
1	1000.018	1000.012	999.999	1000.020	1000.012	0.021
2	999.998	1000.008	999.997	999.996	1000.000	0.012
3	999.990	999.986	1000.010	999.995	999.998	0.026
4	1000.016	1000.021	1000.015	999.998	1000.013	0.023
5	999.996	1000.009	1000.013	1000.016	1000.009	0.020
6	999.998	999.999	1000.015	1000.002	1000.004	0.017
7	1000.017	1000.009	999.999	1000.023	1000.012	0.024
8	999.991	999.996	1000.018	1000.008	1000.003	0.027
9	1000.007	1000.013	1000.008	1000.011	1000.010	0.006
10	1000.008	999.997	999.999	1000.021	1000.006	0.024
11	1000.013	1000.006	1000.008	1000.025	1000.013	0.019
12	999.999	999.998	1000.002	1000.011	1000.003	0.013
13	1000.015	1000.017	1000.003	999.999	1000.009	0.018
14	999.997	1000.001	1000.023	1000.008	1000.007	0.026
15	1000.013	1000.009	1000.025	1000.023	1000.018	0.016
16	999.999	999.999	1000.006	1000.006	1000.003	0.007
17	1000.021	1000.023	1000.027	1000.007	1000.020	0.020
18	999.999	1000.009	1000.017	1000.023	1000.012	0.024
19	1000.014	1000.007	1000.020	1000.021	1000.016	0.014
20	999.999	1000.009	1000.012	1000.011	1000.008	0.013
					Grand avg = 1000.009	$\bar{R} = 0.018$

Range (*R*) Chart

$$\text{Central line (CL)} = 0.018$$

$$\text{Upper control limit (ULC)} = 2.282 \times 0.018 = 0.041$$

$$\text{Lower control line (LCL)} = 0.0 \times 0.018 = 0$$

The values for various subgroups plotted on the $\bar{x}$ and R control charts are given in Figure 9.1.

It has been observed that all the points in the $\bar{x}$ and R control charts are within the upper and lower control limits, and thus the measurement process

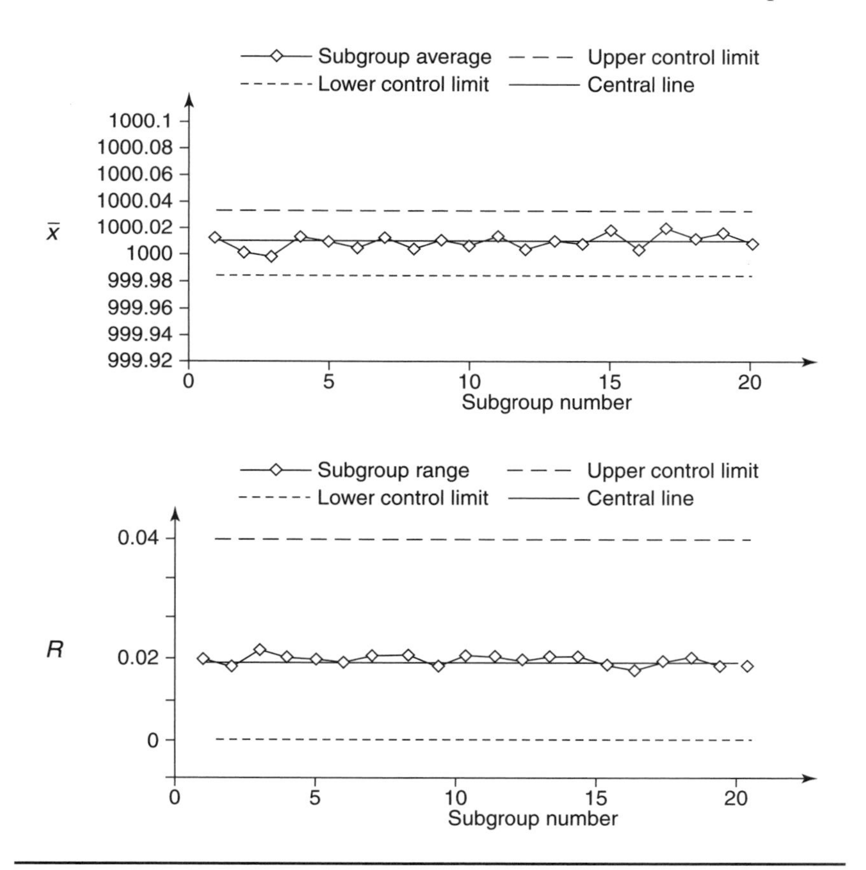

Figure 9.1 $\bar{X}$ and R control charts (check standard nominal value = 1000 ohms).

is in a state of statistical control. If any points are found to be out of the control limits, assignable causes are assumed to exist and they must be investigated and eliminated. The control chart is redrawn after discarding these outlier data.

Interpretation of Control Charts and Uncertainty of Measurement of Measurement Process

Each time a subgroup of observations on the check standard is made in accordance with an approved calibration procedure, the average $\bar{x}$ and range R of the observations is calculated. The values are plotted on the $\bar{x}$ and R charts. As long as the plotted values are within the control limits, the measurement process is expected to be in a state of statistical control, and the results reported by the measurement systems are relied upon and accepted. If either or both the points in the $\bar{x}$ and R charts are found to be out of the control limits, the process is assumed to have gone out of statistical control, implying the presence of an assignable cause. In such a situation the measurement system should not be used until the assignable cause has been detected and eliminated. At times the control limits are redrawn based on the accumulation of fresh data.

An out-of-control situation is an indication that something has gone wrong with the measurement system, and the cause of this has to be identified and corrected. The measurement system does not consist only of the measuring equipment but also includes the operator, environment, and the check standard itself.

An out-of-control condition in the $\bar{x}$ chart indicates a shift in the population average and could be due to measuring equipment going out of calibration or some problem with the check standard itself. Similarly, an out-of-control condition in the R chart indicates an increase in the dispersion of the repeat observations, which could be attributed to inadequate control of the environment or incompetence of the operator involved in the measurements.

The upper and lower control limits are drawn at $\pm 3\sigma$. The process is in statistical control if the points in the control chart are distributed randomly. Sometimes the points lie within the upper and lower control limits, but instead of being randomly distributed they show a trend. Such situations indicate that the process is out of control. Some of these situations are as follows:

- Two out of three consecutive points between 2σ and 3σ
- Four out of five successive points between 1σ and 3σ on either side
- Six points in a row steadily increasing or decreasing
- Eight points in a row on both sides of the central line with none within $\pm 1\sigma$

- Nine points in a row within $\pm 1\sigma$ or beyond on one side of the central line
- Fourteen points in a row alternating up and down
- Fifteen points in a row within $\pm 1\sigma$ above or below the central line

Uncertainty of the Measurement Process

The grand average of $\bar{x}$ is the value assigned to the check standard by the measurement process. The uncertainty of the measurement process is represented by $A_2\bar{R}$, which corresponds to a 99.73% confidence level. The standard deviation of the mean $\bar{x}$, which is represented by $\sigma\bar{x}$, is equal to $A_2\bar{R} / 3$. Thus, $\sigma\bar{x}$ is the standard uncertainty of the measurement process. The control charts for the measurement process are maintained on a continuing basis. As more and more data are available, the grand average of $\bar{x}$ and average range $\bar{R}$ are evaluated afresh, and the central line and upper and lower control limits are redrawn. In the long run, the measurement process matures, and the values of grand average and average range stabilize, as do the upper and lower control limits. The grand average represents the limiting mean envisaged by Churchill Eisenhart.

The check standard is also calibrated independently and more accurately by a calibration laboratory, preferably a national metrology laboratory. This ensures that there is no bias in the check standard and that its assigned value is within acceptable limits. The periodic calibration of the check standard enhances its value as a control tool for the measurement process. If μ is the value assigned to the check standard after calibration, the difference $\bar{x} - \mu$ should be within acceptable limits. The periodic calibration of the check standard also ensures the traceability of the measurement processes being controlled by the check standard.

The application of $\bar{x}$ and R charts in a measurement process helps in the monitoring and control of uncertainty of measurement and helps to ensure its traceability. The power of SPC as an effective tool is encouraging more and more metrologists, particularly in calibration laboratories, to use it. SPC is therefore no longer confined to the production process only.

PROFICIENCY TESTING

It is expected that measurements made on a specific product characteristic in various laboratories should produce identical test results. Similarly, a homogenous sample when analyzed in various laboratories should produce identical analytical results. If results reported by one laboratory are not similar to results reported by other laboratories, that implies that the specific laboratory is not technically competent to carry out that type of testing and analysis. This exercise of testing or calibration of the same or similar items in various laboratories to intercompare their reported results is called interlaboratory comparison. Such comparison has been defined as the "organization, performance and evaluation of tests on the same or similar test items by two or more laboratories in accordance with predetermined conditions." There are a number of purposes for conducting interlaboratory comparison. These include determining the performance of an individual laboratory, judging the adequacy of a new measurement method, validation of reference values, and so on.

Proficiency testing is the use of interlaboratory comparison results for determining a laboratory's testing or calibration performance. It has been defined as "determination of laboratory testing performance by means of interlaboratory comparison." The measurement results reported by various laboratories are statistically analyzed and compared against reference values. Based on this analysis, a decision is made as to whether a laboratory is technically competent to carry out specific types of tests or calibration. Participation in proficiency testing programs provides testing and calibration laboratories an objective means of assessing and demonstrating the reliability of the results they are producing. Thus, participation in a proficiency testing program is an important measurement assurance activity for testing and calibration laboratories.

Proficiency testing schemes are operated on behalf of laboratory accreditation bodies, and regular participation in them is a mandatory requirement for maintaining accreditation. In addition to this independent proficiency testing, operators also conduct proficiency testing schemes in which laboratories participate voluntarily. Proficiency testing schemes can be classified as follows.

Measurement Comparison Scheme

In the measurement comparison scheme, the item to be measured or calibrated is sent to various participating laboratories successively. This scheme is mostly applicable to calibration laboratories when the measurement artifact could be a reference standard such as a resistance standard, a gage block, or an instrument. The reference values are assigned to the test item by a reference laboratory, which could be a national metrology laboratory. The values assigned to the measurement artifact by various participating laboratories are compared against the reference value, and the performance is judged against specific criteria. The performance criteria take into account the reference value and its uncertainty as well as the value reported by the laboratory and its uncertainty. Each time the artifact is moved to a successive laboratory, the artifact is measured in the reference laboratory to ensure that no significant change has occurred in the assigned value. Since the movement of the artifact is in sequence, completing this type of proficiency testing program takes a lot of time, sometimes as long as a year. In such a situation problems may arise because of the stability of the artifact due to frequent movement. The laboratories are also given feedback on their individual performance during the implementation of the scheme rather than waiting for all the participating laboratories to report their results.

The performance criteria used to judge the performance of an individual laboratory participating in the measurement comparison scheme are based on a number called the En number. If *X*ref is the reference value, *U*ref is the uncertainty of the reference value, and *X*lab and *U*lab are the values reported by a laboratory and its uncertainty, then the En number is calculated as follows:

$$En = (Xlab - Xref) / (Ulab^2 + Uref^2)^{1/2}$$

In this case $(Ulab^2 + Uref^2)^{1/2}$ is the total uncertainty when a reference standard is being measured in the participating laboratory. The value of $Xlab - Xref$ gives the deviation of the participating laboratory with respect to the reference value. The En value will be lesser if the value reported by the laboratory is closer to the reference value. The decision criteria for performance of the laboratory is as follows:

$$En \leq 1 = \text{satisfactory}$$

$$En > 1 = \text{unsatisfactory}$$

Interlaboratory Testing Scheme

In the interlaboratory testing scheme randomly picked up samples from a source of material are sent to various participating laboratories for concurrent testing. Such a scheme is generally used in testing laboratories used for testing food items, body fluids, soils, environmental materials, and so on. The essential condition for this proficiency testing scheme is that the material for which samples have been picked up should be sufficiently homogenous so that variability in the test results cannot be attributed to nonhomogeneity of test items. The laboratory, after completing the tests, reports the results to the body that is operating the interlaboratory testing scheme. In this scheme the result reported by a participating laboratory is also compared against a reference value, which may be a known value with results determined by specific test item formulation, a value determined by comparison with a standard reference material, or a consensus value obtained from expert laboratories. The reference value could also be a consensus value from participating laboratories (for example, the mean of results reported by various participating laboratories after removing outliers). The uncertainty of the reference value should also be evaluated.

The performance criteria of the laboratory are based on the reference value and an appropriate measure of variability which is selected to meet the requirements of the scheme. The value of s is called the standard deviation for proficiency assessment, which is a measure of variation of laboratory bias. Its value depends upon repeatability standard deviation and reproducibility standard deviation. The value of s is evaluated based on interlaboratory comparison studies. To arrive at an accurate conclusion based on a proficiency testing program, it is essential that the uncertainty of the reference value be very small compared with the standard deviation of the proficiency assessment.

The performance criterion in the interlaboratory testing scheme is the z-score, which is defined as $z = (X\text{lab} - X\text{ref}) / s$. Basically the z-score tells us how far the result reported by the participating laboratory is away in terms of number from an appropriately selected standard deviation s.

The decision criteria for the performance of the laboratories is as follows:

$$|z| \leq 2 = \text{satisfactory}$$

$$2 < |z| < 3 = \text{questionable}$$

$$|z| \geq 3 = \text{unsatisfactory}$$

The international standard ISO/IEC Guide 43-1, *Proficiency testing by interlaboratory comparisons,* specifies the norms for proficiency testing operators and participating laboratories. These include confidentiality and ethical considerations as well.

There are a number of organizations that provide proficiency testing services to testing and calibration laboratories to help them meet the growing demand for proof of reliability of measurements in their operations. Some of the laboratory accreditation bodies stipulate a laboratory's successful participation in proficiency testing as a precondition for accreditation. The requirement for such services is therefore increasing. It is also essential that bodies providing proficiency testing services should be unbiased and neutral, thus having no conflict of interest. The International Laboratory Accreditation Co-operation (ILAC) is working out norms for operation of these bodies. In the United States an organization called the National Association for Proficiency Testing (NAPT) provides much-needed proficiency testing services to the measurement community. It is a nonprofit association with the goal of championing the metrology community's quest to achieve excellence in measurement. The NAPT provides comprehensive support and guidance to metrology laboratories and organizations in determining their measurement competence.

INTERNATIONAL STANDARDS ON QUALITY MANAGEMENT AND MEASUREMENT ASSURANCE

The quality assurance activity focuses mainly on intended product. Similarly, measurement assurance focuses mainly on the intended product of the measurement process, that is, the measurement result. The international standard ISO 9000:2000 uses the term *measurement control system,* which has a meaning identical to that of measurement assurance. A measurement control system has been defined as a set of interrelated or interacting elements necessary to achieve metrological confirmation and continual control of a measurement process. Similarly, the international standard ISO 9004:2000, which is a guideline document, has specified how organizations are to comply with the requirements for the control of measuring and monitoring devices as per ISO 9001:2000. This guideline standard in turn refers to ISO 10012 for further guidance in respect to requirements for quality of measurements. The standard ISO 10012, *Quality assurance requirements for measuring equipment,* has two parts, viz.:

Part 1 : Metrological confirmation system for measuring equipment

Part 2: Guidelines for control of measurement processes

Thus, compliance to the two parts of ISO 10012 forms part of a measurement assurance program.

The concepts pertaining to measurement assurance described in this chapter are applicable to all metrological applications in industry and in testing and calibration laboratories. The international standard ISO/IEC 17025 has also recommended the use of statistical techniques for quality control of test and calibration results.

It is clear from this requirement that statistical control charts and trend charts are effective quality control techniques that can be used in testing and calibration laboratories as measurement assurance tools. The standard also recommends participation in interlaboratory comparison and proficiency testing programs as a measurement assurance activity.

TRACEABILITY, UNCERTAINTY, AND MEASUREMENT ASSURANCE

The objective of establishing a measurement assurance program in an organization is to ensure traceable measurement results with an acceptable value of uncertainty of measurement so that a measurement result can be used for a specific application. To reiterate the need for measurement quality assurance, scientists at the National Institute of Standards and Technology (NIST) in the United States crafted a new definition of traceability, as follows:

> Traceability to designated standards (national, international or well-characterized reference standards based upon fundamental constants of nature) is an attribute of some measurements. Measurements have traceability to the designated standards if and only if scientifically rigorous evidence is produced on a continuing basis to show that the measurement process is producing measurement results (data) for which the total measurement uncertainty relative to a national or other designated standard is qualified.

The emphasis in that definition is on evidence of known measurement uncertainty relative to a national standard. Such evidence can be provided by designing and implementing an adequate measurement assurance program.

The NIST also has defined the term *measurement assurance program,* or *MAP,* as follows:

> A MAP is a quality assurance programme for a measurement process that qualifies the total uncertainty of measurements with respect to a national or other designated standard and demonstrates that the total uncertainty is sufficiently small to meet the user's requirements.

A measurement assurance program thus includes but is not limited to the following aspects of a measurement system:

- Traceability
- Calibration
- Metrological confirmation
- Uncertainty of measurements
- Statistical control of the measurement process
- Repeatability and reproducibility (R&R)
- Proficiency testing

However, the emphasis of a measurement assurance program is also on the demonstrability of measurement assurance–related activities through objective evidence.

REFERENCES

Ellis, Charles J. 2000. Profile: National Association of Proficiency Testing. *CALLAB, The International Journal of Metrology* 7, no. 2:26–27.

Garner, Ernest L., and Stanley D. Resberry. 1993. What is new in traceability? *Journal of Testing and Evaluation,* April, 8–9.

Grahanen, Christopher. 2000. The how's and why's of proficiency testing. *CALLAB, The International Journal of Metrology* 7, no. 2:20–25.

International Organization for Standardization (ISO). 1991. ISO 8258:1991, *Shewhart control charts.* Geneva, Switzerland: ISO

———. 1992. ISO 10012-1:1992, *Quality assurance requirements for measuring equipment—Part 1: Metrological confirmation system for measuring equipment,* Geneva, Switzerland: ISO

———. 1997. ISO 10012-2:1997, *Quality assurance for measuring equipment—Part 2: Guidelines for control of measurement processes.* Geneva, Switzerland, ISO

———. 2000. ISO 9000:2000, *Quality management system—Fundamentals and vocabulary,* Geneva, Switzerland: ISO

———. 2000. ISO 9004:2000 *Quality management system—Guidelines for performance improvements.* Geneva, Switzerland: ISO

International Organization for Standardization (ISO)/International Electrotechnical Commission (IEC), ISO. 1996. ISO/IEC Guide 43-1:1996, *Proficiency testing by interlaboratory comparison—Part 1: Development and operation of proficiency testing schemes.* Geneva, Switzerland.

———. 1999. ISO/IEC 17025:1999, *General requirements for the competence of testing and calibration laboratories.* Geneva, Switzerland.

National Bureau of Standards (U.S.). May 1984. Special Publication 676-I, *Measurement assurance programs—Part 1: General instructions.* Washington, DC: U.S. Department of Commerce.

Schumacher, Rolf B. F. 1987. *Measurement uncertainty—Measurement assurance handbook.* San Clemente, California: Coast Quality Metrology System, Inc. See pp. 6-1 to 6-45 and 7-1 to 7-19.

10

Measurement System Capability Requirements in Industries

THE NEED FOR CAPABLE INDUSTRIAL MEASUREMENT SYSTEMS

The international standard ISO 9001:2000 specifies requirements for a quality management system in manufacturing and service organizations. Compliance with this standard demonstrates an organization's ability to consistently provide products that meet customers' and applicable regulatory requirements, and that thus enhance customer satisfaction. The standard stipulates that the organization shall determine the monitoring and measurements to be undertaken and the monitoring and measuring devices needed to provide evidence of conformity of products to determined requirements. It shall establish processes to ensure that monitoring and measurement can be carried out and are carried out in a manner that is consistent with the monitoring and measurement requirements.

This requirement basically pertains to conformance of products to their stipulated requirements. Products are tested for compliance at various stages of their life cycle.

Measurements are made in manufacturing and service organizations to facilitate decisions in connection with the following activities:

- Statistical process control (SPC) of the manufacturing process.
- Compliance of products during various stages of manufacturing. A product is inspected at various stages and processed further only after it has met its specification requirements.
- Final inspection of the end product before it is dispatched to the user.
- Inspection of raw material and incoming components before they are accepted into the production line.

The standard also stipulates that the organization shall apply suitable methods for monitoring and, where applicable, measurement of the quality management system processes. These methods shall demonstrate the ability of the processes to achieve planned results. This requirement is generic and is applicable to all processes that form part of a quality management system. The statistical control of a manufacturing process or measurement process falls in this category.

The measurements may be made using simple mechanical devices or a complex electronic system. It is often taken for granted that a measurement system is perfect and that the measurement results are correct. Decisions are made without considering the variability of measurement results and their possible impact on the correctness of those decisions. A perfect measurement system would accept all products which have characteristics within the specification limits—that is, the upper specification limit (USL) and lower specification limit (LSL)—and reject all other products. However, a perfect measurement system does not exist.

There is always an uncertainty U about a measurement result. That uncertainty affects compliance decisions for certain product characteristics. Figure 10.1 depicts the variation of probability of acceptance of various products with specified characteristics for a perfect measurement system and a practical measurement system.

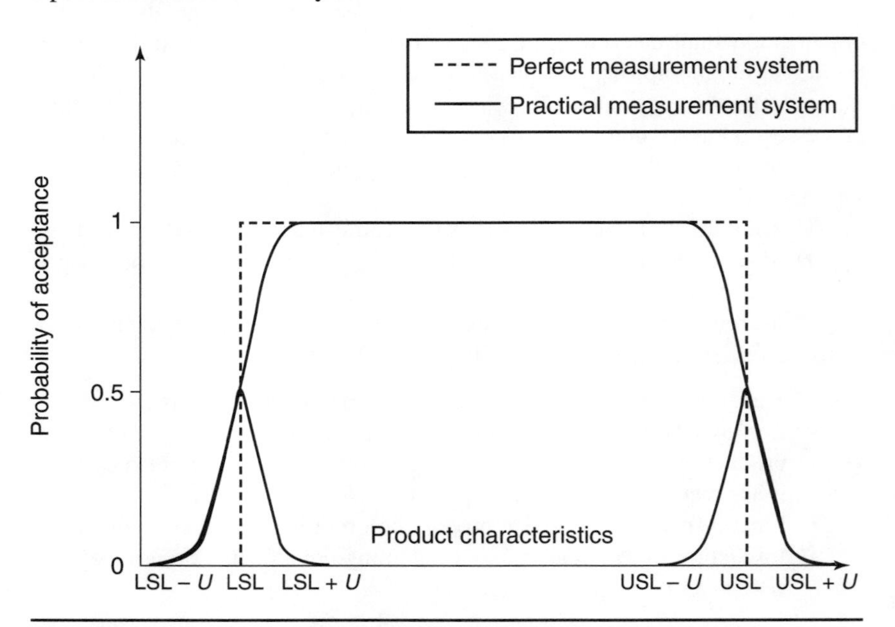

Figure 10.1 Probability of acceptance and compliance decision.

The normal curve around LSL and USL depict measurement uncertainty at those points.

The following situations arise in a practical measurement system:

- **Case 1:** If the product characteristics value is between LSL + U and USL – U, the probability of acceptance is 1, and all products will be accepted as desired.
- **Case 2:** If the product characteristics value is more than USL + U and less than LSL – U, the probability of acceptance is zero, and all products will be rejected as desired.
- **Case 3:** Ideally products with characteristics between LSL and LSL + U and, similarly, between USL and USL – U should be accepted, but due to measurement system limitations some of them could be rejected. Similarly, products with characteristics between LSL – U and LSL and also between USL and USL + U should be rejected, but due to measurement system limitations some of them might be accepted. Thus, there is a gray zone where a noncomplying product could be accepted and a complying product could be rejected. Thus, an organization runs the risk of making the wrong decision in terms of accepting a noncomplying product and vice versa. That risk can be minimized if the measurement system variability, which is equal to $2U$, is small compared with USL – LSL.

Organizations have to ensure that the risk is within acceptable limits. Advances in production technology have now made it possible to manufacture engineering products to very close tolerances. The application of SPC techniques to control such manufacturing processes places a greater demand on the quality of measurement data. The quality of a measurement result is described by its uncertainty, which in turn depends on the unknown bias and the precision of the measurement result. For a measurement system, therefore, it is essential that its uncertainty is known and is consistent with the required measurement capability.

A measurement system is considered to be capable if the measurement result produced by that system meets the uncertainty requirements stipulated for a measurement result. The capability requirement should be met in all types of environmental conditions under which measurements are being made or are likely to be made. The capability of a measurement system therefore needs to be quantified, evaluated, and verified before the system is put to use. Its capability also needs to be verified at periodic intervals for its users to have continued faith in that capability. The initial and periodic verifications are done through measurement system capability studies.

CAPABILITY REQUIREMENTS FOR A MEASUREMENT SYSTEM

A measurement system can be considered capable of meeting a measurement requirement if it meets the following criteria:

- The measurement process is in a state of statistical control. This implies that variations present in the measurement system are due only to common or natural causes and not to assignable causes. This aspect has been explained in chapter 2.
- In the case of a manufacturing process, the extent of its variability is indicated by the process capability index. To control the process, the variability of the measurement process must be small compared with the variability of the manufacturing process. If this is not so, variations in the product characteristics will be masked by variations in the measurement process. In such a case, a decision that the manufacturing process has gone out of control could be made purely on the basis of disproportionate variation in the measurement process, even though the process is actually within control. This may lead to unnecessary adjustment of the process and avoidable loss of money and time. A similar situation can arise when making a compliance decision about a product based on its specification limits. If the measurement process variability is comparable to the specification limits, a complying product could be classified as noncomplying, and vice versa. This is particularly true if the product characteristic lies near the upper or lower specification limit. To minimize the risk of making such a wrong decision it is essential that the variability of the measurement process be small compared with the variability of the characteristic that is being measured. The requirement for the ratio of measurement system variability to the variability of the product characteristics depends on the criticality of the decision being made based on the measurement result.
- The measuring equipment or gage should have no bias and must be validly calibrated so that the measurement results are traceable to national standards of measurement.
- The measurement process should be capable of detecting and indicating small changes in the measured characteristics.

MEASUREMENT SYSTEM ASSESSMENT AND CAPABILITY INDICES

Before a measurement system is put into operation for a specific measurement application, it has to be assessed for meeting the capability requirements listed in the previous section. A number of methods can be used for assessing a measurement system. However, a detailed and exhaustive method is specified in the reference manual *Measurement Systems Analysis* (*MSA,* for short). This guidance document for assessing the capability of a measurement system was published jointly by the "Big Three" U.S. automobile companies, viz., Chrysler, Ford, and General Motors. The manual is meant to be used by the suppliers of these companies to assess the capability of measurement systems in their manufacturing processes. This is an essential requirement for compliance to QS-9000. The QS-9000 standard stipulates quality system requirements; the Big Three have published it for implementation by their suppliers. This standard is based primarily on ISO 9001 but includes a number of additional requirements which lay the thrust on quality improvement.

MSA stipulates that a measurement system is to be assessed in phases. The phase 1 study is carried out before the measurement system is put into operation. One objective is to ensure that the system meets all the aforementioned capability requirements. Another objective is to ensure that the measurement system is robust and that its capability will not be degraded when it interacts with the various environmental conditions under which measurements are likely to be made. When the measurement system meets all the capability requirements, it is considered acceptable; otherwise the system has to be modified, or a new measurement system which meets the capability requirements has to be put in place. Once the measurement system has been accepted, its users have to periodically ensure that it continues to meet the capability requirements. This is the second phase of the measurement system assessment.

A measurement system's capability is characterized by the following parameters:

- **Bias:** Bias is an indication of the inaccuracy of a measurement system. It is the difference between the average of a number of repeat measurements of the same quantity and the accepted reference value. An accepted reference value is an agreed-upon value for that quantity which is determined by averaging several measurements taken at a higher-level metrology laboratory through a calibration process.
- **Repeatability:** Repeatability is linked with the ability of a measuring instrument to obtain the same measurement result repeatedly and also with the operator's ability to use the measuring instrument exactly the same way every time. It is due to inherent variability in the behavior of a gage and positional variation of the part in the gage. When assessing the repeatability of a measurement system, one must ensure that the operator is consistent and is using the measuring equipment the same way every time. Repeatability is thus defined as variation in repeat measurements obtained with one measuring instrument by one appraiser while measuring identical characteristics of the same part repeatedly. It is determined by measuring the same part several

times with the same measurement system. Repeatability is also called equipment variation (EV). The spread of the resulting distribution is a measure of repeatability as shown in Figure 10.2.

- **Reproducibility:** Reproducibility is the variation in repeat measurement results if the measurements are made by various operators using the same measuring equipment and the same method. A measurement result is an average of a number of repeat observations. Reproducibility is thus defined as the variation in the averages of repeat measurements made by different appraisers using the same measuring instrument on identical characteristics of the same part, as shown in Figure 10.3. Reproducibility, which is also called appraiser variation (AV), is determined based on the repeat measurement data taken on the same part by the same measuring instruments by various appraisers.
- **Repeatability and reproducibility (R&R):** These are the indices of the precision of a measurement system under two different conditions of operation—one pertaining to variability due to measuring equipment and the appraiser, and the other pertaining to variability due to different appraisers. The combination of these variabilities (that is, repeatability and reproducibility, or R&R) is the total variability of the measurement system.
- **Statistical stability:** Statistical stability ensures that the measurement process is under statistical control and that there are no long-term and short-term drifts in the measurement system. Statistical stability is the term applied to all aspects of a measurement process including its drift, bias, repeatability, and reproducibility. Statistical stability is a more general term which characterizes the performance of a measurement process in the present and also allows us to predict the performance of the process in the future. The measures of repeatability and reproducibility show the state of the measurement process during the period the R&R study was carried out; they do not ensure the statistical stability of the measurement process. Rather, unless a measurement

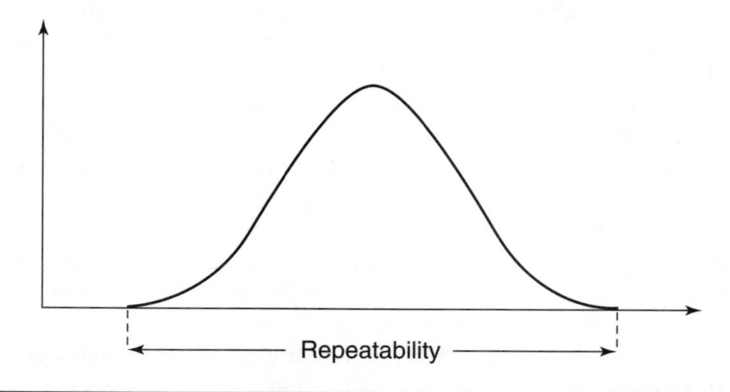

Figure 10.2 Repeatability of a measurement system.

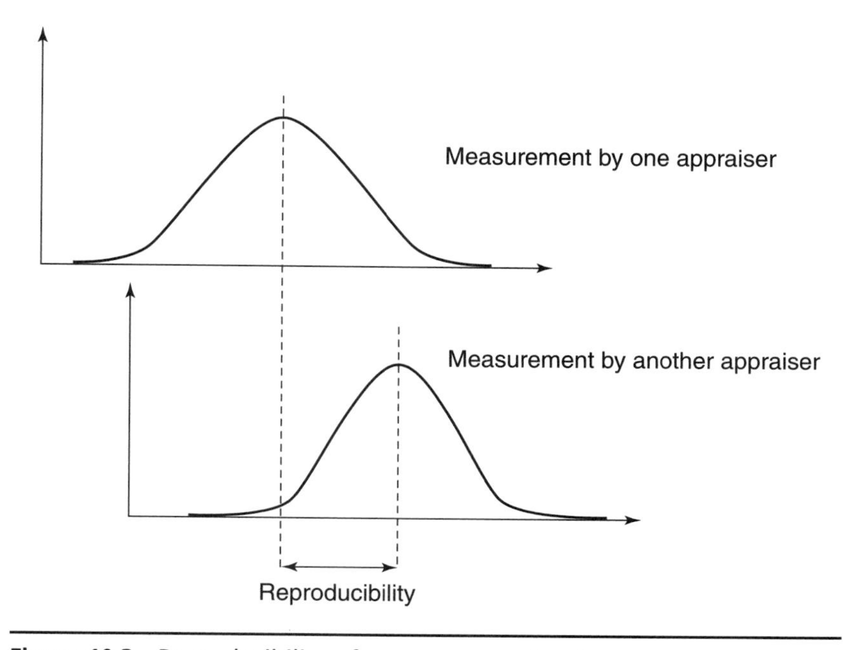

Figure 10.3 Reproducibility of a measurement system.

process is statistically stable, the R&R study may result in misleading conclusions. The statistical stability of a measurement process is determined and maintained through the use of SPC techniques. Average ($\overline{x}$) and range (R) control charts provide a means of separating variations due to common causes from variations due to special causes.

Knowledge about the statistical stability of a measurement process is obtained by studying variations in measurements obtained with a measurement system on the same part over a period of time. The part is used as a standard; it is also called a check standard. Based on knowledge of the measurement process's variability in the past, its future performance can be predicted.

GAGE REPEATABILITY AND REPRODUCIBILITY (R&R) ASSESSMENT

When a measurement system is to be used for the first time for a specific measurement application, its capability is assessed as follows:

- One must ensure that the measurement system does not have a bias. Measurements made by the equipment are compared against those made on a reference standard whose value is known and is traceable to a national standard of measurement, thus ensuring that the measurement system does not have a bias.

- One must ensure that the measurement process is under statistical control.
- The repeatability and reproducibility (R&R) study on the measurement system is carried out by having various appraisers take repeat measurements on the same components using the same equipment. The components chosen should represent the total parts variability. During this study, range charts are used to ensure that the appraisers are consistent in their method of measurements. Average control charts are used to ensure that the variability of the measurement system is much less than the variability of the manufacturing process and less than the specification limits of the product.

Thus, a gage R&R study starts with the collection of statistically valid data. The repeatability and reproducibility of a measurement system is evaluated by analysis of data gathered from repeat measurement observations. A number of parts representing the process variability are collected. The study is conducted with two or three appraisers at a time. Each appraiser measures the parameter of each part a number of times. If there are 10 parts, and three repeat measurements are made on each part by each appraiser, each appraiser will be making 30 observations during the assessment. The measurements are made in random order. The appraisers should carry out the measurements as a routine activity without being too conscious about the study.

The measure of variability of any statistical data is their standard deviation, so the index of repeatability is also the standard deviation of the repeat measurements taken by the measurement system on the same part. This is called the repeatability standard deviation. The set of these repeat measurements is called a subgroup. The variability in the data within the subgroup is the measure of repeatability. The range of the data in the subgroup and its standard deviation are interrelated, according to the relationship given by $s = \overline{R} / d_2$, where s is the estimate of the standard deviation of the data and $\overline{R}$ is the average range. If a number of subgroups of measurement data are obtained, each subgroup will have its range. The average of all these ranges is $\overline{R}$. The factor d_2 is a constant, and its value can be obtained from Statistical Table 1 in Appendix D for various values of number of trials, number of appraisers, and number of parts used for R&R studies. Thus, knowing the value of $\overline{R}$ and d_2, the repeatability standard deviation can be evaluated.

The index of variability due to reproducibility is called the reproducibility standard deviation. This is also evaluated by the knowledge of the range of the averages and the factor d_2. The range of the averages of various subgroups is calculated as the difference of the maximum average value and the minimum average value. The reproducibility standard deviation can be obtained from the range of averages R_o and d_2 factor, which depends upon the number of appraisers taking part in the R&R study.

$$\text{Reproducibility standard deviation} = R_o / d_2$$

The value of d_2 can be obtained from Table 2 in Appendix D for reproducibility calculation.

Repeatability Assessment

To judge the variability due to appraisers, range control charts are drawn for each appraiser as follows:

- The repeat measurements of a number of trials on each part by each appraiser are considered as one subgroup.
- The range for each subgroup is calculated as the difference between the maximum observation and the minimum observation in each subgroup.
- The average of all these ranges for each part and also for each appraiser is calculated. This average is represented as $\overline{R}$.
- The values of ranges as obtained by an appraiser for each part are drawn in a range control chart. Such charts are drawn for each appraiser.
- Based on the value of the average range $\overline{R}$, the upper and lower control limits for the range charts are drawn as follows:
  - Upper control limit (UCL) for range chart = $D_4 \times \overline{R}$.
  - Lower control limit (LCL) for range chart = $D_3 \times \overline{R}$.
  - The values of D_3 and D_4 depend on the number of observations in the subgroup and are given in the table in Appendix C.
- The range control chart is studied for the position of various range points with reference to upper and lower control limits. If for a specific appraiser all the points are within the control limits, that implies that the appraiser is consistent in taking measurements. If the points on the range control charts for each appraiser are within the control limits, that implies that there is a consistency among the appraisers also and that they are measuring the part the same way. If some points are out of the control limits in the range control chart for one appraiser but within control for the other appraisers, that implies that the measurements taken by that appraiser differ from the measurements taken by the other appraisers. The reasons for these out-of-control points should be investigated. This indicates the presence of an assignable cause which could be the incompetence of that appraiser. If all appraisers have some out-of-control points, it implies that the measurement system is not robust

enough and is sensitive to the way measurements are made by the appraisers. The measurement system needs to be improved or changed altogether before it is put into operation for a specific measurement application.

- If the range charts are under statistical control for all the appraisers, the repeatability standard deviation is obtained from the value of $\overline{R}$ and d_2. The value of d_2 is obtained from Table 1 in Appendix D for repeatability calculations.
- The repeatability index is obtained by multiplying the repeatability standard deviation by 5.15. The multiplier 5.15 gives the interval corresponding to a 99% confidence level.

Reproducibility Assessment

For the measurement results reported by various appraisers to be acceptable, the variations in them must be within acceptable statistical limits. This implies that the variability among appraisers is consistent. If the variability is inconsistent, the overall averages reported by the various appraisers will differ, and this difference will appear as a bias in the result reported by one appraiser with reference to other appraisers. The consistency of the variability among the appraisers can be judged using an average control chart.

The reproducibility standard deviation σ_o can be estimated by determining the overall average for each appraiser and then determining the range of those averages. The reproducibility index is obtained by multiplying the reproducibility standard deviation by 5.15.

To determine the adequacy of the measurement system to detect part-to-part variation, control limits for part average control charts are drawn based on the average range $\overline{R}$ from which the repeatability or the equipment variation (EV) has been calculated. The requirement is that the equipment variation has to be small compared with the part variation. If this is so, the majority of the points on the average control chart (with the upper and lower control limits drawn based on the repeatability range $\overline{R}$) should remain outside these control limits.

EXAMPLE OF AN R & R ASSESSMENT

A dial gage is being used to measure the diameter of a rod. The nominal value of the diameter is 6.30 ±0.25 millimeters (mm). Measurements of the diameter are made by three appraisers A, B, and C. Five samples of the rods are taken. Each appraiser measures the diameter of all the five samples selected randomly. Three measurements are taken on each sample by each appraiser, and the observations are recorded in Table 10.1. We wish to find

Table 10.1 Measurement data on the diameter of five samples.

Sample no.	Appraiser A					Appraiser B					Appraiser C				
	Trial 1	Trial 2	Trial 3	Average	Range	Trial 1	Trial 2	Trial 3	Average	Range	Trial 1	Trial 2	Trial 3	Average	Range
1	6.340	6.340	6.338	6.339	0.002	6.352	6.349	6.351	6.351	0.003	6.330	6.332	6.336	6.333	0.006
2	6.402	6.414	6.405	6.407	0.012	6.429	6.422	6.422	6.424	0.007	6.404	6.414	6.408	6.409	0.010
3	6.491	6.492	6.497	6.493	0.006	6.519	6.511	6.512	6.514	0.008	6.492	6.488	6.494	6.491	0.006
4	6.574	6.565	6.565	6.568	0.009	6.600	6.592	6.598	6.597	0.008	6.574	6.572	6.572	6.573	0.002
5	6.726	6.720	6.731	6.726	0.011	6.741	6.738	6.739	6.739	0.003	6.725	6.718	6.718	6.720	0.007
Averages				6.5066	0.0080				6.525	0.0058				6.5052	0.0062

the repeatability index and reproducibility index for the dial gage and operator combinations. We also want to establish whether the measurement system is consistent.

Having tabulated the values of repeat measurement observations for all the five samples and all the three trials, the averages and ranges for each subgroup for all the samples and all the appraisers are calculated and entered in Table 10.2. The next step in the exercise is to find out whether all the appraisers are consistent in the use of the dial gage. This is done by plotting range control charts for all the appraisers for the various parts.

To draw the upper and lower control limits on the range charts, the value of the average range for all the appraisers is calculated.

$$\text{Average range for appraiser A} = \frac{0.002 + 0.012 + 0.006 + 0.009 + 0.011}{5} = 0.0080$$

$$\text{Average range for appraiser B} = \frac{0.003 + 0.007 + 0.008 + 0.008 + 0.003}{5} = 0.0058$$

$$\text{Average range for appraiser C} = \frac{0.006 + 0.010 + 0.006 + 0.002 + 0.007}{5} = 0.0062$$

$$\text{Average range for all appraisers } \overline{R} = \frac{0.0080 + 0.0058 + 0.0062}{362} = 0.0067$$

Thus, the control limits for the range chart are calculated thus:

$$\text{Upper control limit} = \overline{R} \times D_4 = 0.0067 \times 2.575 = 0.017$$

$$\text{Lower control limit} = \overline{R} \times D_3 = 0.0067 \times 0.0 = 0.00$$

The values of D_4 and D_3 have been obtained from Appendix C for three observations in each subgroup, which gives $D_4 = 2.575$ and $D_3 = 0.0$. The range chart for each appraiser is plotted as shown in Figure 10.4 based on the preceding data.

From the range control chart it is clear that the ranges for each subgroup, for each part, and for all the appraisers are in control. This implies that there is consistency among the various appraisers individually and also among

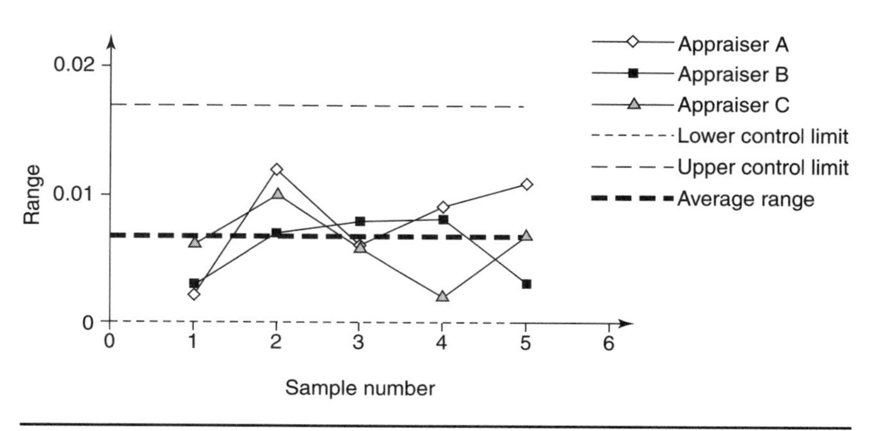

Figure 10.4 Range control chart for various appraisers.

themselves as the upper and lower control limits for range have been drawn based on the average range for all the appraisers. Thus, it has been established that each appraiser is using the gage correctly. If any point on the range control chart is found to be out of control, it is investigated for an assignable cause, which could be the incompetence of an appraiser in making measurements.

Repeatability Index

To obtain the repeatability index, first the repeatability standard deviation is determined from the value of $\overline{R}$ and d_2.

$$\text{Repeatability standard deviation} = \overline{R} \,/\, d_2$$

$$\overline{R} = 0.0067$$

The factor d_2 is obtained from the values given in Table 1 of Appendix D. The value of d_2 depends upon two factors. One factor is the number of trials represented by m. The values of d_2 are given for a number of trials equal to 2, 3, and 4. Another factor is the number of parts times the number of appraisers, and that is represented by g. In any study a minimum of five parts and two appraisers should participate. Thus, the g values given in the table are given for $g = 10$ to 15. For values of $g > 15$, the value of d_2 is the same for the same number of trials.

For the preceding example, $m = 3$, and $g = 5$ parts $\times$ 3 appraisers, and hence $g = 15$. The value of d_2 is thus equal to 1.71.

Repeatability standard deviation = 0.0067 / 1.71 = 0.0039.

σ_e = 0.0039; σ_e represents the standard deviation due to equipment.

Repeatability = 5.15 × 0.0039 = 0.020 mm.

The multiplying factor 5.15 represents the 99% confidence level of the normal distribution, corresponding to $k = 2.575$ and a spread of ±2.575 from the mean (that is, 2 × 2.575 = 5.15). A repeatability figure of 0.020 mm implies that 99% of the measurements will have a variability within 0.020 mm if the measurements are taken with the same gage and by appraisers who are consistent.

Equipment variation (EV) = 0.020 mm

Reproducibility Index

Overall average for all five parts as reported by appraiser A

$$= \frac{6.339 + 6.407 + 6.493 + 6.568 + 6.726}{5} = 6.5066$$

Overall average for all five parts as reported by appraiser B

$$= \frac{6.351 + 6.424 + 6.514 + 6.597 + 6.739}{5} = 6.525$$

Overall average for all five parts as reported by appraiser C

$$= \frac{6.333 + 6.409 + 6.491 + 6.573 + 6.720}{5} = 6.5052$$

Range R_o of the results as reported by various appraisers

$$= 6.5250 - 6.5052 = 0.0198$$

R_o is obtained by subtracting the lowest average from the highest average of various appraisers.

Reproducibility standard deviation $\sigma_o = R_o / d_2$

Since range R_o is determined from three averages, the value of d_2 can be obtained from Table 2 of Appendix D; for three appraisers, $d_2 = 1.91$.

$$\text{Reproducibility standard deviation } \sigma_o = 0.0198 / 1.91 = 0.0104,$$

where σ_o represents the standard deviation due to operator.

Hence,

$$\begin{aligned} \text{Reproducibility} &= 5.15 \times \sigma_o \text{ for 99\% of the measurement} \\ &= 5.15 \times 0.0104 \\ &= 0.0534 \text{ mm} \end{aligned}$$

The reproducibility standard deviation figure derived from the preceding calculations represents the variability due to the appraisers and also due to the gage itself. To find the variance due to the appraiser only, the variance due to the gage has to be subtracted from the square of the reproducibility standard deviation in accordance with the law of combining variances.

The factor σ_e is the standard deviation for the repeat observations due to variability of the gage, and hence its variance will be σ_e^2. If the result is expressed as an average, the variance of the average due to the variability of the gage will be equal to $\sigma_e^2 / n \cdot r$; where n is the number of parts that are being measured and r is the number of trials. For each appraiser the overall average has been drawn based on $n \cdot r$ observations.

Thus, variance due to the appraiser only is given by

$$\begin{aligned} \sigma_o^2 &= (R_o / d_2)^2 - \sigma_e^2 / n \cdot r \\ &= (0.0104)^2 - (0.0039)^2 / 15 = (0.000108) - (0.000001) \end{aligned}$$

Hence, the appraiser standard deviation $\sigma_o = 0.0103$. Thus,

$$\text{Reproducibility} = 5.15 \times 0.0103 = 0.0530$$

$$\text{Repeatability} = 0.020 \text{ mm}$$

$$\text{Reproducibility} = 0.530 \text{ mm}$$

The variability of the measurement system consists of the variability due to the gage and the variability due to the appraisers, their respective standard deviations being σ_e and σ_o. The measurement system variance σ_m^2 is therefore equal to

$$\sigma_m^2 = \sigma_e^2 + \sigma_o^2 = (0.0039)^2 + (0.0103)^2$$
$$= (0.0000152) + (0.0001061) = 0.0001213.$$

$$\sigma_m = 0.011.$$

Thus, the measurement system variability or R&R = 5.15 × 0.011 = 0.057 mm.

Part-to-Part Variation

The standard deviation of part-to-part variation σ_p is either obtained from the independent process capability studies of the manufacturing process or estimated from the data from the measurement system study presented earlier. To determine the standard deviation on account of part-to-part variation, the average of various parts for all the trials as reported by various appraisers is determined. From the part averages, the range of the part averages can be determined by subtracting the lowest part average from the highest part average.

If σ_p is the standard deviation for part-to-part variation and R_p is the range of part averages, σ_p can be determined by the following relationship:

$$\sigma_p = R_p / d_2$$

For the example presented earlier, the averages for various parts for all appraisers are determined as follows:

$$\text{Average of sample 1 for all appraisers} = \frac{6.339 + 6.351 + 6.333}{3} = 6.341$$

$$\text{Average of sample 2 for all appraisers} = \frac{6.407 + 6.424 + 6.409}{3} = 6.413$$

$$\text{Average of sample 3 for all appraisers} = \frac{6.493 + 6.451 + 6.494}{3} = 6.500$$

$$\text{Average of sample 4 for all appraisers} = \frac{6.568 + 6.597 + 6.573}{3} = 6.579$$

$$\text{Average of sample 5 for all appraisers} = \frac{6.726 + 6.739 + 6.720}{3} = 6.728$$

$$\text{Range } R_p \text{ of sample averages} = 6.728 - 6.341 = 0.387$$

The value of d_2 is obtained from Table 3 in Appendix D for five parts:

$$d_2 = 2.48$$

Thus, $\sigma_p = 0.387 / 2.48 = 0.156$ mm.

Thus, part variation (PV) for 99% of the parts is

$$\text{PV} = 5.15 \times \sigma_p$$
$$= 5.15 \times 0.156 = 0.803 \text{ mm}$$

Measurement System Variation, or Gage R&R

$$\text{Gage R\&R} = 5.15\ \sigma_m.$$

$$\text{R\&R} = \sqrt{(\text{EV})^2 + (\text{AV})^2} \qquad = \sqrt{(0.02)^2 + (0.053)^2}$$
$$= \sqrt{0.\ 0.0004 + 0.002809} = \sqrt{0.00321}$$

$$\text{Gage R\&R} = 0.057 \text{ mm}$$

Total Variation (TV)

The total variation consists of the variation due to the measurement system plus the variation due to part-to-part variation. Thus, the standard deviation pertaining to total variation σ_t can be determined as follows:

$$\sigma_t = \sqrt{(\sigma_m)^2 + \sigma_p)^2} \qquad = \quad \sqrt{(0.011)^2 + (0.156)^2}$$
$$= \sqrt{(0.000121) + 0.024336} \quad = \quad 0.1564$$

$$\text{Total variation} = 5.15 \times \sigma_t \quad = \quad 0.8054 \text{ mm}$$

Part Average Control Chart

To draw the part average control chart, the grand average for all the appraisers, all the parts, and all the trials needs to be determined. Thus,

$$\text{Grand average } G = \frac{6.5066 + 6.5250 + 6.5052}{3} = 6.512$$

Control limits for the average chart are drawn at = $G \pm A_2\overline{R}$, where $G = 6.512$, $\overline{R} = 0.0067$, and A_2 is obtained from the table in Appendix C.

For a subgroup of size 3, the value of A_2 is equal to 1.023.

$$\text{Upper control limit} = 6.512 + 1.023 \times 0.0067 = 6.519$$

Similarly,

$$\text{Lower control limit} = 6.512 - 1.023 \times 0.0067 = 6.505$$

For each appraiser the subgroup averages for various parts reflect part-to-part variation, and they have been plotted on an average control chart. The plots of part averages for various parts for each appraiser are shown in Figure 10.5. The control limits have been determined based on the repeatability average range. Thus, the repeatability variation due to the gage is quite small compared with the part-to-part variation. That is why the majority of points for each appraiser are outside the control limits drawn based on the repeatability index. It implies that the measurement system is capable of detecting any changes in the process variation and hence is capable of controlling the manufacturing process.

If the majority of the points on the average control chart are within the control limits, it implies that the variabilities of the measurement process and the manufacturing process are comparable, and hence the measurement system is not capable of detecting small changes in the manufacturing process and thus is not suitable for the statistical control of the manufacturing process.

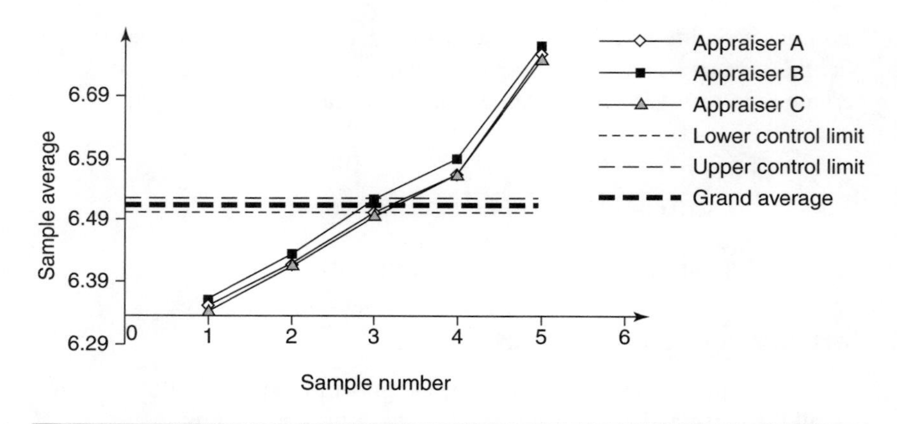

Figure 10.5 Part average control chart.

Acceptability Criterion from Control Charts

For a measurement system to be acceptable its variability has to be small compared with the variability of the parameter being measured, viz., manufacturing process variability or specification tolerance of a product. Range (R) and average ($\bar{x}$) control charts help to determine the capability of a measurement process for a specific measurement application in the following ways:

- R control charts ensure that the appraisers are consistent while making repeat measurements on the same parts with the same measuring equipment or gage.
- $\bar{x}$ control charts help in determining the relative variability of the measurement system and part-to-part variation and thus in assessing the measurement system's capability to detect small changes in the manufacturing process and its capability for statistical control of the manufacturing process.
- If the measurement system is used to decide whether a product complies with the product's specification limits, then the measurement process variability is compared against the specification tolerance in place of part-to-part variation. In this case the measurement process variability has to be small compared with the specification tolerance.

CONDUCTING AN R&R STUDY AND *MEASUREMENT SYSTEMS ANALYSIS*

The capability of measurement system depends on its variability as compared with the part-to-part variability and total variability. *Measurement Systems Analysis (MSA),* the reference manual published jointly by Chrysler, Ford, and General Motors, defines the variability indices due to various components and provides the formulas used for the calculation of these components. These indices are obtained by multiplying the corresponding standard deviation by the factor 5.15, which corresponds to a 99% confidence level. The details are as follows:

1. **Repeatability, or equipment variation (EV):**

$$EV = K_1\bar{R},$$

where $\bar{R}$ is the average of the repeatability range for various parts and appraisers, and K_1 is equal to 5.15 / d_2 for a 99% confidence level. The values of K_1 and d_2 for various values of number of trials and number of parts and number of appraiser multiplication are given in Table 1 of Appendix D.

2. **Reproducibility, or appraiser variation (AV):**

$$AV = \sqrt{K_2 \cdot R_o^{\,2} - \frac{(EV)^2}{n \cdot r}}$$

where R_o is the range of the average for all parts for various appraisers, and K_2 is equal to 5.15 / d_2. For a 99% confidence level, the values of K_2 when the number of appraisers equals 2 , 3, and 4 are given in Table 2 of Appendix D.

3. **Repeatability and reproducibility (R&R):**

$$\text{R\&R} = \{(\text{EV})^2 + (\text{AV})^{2}\}^{0.5}$$

4. **Part variation (PV):**

$$\text{PV} = K_3 \,.\, R_p$$

where R_p is the range of part averages, and the value of K_3 is equal to 5.15 / d_2 for a 99% confidence level. The value of K_3 for various values of the number of parts is given in Table 3 of Appendix D.

5. **Total variation (TV):**

$$\text{TV} = \sqrt{\{(\text{PV})^2 + (\text{R\&R})^2\}}$$

Total variation includes variation due to the manufacturing process as well as variation due to the measurement process.

6. **Percentage of total variation:**

$$\text{Percent equipment variation EV} = 100(\text{EV} / \text{TV})$$

$$\text{Percent appraiser variation AV} = 100(\text{AV} / \text{TV})$$

$$\text{Percent R\&R} = 100(\text{R\&R} / \text{TV})$$

$$\text{Percent part variation PV} = 100(\text{PV} / \text{TV})$$

Using the data from the previous example the percent variations are as follows:

$$\text{Percent EV} = \frac{0.020}{0.8054} \times 100 = 2.48\%$$

$$\text{Percent AV} = \frac{0.053}{0.8054} \times 100 = 6.50\%$$

$$\text{Percent R\&R} = \frac{0.057}{0.8054} \times 100 = 7.08\%$$

$$\text{Percent PV} = \frac{0.803}{0.8054} \times 100 = 99.70\%$$

It is observed that the variability due to the measurement system is only 7.08% of the total variation, and hence the measurement system is acceptable for statistical control of the manufacturing process.

Conducting R&R Studies

The reference manual *Measurement Systems Analysis* has recommended a method for conducting R&R studies and has also suggested a specific format for collecting the data and calculation of various indices. Generally two to three appraisers participate in the study. The collection of data pertaining to repeat measurements and the analysis of data are done according to the following procedure:

- Ten samples, which represent the expected range of the process variation, are collected for the purpose of the study, as the study aims at finding part variations also.
- The samples are identified by marking them with specific identification, but in such a way that the identity of the sample is not visible to the appraiser in order to ensure that the measurement results are not biased.
- Each appraiser makes a number of repeat measurements, generally two or three on each part, and reports the measurement observations.
- The repeat measurements are made over a period of time.
- The parts and the appraisers are picked randomly for taking measurements.
- The gage is calibrated before the studies are conducted, and the gage's accuracy is ensured to be within its acceptable tolerance.
- The repeat measurement data are entered in the format given in Table 10.2. The recommended table format has 12 columns. In the first column are the serial numbers of the rows and the identity of the appraiser. In the next 10 columns are part numbers. The last column indicates the averages for various appraisers/trials. The table has five rows for each appraiser: three rows for recording repeat observations on various parts by various appraisers, a fourth row for averages for various parts for repeat trials, and a fifth row for range of repeat observations on various parts. Thus, for three appraisers there will be a total of 15 rows. An additional five rows are provided for various calculations. Thus, there are a total of 20 rows.

Determination of Various Indices

(i) **Equipment variation (EV):** To determine the repeatability, or equipment variation, first the repeatability standard σ_e is obtained from the knowledge of the average range $\overline{R}$ for all appraisers and all parts. Row 5 of Table 10.2 shows the ranges of repeat measurement data for various parts for appraiser A, and the last column of row 5 also shows the average range for appraiser A (that is, $\overline{R}_a$). Similarly, row 10 and row 15 show the ranges of repeat measurements for various parts for appraisers B and C respectively, and the last column in those rows also gives the average range for appraiser B (that is, $\overline{R}_b$) and appraiser C (that is, $\overline{R}_c$). The average appraiser range $\overline{R}$ is the average of the average ranges of appraisers A, B, and C (that is, $\overline{R}_a$, $\overline{R}_b$, and $\overline{R}_c$).

Thus, $$\overline{R} = \frac{\overline{R}_a + \overline{R}_b + \overline{R}_c}{3}$$

Repeatability, or EV, is determined from the formula given earlier.

(ii) **Appraiser variation (AV):** The reproducibility, or appraiser variation, is an indication of consistency in measurements among various appraisers. This is reflected in variation in the average values for all parts and all trials as reported by various appraisers. The averages of repeat trials for various parts are shown in row 4 for appraiser A, row 9 for appraiser B, and row 14 for appraiser C. The averages for all trials and all parts are also given in the last column of rows 4, 9, and 14 for appraisers A, B, and C respectively, represented by $\overline{x}_a$, $\overline{x}_b$, and $\overline{x}_c$. The reproducibility standard deviation σ_o is calculated based on the range R_o of $\overline{x}_a$, $\overline{x}_b$, and $\overline{x}_c$. R_o is equal to the difference between the maximum value and minimum values of $\overline{x}_a$, $\overline{x}_b$, and $\overline{x}_c$. Having determined the value of R_o, the reproducibility or appraiser variation is determined from the formula given earlier in this section.

(iii) **Repeatability & reproducibility (R&R):** Having calculated EV and AV, the value of R&R can be calculated from the formula given earlier in this section.

(iv) **Part variation (PV):** The averages for various parts based on the measurements by various appraisers in various trials are indicated in row 16. The last column of row 16 also shows the range R_p of various part averages and the grand average G of repeat measurements by all appraisers on all parts in various trials. The part standard deviation σ_p can be calculated from the range R_p of part averages, and the part variation PV can be calculated from the formula given earlier in this section.

(v) **Total variation (TV):** Based on the calculated value of PV and R&R, the total variation can be calculated from the formula given earlier in this section.

(vi) **The percent EV, percent PV, percent R&R, and percent TV:** These percentages of total variation can be calculated from the formulas given earlier. A decision regarding a measurement system's capability to perform a specific measurement application can be made based on the calculated values of various indices.

GUIDELINES FOR R&R ACCEPTABILITY OF A MEASUREMENT SYSTEM

The reference manual *Measurement System Analysis* has suggested the following guidelines for acceptance of gage repeatability and reproducibility (R&R):

- If R&R as a percentage of total variation is less than 10%, the measurement system is acceptable
- If R&R as a percentage of total variation lies between 10% and 30%, the measurement system may be acceptable based upon the importance of the application, the cost of the gage, the cost of repairs, and so on
- If R&R as a percentage of total variation is greater than 30%, the measurement system needs improvement

EXAMPLE OF AN R&R STUDY—DATA COLLECTION AND ANALYSIS

Data for an R&R study for assessing the suitability of a measurement system were collected. The data are presented below and in Table 10.2.

- Parameter: thickness of copper plate (1.75 ±0.15 mm)
- Equipment calipers/dial gages least count 0.001 mm
- Number of appraisers: A, B, and C
- Number of trials: 3
- Number of parts: 10

The following steps are involved in the calculation:

(i) Enter the repeat observations for appraisers A, B, and C for trials 1, 2, and 3 for the parts numbered 1 to 10 in rows 1, 2, 3, 6, 7, 8, 11, 12, and 13.

(ii) Calculate the averages for various parts 1 to 10 for all the three trials, and record them in row 4 for appraiser A, row 9 for appraiser B, and row 14 for appraiser C. Also calculate the grand averages $\overline{x}_a$, $\overline{x}_b$, and $\overline{x}_c$, and record them in the last column of rows 4, 9, and 14 respectively.

Table 10.2 Data collection and analysis for R&R studies.

Appraisers/trial		Part number (Thickness in mm)										
		1	2	3	4	5	6	7	8	9	10	Average
1 A	1	1.722	1.746	1.749	1.764	1.762	1.764	1.764	1.776	1.792	1.802	1.7641
2	2	1.720	1.741	1.742	1.762	1.764	1.768	1.768	1.777	1.793	1.800	1.7635
3	3	1.720	1.742	1.745	1.761	1.761	1.765	1.764	1.775	1.790	1.798	1.7621
4	Average	1.721	1.743	1.745	1.762	1.762	1.766	1.765	1.776	1.792	1.800	$\bar{x}_a = 1.7633$
5	Range	0.002	0.005	0.007	0.003	0.003	0.004	0.004	0.002	0.003	0.004	$\bar{R}_a = 0.0037$
6 B	1	1.721	1.744	1.747	1.763	1.767	1.766	1.765	1.778	1.789	1.805	1.7645
7	2	1.722	1.747	1.744	1.761	1.764	1.765	1.767	1.778	1.787	1.798	1.7633
8	3	1.722	1.743	1.746	1.762	1.765	1.763	1.765	1.776	1.789	1.801	1.7632
9	Average	1.722	1.745	1.746	1.762	1.766	1.765	1.766	1.777	1.788	1.801	$\bar{x}_b = 1.7638$
10	Range	0.001	0.004	0.003	0.002	0.003	0.003	0.002	0.002	0.002	0.007	$\bar{R}_b = 0.0029$
11 C	1	1.723	1.746	1.747	1.763	1.766	1.766	1.769	1.778	1.788	1.801	1.7647
12	2	1.721	1.744	1.748	1.761	1.765	1.766	1.766	1.777	1.789	1.800	1.7637
13	3	1.719	1.746	1.744	1.760	1.763	1.763	1.766	1.777	1.790	1.800	1.7628
14	Average	1.721	1.745	1.746	1.761	1.765	1.765	1.767	1.777	1.789	1.800	$\bar{x}_c = 1.7636$
15	Range	0.004	0.002	0.004	0.003	0.003	0.003	0.003	0.001	0.002	0.001	$\bar{R}_c = 0.0026$
16	Part	1.7211	1.7422	1.7458	1.7619	1.7641	1.7651	1.7660	1.7769	1.7897	1.8006	$G = 1.7635$
I	Average $\bar{x}_p$											$R_p = 0.0795$
17		Average range $\bar{R} = \frac{R_a + R_b + R_c}{3} = \frac{0.0037 + 0.0029 + 0.0026}{3}$										$\bar{R} = 0.0031$
18		Appraiser range $R_o = \bar{x}_b - \bar{x}_a = 1.7638 - 1.7633 = 0.0005$										0.0005
19		Upper control limits for repeatability control chart $D_4\bar{R} = 2.576 \times 0.0031$										0.008
20		Lower control limits for repeatability control chart $D_3\bar{R} = 0 \times 0.0031$										0.000

(iii) Similarly, calculate the ranges of repeat observations for various parts and record them in row 5 for appraiser A, row 10 for appraiser B, and row 15 for appraiser C. Also, calculate the averages of the ranges for various parts R_a, R_b, and R_c, and record them in the last column of rows 5, 10, and 15 respectively.

(iv) Calculate the averages of the observations of all appraisers and all trials for each part 1 to 10, and record them in row 16. Also, calculate the grand average for all the parts and record it in the last column of row 16.

(v) Calculate the average range $\overline{R}$, the average of R_a, R_b, and R_c, and record it in the last column of row 17. $\overline{R} = 0.0031$. This range indicates the variability due to the gage.

(vi) To draw a range control chart calculate the upper and lower control limits thus: UCLR = $D_4 \times \overline{R}$, and LCLR = $D_3 \times \overline{R}$. The values of D_3 and $D_4$are obtained from Appendix B for three observations in the subgroup D_3= 0 and D_4= 2.575.

$$\text{UCLR} = 2.575 \times 0.0031 = 0.008$$

$$\text{LCRL} = 0 \times 0.0031 = 0.000$$

(vii) Compare the ranges of repeat observations for various parts given in rows 5, 10, and 15 for appraisers A, B, and C. It is observed that all the ranges for trials on various parts for appraiser A as given in row 5 are within the control limits for the range chart (that is, all values are less than the upper control limit of 0.008). This implies that the measurement process is consistent as far as appraiser A is concerned. Similar conclusions can be drawn with regard to appraiser B and appraiser C, as all the range values given in row 10 and row 15 are also within the control limits for the range chart.

(viii) Calculate the repeatability, or equipment variation (EV):

$$\text{EV} = \overline{R} \times K_1$$

The value of K_1 is obtained from Table 1 in Appendix C. Number of trials m = 3; number of parts = 10; and number of appraisers = 3. So g = number of parts × number of appraisals = 30.

$$\text{For } m = 3 \text{ and } g = 30,\ K_1 = 3.04$$

$$\overline{R} = 0.0031$$

$$\text{EV} = 0.0031 \times 3.04 = 0.0094 \text{ mm}$$

(ix) Calculate the reproducibility, or appraiser variation (AV):

$$AV = \sqrt{(K_2 \cdot R_o - \frac{(EV)^2}{n \cdot r}}$$

The reproducibility range R_o is obtained from the study of the averages of all parts by various appraisers (that is, $\bar{x}_a$, $\bar{x}_b$, and $\bar{x}_c$) given in the last column of rows 4, 9, and 14. Here, $\bar{x}_b = 1.7638$ is the maximum value, and $\bar{x}_a = 1.7633$ is the minimum value. Thus, $R_o = \bar{x}_b - \bar{x}_a$ (that is, $1.7638 - 1.7633 = 0.0005$). This value is recorded in the last column of row 18.

The value of K_2 is obtained from Table 2 of Appendix C. For three appraisers, $K_2 = 2.70$.

$$\text{AV} = 0.005 \times 2.70 = 0.00135 \text{ mm}$$

Since the variation among appraisers is less than the equipment variation, the effect of EV on AV can be ignored and $(\text{EV})^2/n \cdot r$ is not taken into consideration while calculating AV.

(x) Calculate R&R:

$$\text{R\&R} = \sqrt{(\text{EV})^2 + (\text{AV})^2} = \sqrt{\{(0.0094)^2 + (0.00135)^2\}}$$

$$= \sqrt{(8.836 \times 10^{-5} + 1.8225 \times 10^{-6})}$$

$$\text{R\&R} = 0.0095 \text{ mm}$$

(xi) Calculate the part variation (PV):
Part variation is calculated from the range of the part averages as reported in row 16. The maximum average is for part 10 (that is, 1.8006), and the minimum average is for part 1 (that is, 1.7211). Thus, the part range $R_p = 1.8006 - 1.7211 = 0.0795$.

$$\text{Part variation PV} = R_p \cdot K_3$$

The value of K_3 is obtained from Table 3 of Appendix C. For 10 parts, $K_3 = 1.62$.
So part variation PV $= R_p \cdot K_3 = 0.0795 \times 1.62 = 0.1288$.

(xii) Calculate the total variation TV:

$$\text{TV} = \sqrt{\{(\text{PV})^2 + (\text{R\&R})^2\}}$$

$$= \sqrt{\{(0.1288)^2 + (0.0095)^2\}}$$

$$= \sqrt{(0.016589 + 0.000009)}$$

$$= 0.1291$$

(xiii) Calculate the percent total variation:

$$\text{Percent EV} = \frac{0.0094}{0.1291} \times 100 = 7.28\%$$

$$\text{Percent AV} = \frac{0.00135}{0.1291} \times 100 = 1.06\%$$

$$\text{Percent R\&R} = \frac{0.0095}{0.1291} \times 100 = 7.36\%$$

$$\text{Percent PV} = \frac{0.1288}{0.1291} \times 100 = 99.77\%$$

Since the R&R variability is only 7.36% of the total variability, the measurement system is suitable for control of the manufacturing process.

GAGE R&R STUDY LIMITATIONS AND STATISTICAL CONTROL OF THE MEASUREMENT SYSTEM

R&R studies are aimed primarily at evaluating the capability of a measurement system before it is put into use and subsequently at periodic intervals to ensure that the measurement system continues to be capable. The subsequent verification is performed as a routine part of an organization's normal activities, such as a calibration program or a preventive maintenance program.

Though R&R studies are quite popular with industries, they have certain limitations such as the following:

- Results are derived on a periodic basis. In-between studies are carried out only to evaluate the measurement system for any suspected changes in it. The approach is not based on continuous control of the measurement process.
- The approach is time consuming and hence costly.
- Confidence in the results is affected by the quality of parts chosen.
- The instruments and gages used in R&R studies are calibrated at periodic intervals or prior to the studies. This therefore does not address the issue of what happens when an instrument or gage becomes defective or drifts out of its predetermined uncertainties during the in-between period. In such a situation the manufacturing process could be adjusted erroneously and produce out-of-specifications products. It is no longer acceptable to rely on postproduction inspection to segregate defectives caused by the measurement system.

The answer to the preceding problems lies in statistical control of the measurement process. Statistical process control techniques are used to control and quantify the variability of a measurement system on a continuing basis. SPC also ensures that the variability is consistent and is within the acceptable limits for a specific measurement application. As explained earlier, the most frequently used SPC technique is the use of average ($\bar{x}$) and range (R) charts. One can also ensure that the bias in the measurement system is within a stipulated tolerance by using check standards calibrated at a higher-echelon calibration laboratory. Although statistical control charts are used in R&R studies in a limited way, R&R studies are not substitutes for statistical process control of a measurement system. The application of control charts is becoming quite common in calibration laboratories for controlling the calibration process, and such charts are now becoming popular with manufacturing and service industries for characterizing and controlling the measurement process.

THE PROCESS MEASUREMENT ASSURANCE PROGRAM (PMAP)

Jerry L. Everhart has recommended the use of a process measurement assurance program, or PMAP, for controlling a measurement system on a continuous basis. The objective of the PMAP is to determine, monitor, continually control, and improve the measurement capability of the measuring system and its operating environment. Unlike calibration, which is done in a controlled environment, it ensures the calibration of a measurement system in its operating environment. Also, unlike R&R studies, which are done on a periodic basis, a PMAP is done on a continuing basis.

The PMAP as recommended by Everhart is based on the evaluation of a production operator's capability to make valid measurements in a production environment. The evaluation is carried out against the baseline or reference values that have been provided by measurement experts and worked out in a production environment. The PMAP also helps in estimating the uncertainty that may be present in each product value that is being measured. The following paragraphs outline the PMAP procedure.

To establish the reference values, measurements are made on a control standard, which plays the same role as that of a check standard in calibration laboratories. The control standard is chosen or manufactured to represent the specific features of the product that will be measured with the measurement system. It is essential that the control standard be stable with reference to the actual production environment. Initial measurements are carried out on the control standard by measurement experts, generally a person from a standards and calibration laboratory. These measurements are called control measurements. About 20 to 25 repeat measurements are made at various times on the control standard with the measurement system that is intended to be controlled.

The mean ($\bar{x}$) and standard deviation s of the repeat measurements are calculated, and the reference limits for control are established as the mean ($\bar{x}$) and the upper and lower control limits at three standard deviations from the mean, corresponding to a 99.97% confidence level. The standard certified value for the control standard is also obtained from a higher-echelon independent calibration laboratory. The PMAP control chart and reference values are shown in Figure 10.6.

Based on the PMAP control chart the random variability and systematic error with reference to the certified value of the measurement system are known, and they are shown in the figure. At each calibration the reference values are reestablished, and in between calibrations the measurement experts continue collecting measurement data on the control standard. The reference limits are valid until new reference limits are reestablished at the time of the next calibration. With the reference limits established, the control standard is measured by production personnel using the same procedure and operations that will be used to measure the product. This control measurement is made and recorded prior to manufacturing and measuring the product. This could

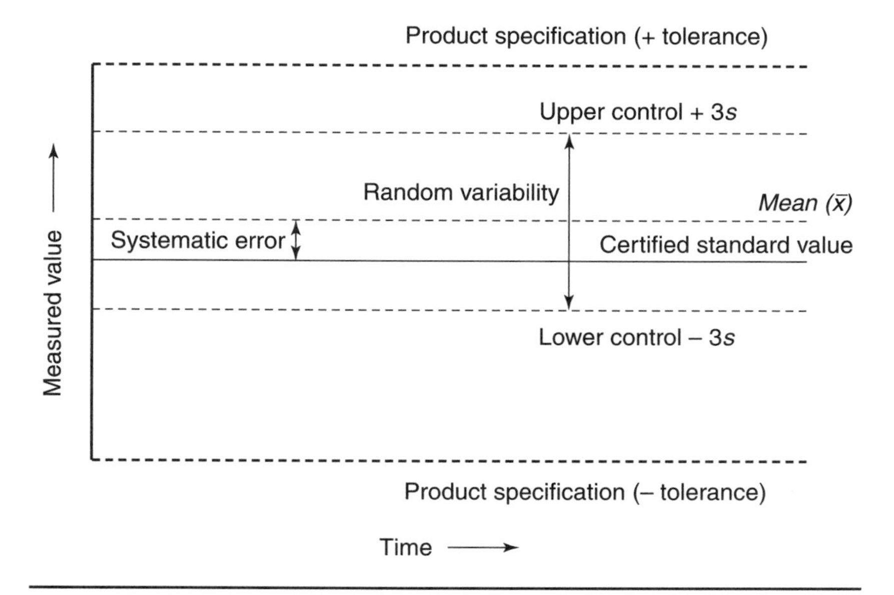

Figure 10.6 PMAP control chart and reference values.

be done at the beginning of each working day or each production shift. The control measurements are made using the total measurement system, including the computer and its software, as if the actual production item were being measured.

The control measurement results obtained by production personnel are compared with the preestablished reference limits to ensure that the measurement system is still in calibration prior to measuring the product. The control measurement of the standard is repeated at the end of the production period. The control measurement results obtained by production personnel should be within the reference limits; thus, they ensure the measurement system's proper performance before the product leaves the organization. There will never be a chance that nonconformance reports regarding the product will be obtained after it has reached the customer. If the control measurements are out of the reference limits, immediate corrective action is taken.

The PMAP control measurements are recorded according to the date and time of each measurement. Based on the control measurement data made by production personnel, the upper and lower control limits can be determined, which will indicate the uncertainty of the assigned value of the product for that specific production time period. This information can also be used to set the design and specification limits for the product and in a limited way to control the manufacturing process. Thus, the PMAP is based on the concept of parallel control of the product and control of the measurement system; it provides total and continuous control of the final product, which results in a higher-quality product at a lower cost.

QUEST FOR EXCELLENCE: SIX SIGMA AND THE ROLE OF MEASUREMENTS

Due to an ever-changing world market and increasing global competition, manufacturing and service organizations are making major changes in their business performance and customer orientation. Their thrust is now on achieving excellence in all operations of the organization with a focus on the quality of products and services to achieve the satisfaction of customers, external as well as internal. To achieve excellence organizations must believe in quality and consistently pursue quality improvements in all their operations. They can achieve such improvements through an organization-level focus on customer satisfaction, leadership, the involvement and empowerment of people, a prevention-oriented approach, and more.

Quality gurus have propagated many concepts by which service and manufacturing organizations can achieve excellence, and those concepts are being successfully practiced. Measurement processes play an important role in the implementation of these concepts and techniques. Six Sigma is one such technique wherein measurement processes play an important role.

Six Sigma Implementation Program

Sigma is a unit of standard deviation and a measure of the variability of a process or product characteristics. This variability is viewed in the context of upper and lower specification limits of the product. *Three sigma* corresponds to 66,807 defects per million; *four sigma* corresponds to 6210 defects per million; *five sigma* corresponds to 233 defects per million; and *six sigma* corresponds to 3.4 defects per million. Thus, sigma is a metric which really refers to the extent to which a process is capable of deviating from preset specifications without causing errors.

Six Sigma is a quality improvement technique invented at Motorola which aims at reducing defects to the magic figure of 3.4 defects per million or less. Six Sigma is thus a powerful tool used to enhance quality levels in every process, not just on the shop floor but in all other organizational processes such as accounting, customer service, supply chain management, advertising, and so on. It is believed that all processes can deviate from the ideal level and can cost the organization in terms of time, labor, and material. Six Sigma as a quality improvement tool has a direct link with profitability by reducing the cost of poor quality which the organization has to incur through rework, rejection, and lost customers. It is a proven tool to ensure customer satisfaction, enhance profits, and eliminate flaws.

The implementation of the Six Sigma technique calls for a clear mathematical interpretation of Six Sigma. The output of any organizational process, such as production from an assembly line, bills prepared by the accounts department, or vendor ratings by the supplier management department, can always be analyzed for number of defects and defectives generated by the process. For each process certain parameters called critical-to-quality (CTQ) parameters are defined. These parameters are critical to the quality of the process output and have a direct bearing on customer satisfaction. In the Six Sigma technique the output of the process is measured on each of these CTQ parameters.

Six Sigma thus focuses on two approaches. One is to make improvements in the process so that the chances of defects are lowered, and another is to make the design robust so that minor variations in the component tolerances can be accommodated in the design. Interestingly the origin of the Six Sigma technique is attributed to a study carried out at Motorola. The study revealed that a positive statistical correlation exists between the field life of a product and the number of flaws that have been spotted and corrected while the product is being manufactured. Thus, the best way to prevent field failures is to ensure that processes do not produce defectives at all. The approach is to set high targets for all processes in terms of CTQ parameters. Then each process is broken down into smaller sequences, and each smaller sequence is examined for its potential for error and elimination of those error potentials. Thus, breaking down the process and tracking down the root cause of defects in a subprocess is a key to the Six Sigma technique. Six Sigma is applied in the following five steps:

- **Define:** The first step in the application of the Six Sigma technique is identifying the process to which the technique needs to be applied and establishing the targets.
- **Measure:** The second stage identifies the metrics that will be improved using Six Sigma. These metrics could be product based, such as certain characteristics of a product, or process based, such as number of defects per million in account records or checks issued. Critical-to-quality characteristics that have the greatest impact on customer satisfaction are identified. The targets are set for achieving defects per million for each of the identified CTQ parameters.
- **Analyze:** The gap between the existing state of quality (the existing CTQ parameters) and the new goals as well as the methodology for achieving them are analyzed. Various tools and techniques for quality improvement such as benchmarking, cause-and-effect diagrams, Pareto analysis, and so forth are used. This is a key component of a defect-reducing program. The aim is to identify the causes of defects in each subprocess so that they can be eliminated by redesign of the product or the process.
- **Improve:** The methods agreed upon during the analysis phase are implemented during the improvement phase. The effect of key process variables on the CTQ parameters is quantified. The maximum acceptable range of the specification is decided. A decision is also made as to whether the Six Sigma capability will be achieved through enlarging design width to accommodate greater variability in the output or through process improvement. During the improvement phase the output needs to be measured continuously to assess the improvements in the CTQ parameters.
- **Control:** In the control phase of Six Sigma implementation, the new process or design conditions arrived at based on analysis and improvement are documented and made permanent in the system. The process is again reviewed after a settling-in period to ensure that the improvements are being sustained.

Role of Measurements in Six Sigma

Measurements are made on products and process parameters during all stages of a Six Sigma implementation program. However, the word *measure,* which is the second phase of Six Sigma, means more than just taking measurements. It includes identification of key process variables, identification of CTQ characteristics, failure mode and effects analysis, and most important, quantification of the effectiveness of the current measurement system.

The *measure* phase of Six Sigma is much more than measuring a value and assigning a number; this along with quantifying the capability of a measurement system is very crucial to implementing Six Sigma techniques. Six Sigma is basically a quality improvement tool like many other classic tools, and the measurement system plays an important role in it. A product or process can be improved only if the characteristics of the product or process can be measured accurately. A clear link exists between quality improvement and quality of measurement results.

REFERENCES

ANSI/ISO/ASQ Q9001-2000. *Quality management systems—Requirements.* Milwaukee, WI: ASQ Quality Press.

Brown, Bradley J. 1993. Measurement system usability: A system for determination of plantwide measurement system capability. *Proceedings of Asian Quality Conference 1993.*

Chrysler Corporation, Ford Motor Company, and General Motors Corporation. 1995. *Measurement systems analysis,* 2d ed. Southfield Michigan, Automative Industry Action Group.

Ermer, Donald S. 1997. *Reliable data is an important commodity. The Standard* (winter): 15–29.

Everhart, Jerry L. 1997. Developing a process measurement assurance program (PMAP). *CALLAB, The International Journal of Metrology* 4, no. 1:22–28.

Hradesky, John L. 1988. *Productivity and quality improvement.* New York: McGraw-Hill.

International Organization for Standardization (ISO). 1994. ISO 5725-1:1994, *Accuracy (trueness and precision of measurement methods and results)—Part 1: General principles and definitions.* Geneva, Switzerland: ISO

Lahiri, Jaideep. The enigma of Six Sigma. *Business Today* (New Delhi), *Millennium Issue.*

Pearson, Thomas A. 2001. Measure for Six Sigma success, *Quality Progress* 34, no. 2:35–40.

Taylor, Wayne A. *Optimization and variation reduction in quality.* New York: McGraw-Hill.

11

Measurement Uncertainty Considerations in Testing Laboratories

REQUIREMENTS CONCERNING THE UNCERTAINTY OF TEST RESULTS

International standard ISO/IEC 17025:1999 specifies the general requirements for the competence of testing and calibration laboratories, covering both quality management and technical requirements. Laboratory accreditation bodies, regulatory authorities, and users of the testing laboratories use this standard as a basis for assessing and recognizing the competence of testing laboratories.

The requirement for a statement of the uncertainty of test results reported by testing laboratories has been a gray area for quite some time. The reporting of the uncertainty of calibration results has been a well-accepted requirement by calibration laboratories as well as by their users. International standard ISO/IEC Guide 25:1990 on the technical competence of testing and calibration laboratories has now been superseded by ISO/IEC 17025:1999. Guide 25 specifically stipulated that the certificate or report should include a statement of the estimated uncertainty of calibration or test results where relevant. The reporting of uncertainty of calibration results was a well accepted practice but this was not so with test results. Thus, it was mandatory that a calibration result be reported along with its uncertainty, whereas this was not emphasized for a test result. It was left to the testing laboratory and its clients to decide whether the reporting of the uncertainty of a test result was relevant or not. There is still nonclarity among testing laboratories about the relevance of the uncertainty of a test result and the method of its evaluation and reporting.

The user of a test result can use it effectively only if the uncertainty of the test result is reported. In order to make a decision on the compliance or noncompliance of a product with its specifications, one needs information on test results and the product's uncertainty. How this information is to be used for compliance decisions has been explained in a guideline document published by the International Laboratory Accreditation Cooperation (ILAC), details of which are given in subsequent parts of this chapter. The new standard has therefore stipulated that testing laboratories shall have and apply procedures for estimating uncertainty of measurement.

From the above, it is clear that an accredited testing laboratory or a testing laboratory otherwise in compliance with the ISO/IEC 17025 standard should have a procedure for evaluating the uncertainty of test results and should have evidence that the uncertainty is evaluated according to the established procedure. However, the reporting of the uncertainty of the test result in the test report is not mandatory. The standard stipulates that, where applicable, a statement of the estimated uncertainty of measurement shall be included in the test report for the interpretation of the test result. The applicability of reporting the uncertainty statement is defined in the standard for the following situations:

- When the uncertainty is relevant to the validity or application of the test results
- When the client desires an uncertainty statement
- When the uncertainty affects conformance to specification limits

FACTORS AFFECTING THE MEASUREMENT UNCERTAINTY OF TEST RESULTS

There is a large spectrum of technological areas in which products are tested in testing laboratories. The tests are conducted to evaluate the characteristics and properties of products and materials. Thus, there are testing laboratories operating in various technological areas.

Some representative types of testing in technological areas are as follows:

- Electrical
- Electronic
- Mechanical
- Chemical and analytical
- Nondestructive
- Material
- Clinical and pathological
- Food
- Soil, water, and other environmental factors
- Forensic science

- Microbiological and genetic engineering
- Metallurgical

There is an endless list of technological areas in which products and materials are being tested. Recent advances in new areas of technology are resulting in a need for new types of testing techniques.

Irrespective of the area of testing, the principal factors that contribute to the uncertainty of test results remain the same. The factors are as follows:

- The test method, including sampling, handling, transport, storage, and preparation of items to be tested
- The measuring equipment and its calibration
- The environmental conditions under which tests are conducted
- The operator

These factors influence the testing operations of a testing laboratory and the value of its test results. Laboratories should therefore ensure that the factors are controlled so that the uncertainty of test results is within acceptable limits.

GUIDELINE DOCUMENTS ON EVALUATION OF THE MEASUREMENT UNCERTAINTY OF TEST RESULTS

The procedure for evaluating the uncertainty of test results by testing laboratories is primarily based on the following two documents published by the International Organization for Standardization (ISO):

- *Guide to the Expression of Uncertainty in Measurement (GUM),* 1995
- ISO 5725:1994, *Accuracy (trueness and precision) of measurement methods and results*

GUM and Uncertainty of Test Results

The applicability scope of *GUM* states that the general rules for evaluating and expressing uncertainty in measurements as given in *GUM* can be applied at all levels of accuracy and in many fields, from the shop floor to fundamental research. Thus, the uncertainty of measurement results associated with the testing of products and materials can also be evaluated by the principles given in *GUM*. This applies to all types of testing laboratories, including analytical laboratories. The basic features of the approach suggested in *GUM* are as follows:

- Various uncertainty components should be expressed as standard uncertainty which is characterized by standard deviation. These components can be evaluated by the type-A or type-B method of evaluation.

- Combined standard uncertainty should be obtained as per the law of propagation of uncertainty for combining various standard uncertainty components.
- Expanded uncertainty should be used if the measure of uncertainty is to be defined as an interval about the measurement result. This is obtained by multiplying the combined standard uncertainty by a coverage factor, whose value lies between two and three.

ISO 5725:1994 and Uncertainty of Test Results

The international standard ISO 5725:1994 deals with the accuracy of measurement methods and measurement results and consists of six parts. The accuracy of a measurement result consists of two components. One pertains to the trueness and the other to the precision of the result. Trueness refers to the closeness of the result to the accepted reference value, whereas precision refers to the closeness of repeat measurement observations of the same quantity among themselves. One of the important factors that determine the uncertainty of a test result is the uncertainty of the measurement method. Thus, standardized test methods against which tests have to be carried out are emphasized as a necessary requirement in this standard.

The standard stipulates that the precision and trueness of a standardized test method should be known beforehand. This task is carried out by a panel of experts familiar with the measurement method and its application. The panel should have at least one member with experience in statistical design and analysis of experimental data. The standard test method should be a written document. Generally, standard measurement methods are published by professional bodies charged with the responsibility of developing and standardizing the measurement methods. Two examples of such professional organizations are the following:

- The American Society for Testing and Materials (ASTM) has published large numbers of standard practices, standard guides, and standard test methods. These documents are used globally by testing laboratories in their specific areas of operation.
- The Technical Association of Pulp and Paper Industries (TAPPI) has published a large number of standard test methods for use by testing laboratories in the area of pulp and paper testing.

Standard test methods have also been published by a number of national and international standardization bodies, such as the ISO and the International Electrotechnical Commission (IEC).

UNCERTAINTY OF STANDARD TEST METHODS

ASTM standard E177-90 (reapproved in 1996) on the use of the terms *precision* and *bias* in ASTM test methods identifies three distinct stages of quantitative test methods:

- Direct measurement or observations of dimensions, properties, and so on
- Arithmetic combination of observed values to obtain a single determination
- Arithmetic combination of a number of determinations to obtain the test result obtained from the test method

In a test method, certain parameters are measured using measuring equipment or a measurement setup. From the measurement values, a single value for the test result is determined by calculation based on a determination equation or functional relationship between the final test parameter and related parameters that have been measured. The test result may consist of a single determination or more. Generally, there are two types of tolerances or uncertainties specified in a test method:

- **Measurement tolerance:** A measurement value is the numerical result of quantifying a particular property. The measurement tolerance represents the exactness with which a measurement is to be made and recorded. There are two types of measurement values and measurement tolerances specified in a test method. The first type is those tolerances which are specified in the test method, such as the dimensions and dimensional tolerances of a specimen which will be subjected to testing. The second type of tolerance is those whose magnitudes are found by testing, such as the mass of a specimen.
- **Determination tolerance:** A single-value test result is obtained based on a determination equation. The determination tolerance or end result uncertainty is evaluated based on the propagation of uncertainty formula given in *GUM*.

The bias and precision of a standard test method should be known beforehand. They are established based on interlaboratory collaborative experiments. It is also essential that the standardized test method should be robust; that is, a small variation in the procedure should not produce an unexpectedly large change in the test result. In an interlaboratory collaborative experiment for the validation of a standard test method, a number of competent testing laboratories participate. A number of samples representing various levels of the characteristics intended to be measured are measured in a number of competent laboratories using the test method. The laboratories take repeat measurements on these samples, and these repeat measurement data are analyzed to evaluate the bias and precision of the test method. The whole exercise is conducted under the guidance of an expert panel.

The parameters that characterize the accuracy of a test method are its precision and bias. Precision has two components, which are the repeatability and the reproducibility of the test method. *Repeatability* characterizes within-laboratory precision and is defined as closeness of agreement between test results obtained in repeatability conditions—the same test method, the same laboratory, the same operator, with the same equipment in the shortest practical period of time. *Reproducibility* characterizes between-laboratory precision and is a measure of the closeness of test results obtained in different laboratories using the same method. The *bias* of a measurement method is defined as the difference between the average of the test results obtained from all laboratories using that method and the accepted reference value.

Numerical Indices of the Precision and Bias of a Test Method

A test result is assumed to have three components. Mathematically, the test result y can be expressed as the sum of these three components: $y = m + B + e$, where m is the mean of the repeat observations, B is the bias of the laboratory, which is constant for a specific laboratory but varies from one laboratory to another, and e is the random error of the test result. The statistical distribution of e is generally a normal distribution with zero meanand a certain amount of variance.

The variance of e within a single laboratory is called the within-laboratory variance σ_w^2. Since σ_w^2 has different values for different laboratories, the average of these within-laboratory variances is called the repeatability variance and is represented by σ_r^2. Its square root is called the repeatability standard deviation (thus, σ_r^2 is the average of σ_w^2).

The variance of laboratory bias B is called the between-laboratory variance σ_L^2. The reproducibility variance σ_R^2 is the sum of the between-laboratory variance σ_L^2 and the within-laboratory variance σ_r^2; thus, $\sigma_R^2 = \sigma_L^2 + \sigma_r^2$. The quantity σ_R is called the reproducibility standard deviation.

The values of σ_r, σ_L, and σ_R are evaluated based on interlaboratory collaborative experiments. Suppose it is intended to evaluate these parameters for a specific material at a specific level involving a number of laboratories. Say p number of laboratories participate in the interlaboratory collaborative program. Each laboratory makes n repeat measurements of the characteristics and reports them. Of these laboratories, a specific laboratory takes n repeat observations, say $x_1, x_2 \dots x_n$, on the sample. The measurement result for this laboratory will be $\bar{x}$, the average of the repeat observations. The variance of these observations for the laboratory is calculated and is given by σ_w^2. This variance is calculated for all the laboratories based on their repeat measurement data. Since the different laboratories will have different variances, their mean gives the within-laboratory variance, or repeatability variance, σ_r^2.

The estimate of σ_L^2 is obtained based on the averages m of the data reported by p laboratories. The grand average of the averages m reported by the various laboratories gives the result as obtained by the test method. The variance of the averages reported by the various laboratories is calculated, and it gives the value of the between-laboratory variance σ_L^2. Based on the calculated values of between-laboratory variance σ_L^2 and within-laboratory variance σ_r^2, the reproducibility variance σ_R^2 is obtained. The detailed methods of evaluating the repeatability standard deviation and reproducibility standard deviation of standard test methods are stipulated in parts 2 to 5 of the international standard ISO 5725:1994. This standard gives detailed procedures not only on interlaboratory collaborative experiments but also on the quality of the data and the criteria for statistically identifying and rejecting outlying observations.

Repeatability and Reproducibility Indices

Repeatability and reproducibility (R&R) figures are used extensively in quantifying the quality of measurement methods and measurement results. The R&R indices depicting the capability of a measurement system in industrial metrology have been defined in chapter 10. In the case of testing laboratories, repeatability and reproducibility indices are used as a measure of the differences in repeat measurements within a laboratory or between laboratories. In essence, both are linked with the quality of the test method and test result.

The repeatability standard deviation σ_r gives an indication of the spread of repeat measurements. For example, one can say that 99.73% of repeat observations will lie within an interval of $\pm 3\sigma_r$ from the mean. But in a testing laboratory, more than the spread of the repeat observations, one is interested in knowing whether the difference between two observations is or is not significant enough to determine their acceptability. According to the law of combining variances, the variance of the difference between two repeat observations is equal to the sum of their individual variances, that is, to $2\sigma_r^2$, where σ_r is the repeatability standard deviation. Thus, the difference between two observations under repeatability conditions will have a standard deviation equal to $\sigma_r\sqrt{2}$. Similarly, if the reproducibility standard deviation is σ_R, the difference between two observations reported by two laboratories under reproducibility conditions will have a standard deviation equal to $\sigma_R\sqrt{2}$. In normal statistical practice, for examining the difference between two repeat observations, a parameter called critical difference is used which is k times the standard deviation. The value of k depends on the confidence level associated with the critical difference and on the underlying distribution of the data. For R&R limits, the confidence level is specified as 95% and the distribution is assumed to be normal. Thus, the value of k is taken to be equal to 1.96. The difference between two results will then be expected to lie between $1.96 \times \sqrt{2}$, or 2.77, times the repeatability standard deviation σ_r and 2.77 times the reproducibility standard deviation σ_R, depending upon whether the two results have been obtained in the same laboratory or different laboratories.

The definitions of *repeatability* and *reproducibility* in the context of testing laboratories are as follows:

- **Repeatability (*r*):** A quantity that 95% of the time will not be exceeded by the absolute value of the difference between two single test results made under repeatability conditions. Its value is equal to $2.77\sigma_r$ where σ_r is the repeatability standard deviation evaluated based on interlaboratory collaborative experiment.
- **Reproducibility (*R*):** A quantity that 95% of the time will not be exceeded by the absolute value of the difference between two single test results made on the same material using the same method in two different laboratories. Its value is equal to $2.77\sigma_R$ where σ_R is the reproducibility standard deviation evaluated based on interlaboratory collaborative experiment.

Evaluation of the Bias of a Standard Test Method

The bias of a standard test method is defined as the difference between the average of the test results reported by various laboratories and the accepted reference value. Every laboratory participating in an interlaboratory collaborative experiment reports a test result, which is the average of the repeat observations taken in that laboratory. The grand average of these test results reported by various laboratories gives the value of the test result as per the test method. In order to determine the bias, all the participating laboratories are also required to make repeat measurements on a reference material which represents the characteristics to be measured by the standard test method. The certified value of the reference material is known. The difference between the test result of the test method and the reference value thus gives the bias of the test method. The uncertainty of the value of the reference material is not taken into consideration in the evaluation of the bias. The absolute value of the estimated bias has to be smaller than or equal to half of the uncertainty interval or one-fourth of the desired uncertainty of testing in order for the standard test method to be accepted.

The uncertainty of a standard test method is therefore characterized by the repeatability index *r* and reproducibility index *R* of the measurement method. The bias of a standard test method has to be low, and generally a test method is accepted as a standard test method if its bias meets the criteria given above.

UNCERTAINTY EVALUATION DURING THE USE OF STANDARD TEST METHODS WHOSE PRECISION AND BIAS ARE KNOWN

The task of evaluating the uncertainty of a test result and judging the acceptability of a test result is relatively easy for testing laboratories following standard test methods. The international standard ISO/IEC 17025:1999 clarifies this aspect as a note which states that "in those cases where a well-recognized test method specifies limits to the values of the major sources of uncertainty of measurement and specifies the form of presentation of calculated results, the laboratory is considered to have satisfied this clause pertaining to reporting uncertainty of test results by following the test method and reporting instructions."

Testing laboratories which follow a standard test method in testing products need not evaluate the uncertainty of test results every time they report results, as they are considered to have complied with the uncertainty requirement provided that the standard test method stipulates the precision and bias of the test method. They have only to ensure that the test results as obtained based on the test method comply with the requirements of the bias and precision as stipulated in the standard test method. The standard test methods published by the ASTM and TAPPI stipulate the precision and bias as well as how the test results have to be reported. An example of this is given in ASTM standard D2699 on the test method for measurement of research octane number of petroleum products and lubricants. The standard test method stipulates the following for precision and bias.

- **Precision:** The precision of the test as determined by this test method by the statistical examination of interlaboratory replicate test results at the most commonly used octane level is as follows:
  - **Repeatability:** At the 90.0 and 95.0 octane levels, the differences between two test results obtained by the same operator with the same apparatus under constant operating conditions on identical test

material would, in the long run, in the normal and correct operation of the test method, exceed the values in the following table only in 1 case in 20.

— **Reproducibility:** At the 90.0 and 95.0 octane levels, the difference between two single and independent results obtained by different operators working in different laboratories on identical test material would, in the long run, in the normal and correct operation of the test method, exceed the values in the following table only in 1 case in 20.

Average research octane level	Repeatability limit octane number	Reproducibility limit octane number
90	0.2	0.7
95	0.2	0.6

These precision limits were calculated from the replicate octane number results obtained by the National Exchange Motorfuel Group (NEG) participating in cooperative testing programs October 1979 through July 1983. These data were analyzed by procedures detailed in the *Manual on Determining Precision Data for ASTM Methods on Petroleum Products and Lubricants.*

- **Bias:** The bias statement is currently being developed.

Another ASTM standard, ASTM D4683-96, which pertains to the standard test method for measuring the viscosity of engine oil at high shear rate and high temperature by a tapered-bearing simulator, specifies the following repeatability and reproducibility figures for the test method. The same definitions for these parameters are given as in the preceding example.

Repeatability: 2.3% of the mean

Reproducibility: 3.65% of the mean

As regards bias, the standard test method specifies that since there is no accepted reference material suitable for determination of the bias of this test method, it cannot be determined.

Another example of R&R indices is given by TAPPI test method T 235 cm 100-2000 for the alkali solubility of pulp at 25°C. It specifies the values for the test method as follows:

Repeatability: 0.24 (10% NaOH solution), 0.20 (18% NaOH solution)

Reproducibility: 2.04 (10% NaOH solution), 1.26 (18% NaOH solution)

Repeatability and reproducibility values were obtained from an interlaboratory study conducted by eight laboratories on four bleached pulp samples

The Repeatability Limit and Control of Uncertainty of a Test Result in the Testing Laboratory

The value of the repeatability limit stipulated in a standard test method can be used to control the uncertainty of a test result. It is presumed that the testing laboratory is carrying out tests according to the standard test method for which the repeatability index r has been established based on interlaboratory collaborative experiment. It is also assumed that the test result is given as a single numerical figure and that the laboratory is performing the tests on a continuing basis. If a complying product is tested according to the method, the laboratory has now to ensure that the range of repeat test results is within the critical difference, which is a multiple of the repeatability standard deviation. If the range exceeds the critical difference, this implies that the variability of the test result has increased. Since the test method is standardized, the testing laboratory will have to look for the reasons for this increase in variability in factors other than the test method and take the necessary corrective action. The reasons could be, for example, inadequate control of environmental factors, training of the operator, or calibration of equipment. The acceptability criteria for test results in such a situation are specified in international standard ISO 5725-6:1994, *Accuracy (trueness and precision) of measurement methods and results, part 6—Use in practice of accuracy values.*

Checking the Acceptability of Test Results and Determining the Final Quoted Result

The acceptability of test results is based on the critical difference represented by the repeatability index. If the difference between two single results is less than the repeatability index specified in the standard test method, it is implied that the testing is being carried out according to the standard test method and that the uncertainty of the test result can be calculated based on the repeatability standard deviation σ_r, which is known from the repeatability limit. Testing laboratories should adhere to the following norms for the acceptability and quotation of the final result as per the guidelines specified in part 6 of ISO 5725.

Single Test Result

When only one test result is obtained, it is not possible to make an immediate statistical test of the acceptability of the test result with respect to a specified repeatability measure. If there is suspicion concerning the correctness of the single test result, a second result should be obtained.

Two Test Results

To judge the acceptability of the precision of test results under repeatability conditions, a minimum of two repeat test results are needed. If the absolute difference between the two results is within the repeatability limit, both the results are accepted and the average of the two results is quoted as the final result in the test report. The critical difference for the two results is equal to the repeatability limit, that is, $2.77\sigma_r$.

If the difference between the two results exceeds the repeatability limit of $2.77\sigma_r$, then depending upon the situation one or two more test results are taken. If obtaining test results is expensive in terms of time and money, then the acceptability of the test results and the final quoted result is decided based on three results: the two earlier results and one additional result which was necessitated because the difference between the two earlier results was found to exceed the repeatability limit. In the case that obtaining test results is not expensive in terms of cost and time, the acceptability of the test results and the final quoted result is based on four test results: the two earlier results and two more results necessitated in view of the above.

- **If it is impossible to obtain a fourth test result:** In such a situation, the range of the three results, that is, the difference between the highest and lowest test results, is compared against the critical range at the 95% confidence level for three results, which is equal to $3.3\sigma_r$. If the range of the three results is less than the critical range $3.3\sigma_r$, the arithmetic mean of the three results should be quoted as the final result. If the range of the three test results is greater than the critical range, the median of the three test results is quoted as the final result. The median is the middle value of the three test results when they are arranged in ascending or descending order.
- **If it is possible to obtain a fourth test result:** In this case, the range of the four test results, that is, the difference between the highest and lowest test results, is compared against the critical range at the 95% confidence level for four results, which is equal to $3.6\sigma_r$. If the range of the four test results is less than the critical range $3.6\sigma_r$, the arithmetic mean of the four test results should be reported as the final result. If the range of the four test results is greater than the critical range $3.6\sigma_r$, the median of the four results is quoted as the final result. The median in the case of four test results is the average of the two middle values.

More Than Two Test Results to Begin With

When there are more than two results to start with, say *n*, the final test result is obtained as follows. The procedure is similar to that stated earlier. The range of these n results, that is, the difference between the highest and the lowest test results, is compared against the critical range for n results at the 95% confidence level. The critical range is obtained by multiplying the repeatability standard deviation σ_r by the critical range factor $f(n)$ at the 95% confidence level. The values of $f(n)$ for various values of n are given in Appendix E. If the range of n values is less than the critical range, as calculated above, the reported result should be the arithmetic mean of n test results. If the range of n values exceeds the critical range, then one of the following options is chosen to determine the final result.

- **Case 1:** If it is expensive to obtain test results, the median of n test results should be reported as the final result.
- **Case 2:** If obtaining test results is not expensive, n more test results are obtained. The range of the total $2n$ test results is compared against the critical range for $2n$ test results. If the range of the $2n$ test results is less than the critical range for $2n$ results, the mean of the $2n$ test results should be quoted as the final result. If this range exceeds the critical range, the median of the $2n$ test results should be quoted as the final result.

The median also indicates the central tendency of the data. If the data are arranged in ascending or descending order, the median is equal to the middle value of an odd number of data. For an even number of data, the median is equal to the average of the two middle values.

If it is observed consistently that the range of the test results exceeds the critical range, it implies that the measurement process used for testing in the laboratory is not capable of meeting the precision requirement of the test method. The reason for this has to be investigated by the testing laboratory and necessary corrective action taken to ensure that the range of test results is within the critical range. This ensures the correct operation of the test method.

Uncertainty of the Test Result

The repeatability limit r is specified in the standard test method. The laboratory has, however, to ensure that the difference between the repeat measurement results obtained on a standard product is within the repeatability limit as stipulated in the standard test method. If this is so, the laboratory can have confidence in its capability to test the parameter according to the test method and thereby to meet the uncertainty requirements of the test method. If, on the other hand, the difference between the test results exceeds the repeatability limit and does so repeatedly, it implies that the laboratory is not capable of undertaking the tests according to the test method and has to take corrective action to improve its measurement process.

To find the uncertainty from the known value of r, the repeatability standard deviation is calculated as $\sigma_r = r / 2.77$. If the difference between two test results is within the critical difference represented by r or the range of more than two test results is less than the critical range consistently, σ_r represents the standard deviation of the repeat measurement data. The value of σ_r can be taken as equal to the combined standard uncertainty. If the result is taken as the arithmetic mean of n number of observations, the combined standard uncertainty is taken as equal to $\sigma_r / \sqrt{n}$. The uncertainty can be expressed as expanded uncertainty, or as an interval around the mean value, according to rules specified in *GUM* by multiplying the combined standard uncertainty by a coverage factor k.

Standard test methods are generally applicable in testing laboratories where chemical analysis or spectral analysis of test samples is carried out. These include laboratories that conduct chemical testing, metallurgical testing, and food analysis, pathology laboratories, forensic science laboratories, and so on, wherein the repeatability limit of a documented test method or test procedure is expected to be specified.

Use of Control Charts for Controlling the Uncertainty of Test Results

ISO 5725-6:1994 recommends the use of average ($\bar{x}$) and range (R) control charts for controlling the uncertainty of test results. The procedure for preparing a control chart and drawing inferences based on it was explained in chapter 9. $\bar{x}$ and R control charts can be applied in testing laboratories where the testing of products is conducted by the laboratory on an ongoing basis. The subgroups can be formed appropriately, for example, the results of the testing carried out each day. The bias of the testing process can be evaluated by the use of standard reference materials.

MEASUREMENT UNCERTAINTY OF TEST RESULTS WHEN THE BIAS AND PRECISION OF THE TEST METHOD ARE NOT SPECIFIED

Repeatability and bias limits specified in test methods are good indicators of the uncertainty of test methods. However, they are generally specified only in test methods involving chemical analysis. In many other areas of testing, the variability in test results cannot be quantified in terms of the repeatability index due to the nonhomogeneity of products and other factors.

The test method generally identifies the requirements for control of various sources of uncertainty. The requirements for the accuracy of measuring equipment used in the test method are specified to ensure that the measurement accuracy of the test process is fit for the purpose. This is necessary to ensure that the uncertainty of the measurement process does not mask the variability of the parameter that is being tested. Some examples of such stipulations in standard test methods follow.

- International standard IEC 61189:1997 deals with the test methods for electrical materials, interconnection structures, and assemblies. Part 1 of this standard specifies general test methods and methodology. This standard stipulates the requirements for estimating systematic and random uncertainty components at the 95% confidence level. It also stipulates the sources of systematic uncertainty as calibration uncertainties, errors in graduation of scale, and so on. The random uncertainty estimation based on repeated measurements of a standard item is also recommended. The standard also specifies how the uncertainty components are to be evaluated using the Student *t* distribution and combined into an overall uncertainty figure. It also suggests uncertainty limits for the measurement of various parameters, such as voltage, current, resistance, and plating thickness.
- IEC publication 770:1984, on the method of evaluating the performance of transmitters for use in industrial process control systems, stipulates the uncertainty of the test method as follows: "The limit of error of the measuring system used for the test shall be stated in the test report and should be smaller than or equal to one-fourth of the stated limits of the transmitter instrument tested."
- Similarly, IEC 731, *Particular requirements for the safety of the dosimeter,* and IEC-601-2-15, part 2, *Particular requirements for the safety of the capacitor discharge X-ray generator,* stipulate the uncertainty requirement for the test instruments and methods as follows: "When demonstrating compliance the uncertainty of the test instrument and method shall not exceed one-third of the applicable tolerance of the quantity being measured."
- Generally, ASTM standard test methods for analytical measurements specify the repeatability and reproducibility and the bias of test methods. However, there are a number of ASTM standard test methods in which the measurement accuracy of various parameters and ultimately the accuracy of the test method are specified. One example is ASTM B193-92, *Standard test method for resistivity of electrical conductor material. Resistivity* is defined as the electrical resistance of a body of unit length and unit cross-sectional area. The method envisages a test specimen in the form of a wire, rod, or bar with a length of at least 300 millimeters and a resistance of at least 0.00001 ohm. The resistivity is

calculated based on measurements of the length, cross-sectional area, and resistance of the test specimen. The standard test method stipulates the following accuracies of measurement:

1. Dimensional measurements to an accuracy of ±0.05%
2. Cross-sectional dimensions of the specimen to be determined by micrometer measurements and sufficient number of measurements to be made to obtain the mean cross section to within ±0.10%
3. Resistance measurements to an accuracy of ±0.15%
4. Specimen to have a uniform cross-sectional area throughout its length within ±0.75%

The test method also states that it provides an accuracy of ±0.30%.

In the above examples, the standard test methods stipulate the specific limits to the values of major sources of uncertainty. If the laboratory is able to give evidence that these accuracy requirements have been met, it can be considered to have satisfied the clause pertaining to the reporting of uncertainty.

Test Result Uncertainty Evaluation Procedure

Products are tested in testing laboratories to find the numerical value of specific characteristics. The international standard ISO/IEC 17025:1999 has made it obligatory for testing laboratories to estimate the uncertainty of test results. It is essential that the measurement uncertainty of a test result is small enough compared with the tolerance of the product characteristic being measured. This enhances the confidence in the test result.

Laboratories generally use well-accepted standard test methods for measuring specific characteristics. If the repeatability index and bias of a standard test method is not specified and major sources of uncertainty are not identified, as is the case with standard test methods in many technological areas, the laboratory itself has to evaluate the total measurement uncertainty of the test result. In order to ensure the fitness of a test result for a specific application, it should be established that the measurement uncertainty of the test result is preferably less than one-fourth, and in no case less than one-third, of the expected variability of the test parameter. For this purpose, the uncertainty of the test result needs to be evaluated. The expected variability of the parameter being measured is generally known to the laboratory itself or to its clients.

The evaluation of the measurement uncertainty of test results is an involved task. It requires a very clear understanding of the scientific and technological principles involved in the testing process and various influence factors that are affecting the testing process. It also calls for good knowledge of statistical analysis of the measurement data and knowledge of the probability distributions associated with various influence factors. However, the general principles for the evaluation of measurement uncertainty remain the same, irrespective of the technological area of testing.

Uncertainty evaluation is done in the following steps:

- The parameter to be measured needs to be defined unambiguously. The environmental conditions in which the testing is to be carried out should be clearly spelled out. These conditions include temperature, humidity, biological sterility, illumination level, and other conditions which may affect the measurement result.
- All factors influencing the measurement process and contributing to the measurement uncertainty must be identified. This step is an important one and should be done by test engineers or analysts with an in-depth knowledge of the testing process and the impact of influence factors on the testing process.
- The first source of uncertainty is the test method itself. The test method has to be robust; that is, the impact of influence factors on the test method should be minimal. A standard method of measurement should not only give details as to how measurements should be made or analysis carried out; it should also be judged as to its fitness for the purpose for which measurement or analysis is being carried out. This is the validation process of the test method. The validation of a test method ensures that the repeatability, reproducibility, and bias of the measurement process are known and are in accordance with requirements. The validation process is used to determine the performance parameters of a test method. It is done through interlaboratory comparison of test results that have been obtained by application of the test method. During the validation process, all the potential sources of errors are identified and their effect on the measurement process is analyzed and investigated to ensure fitness for the purpose of measurement.
- One of the important factors that contribute to measurement uncertainty is the variability among repeat measurements due to sources of random error. This random uncertainty exists in all areas of testing, including analytical measurement. This component of standard uncertainty is evaluated by the method of type-A evaluation as given in *GUM*.
- In certain test methods, there are a number of independent parameters which are measured independently, and the final test result is calculated based on the measured values of these parameters. The mathematical relationship of these parameters to the test results and how the final result is to be evaluated are specified in the test method. The evaluation of the standard uncertainty of all these independent parameters on which repeat measurements are taken is done using type-A evaluation as given in *GUM*.

• The most critical factor is the identification of uncertainty sources that need to be evaluated by type-B evaluation methods. This can be done based on one's experience in similar areas of testing and scientific judgment. Though these factors depend upon the technological area of testing and vary from one area to another, some of the factors that need to be considered are as follows:

- Calibration uncertainty of measuring equipment
- Uncertainty of reference materials and certified reference materials
- Inaccuracy of measuring devices
- Difference in environmental conditions such as temperature or humidity between the operation and calibration of measuring equipment
- Biological sterility
- Sampling variability

Once all uncertainty components that need type-B evaluation have been identified, the statistical probability distribution of these variability components should be known or assumed to be known. This can be found out, for example, from a reference data book or from specifications issued by equipment manufacturers. Sometimes one can assume a specific distribution based on experience. Thus, the standard uncertainty for all the uncertainty components according to the type-B evaluation method of *GUM* is evaluated.

• The test result is obtained based on the mathematical relation given in the test method. The combined standard uncertainty is then evaluated as per the uncertainty propagation formula given in *GUM*. The sensitivity coefficients are calculated as given in *GUM* before evaluation of the combined standard uncertainty.

• Generally, testing is undertaken to determine compliance with specification requirements. Thus, the applicable uncertainty figure will be the expanded or the overall uncertainty. This can be evaluated as explained in *GUM*. Ultimately the expanded uncertainty is compared with the test parameter variability.

A testing laboratory in its day-to-day operation carries out tests on a routine basis. For a specific type of test, the testing process is a repetition of the sequence of operations according to the standard test method. Thus, the uncertainty of the testing process needs to be evaluated when the test method is used for the first time and then subsequently at periodic intervals. A testing laboratory complying with ISO/IEC 17025:1999 has to give objective evidence of the uncertainty evaluation of test results not only once but on a continuing basis at regular periodic intervals.

In certain regulatory product testing situations, compliance with regulatory requirements is judged based on the product-to-product variability of the characteristics being measured. The product variability is taken into consideration by repeating the measurements on samples of the product. Judging compliance with electromagnetic interference requirements is one such example and is given in detail later in this chapter.

Use of Laboratory-Developed and Nonstandard Test Methods

If a laboratory-developed or nonstandard method of testing is used, it is essential to validate the test method. In such situations, one of the components of the test method as stipulated in the standard is a procedure to evaluate the uncertainty of the test method. Evaluation of uncertainty is an essential requirement for the validation of a test method. The uncertainty in such a case should be evaluated according to the procedure specified in *GUM* or through interlaboratory comparison studies in accordance with ISO 5725:1994. Having validated the test method one time before using it is not enough. For consistency in test results, it is essential that the measurement uncertainty be evaluated periodically.

UNCERTAINTY EVALUATION IN CHEMICAL METROLOGY

In analytical chemistry, standard methods of analysis have emphasized the importance of the precision of analytical results. Thus, the traceability of the results has been linked to the standard method rather than a defined standard or SI unit. To give confidence in the quality of analytical results, it is essential that the analytical results for a particular substance be in agreement irrespective of the analytical method used. To what extent this agreement in analytical results is achieved is quantified by the uncertainty of the results. The concept of uncertainty has been recognized by chemists for many years. The publication of *GUM* by the ISO harmonized the methods of uncertainty evaluation in all technological areas of measurement and at all levels of measurement from the shop floor to fundamental research. In view of the need for the harmonization of methods used to evaluate the uncertainty of analytical measurements, a Eurachem/CITAC document entitled *Guide to Quantifying Uncertainty in Analytical Measurement* has been published. Eurachem is a European network of organizations having the objective of establishing a system for the international traceability of chemical measurements. CITAC (Cooperation on International Traceability in Analytical Chemistry) aims to foster collaboration among existing organizations to improve the international comparability of chemical measurements.

The Eurachem/CITAC guide was first published in 1995, and a second edition was published in 2000. The guide shows how the concepts of *GUM* can be applied to chemical measurements. Eurachem/CITAC basically aims at achieving traceability of chemical measurements, and uncertainty is an essential element of traceability. The document gives the following brief guidelines for the benefit of analytical chemists who attempt evaluation of uncertainty in their reported results:

- Identify as many sources of uncertainty as possible. However, there are factors which contribute significantly to uncertainty, and if these major contributors are identified a good estimate of uncertainty can be made. Since the identification of sources and analysis of uncertainty are involved tasks, it is essential that the effort expended not be disproportionate.
- Once the uncertainty for a particular measurement procedure in a particular laboratory has been arrived at, the figure obtained can be used reliably for all subsequent results for that particular procedure and for that particular laboratory. The uncertainty will have to be reevaluated if there is a change in the measurement procedure or measurement setup.
- The procedures introduced by a laboratory to evaluate its measurement uncertainty should be integrated with existing quality assurance measures, which provide valuable information needed for the evaluation of measurement uncertainty.

Like other measurements, analytical measurements are influenced by random errors which give rise to variation in repeat measurements of the measurand. They are also affected by systematic errors which remain constant during the analysis or vary in a predictable way. Inaccuracy of measuring devices gives rise to systematic error, and inadequate control of experimental conditions gives rise to systematic errors that are not constant.

Analytical Method Validation

The fitness of an analytical method for a specific purpose is determined through method validation studies. The evaluation of uncertainty is an integral part of method validation. Such studies produce data on the overall performance of the method and the effect of influence factors on the analytical method. These data are used in ascertaining the validity of the analytical method, which is determined based on in-house validation protocols or interlaboratory comparison studies. The validation of a quantitative analytical method is carried out based on the determination of some or all of the following parameters:

- **Precision:** These measures include the repeatability standard deviation S_r, which can be found out within a laboratory, and the reproducibility standard deviation S_R, which is determined based on interlaboratory comparison studies.
- **Bias:** Bias is usually determined by conducting studies with relevant reference materials, which also helps in establishing traceability.
- **Linearity:** This parameter represents the capability of the analytical method to make measurements in the range of the measurand.
- **Detection limit:** This represents the lower end of the practical operating range of the method.
- **Robustness:** Determination of this parameter establishes the effect of influence factors on the analytical method.

Some analytical measurements produce results which are method dependent; these methods are called empirical methods. In these cases, results are generally reported with reference to the method. The nature of the method—empirical or nonempirical—is an important consideration in uncertainty evaluation. The Eurachem/CITAC guide suggests the following steps in uncertainty evaluation, giving the details of the steps as applicable to chemical analysis:

1. **Specify the measurand.** In many cases of chemical analysis, the measurand is a concentration, that is, the amount of a specific substance in a given volume of liquid. The nature of the analytical method—empirical or nonempirical—should also be part of the measurand definition.
2. **Identify sources of uncertainty.** The guide identifies typical sources of uncertainty in chemical analysis, some of which are as follows:
    - **Sampling:** Due to random variations in different samples
    - **Storage conditions:** Applicable when the sample is required to be stored in specific environmental conditions before analysis
    - **Effect of instruments:** Due to calibration uncertainty of analytical balances, temperature controllers, volume measurements, and so on
    - **Reagent purity:** Associated with the stated degree of reagent purity
    - **Measurement conditions:** Due to the effect of the environment in the laboratory—temperature, humidity, and so on—on the analysis being carried out
    - **Operator effect:** Due to personal bias in measuring; usually consistently high or low
    - **Random effects:** Due to random variations in the analysis

3. **Quantify the standard uncertainty.** The guide suggests two approaches, as follows:

   - Evaluation of the uncertainty arising from each individual source and then combination of the uncertainties
   - Direct determination of the combined uncertainty

   Most of the information needed for the evaluation of uncertainty using either of these approaches is likely to be readily available from method validation studies and from the laboratory's quality assurance and quality control data.

When an uncertainty component is evaluated experimentally based on the dispersion of repeat observations, the standard uncertainty is evaluated by the type-A evaluation method. If it is derived based on previous data, it is evaluated by type-B methods, as explained earlier.

- **Combined standard uncertainty:** This is obtained based on the uncertainty propagation formula after evaluating standard uncertainties and sensitivity coefficients. The mathematical relationship is generally known from the analytical method.
- **Expanded uncertainty:** This is calculated by multiplying the combined standard uncertainty by a coverage factor which can be taken as $k = 2$ or by an appropriate coverage factor from the Student t table.

Example of Uncertainty Evaluation in an Analytical Measurement

Uncertainty of Standard Solution Used as a Calibration Standard

Standard solutions are used in almost every determination measured in analytical laboratories. Routine analytical measurements are relative measurements, and they need a reference standard to provide traceability to SI units. Standard solutions are used as calibration standards for this purpose.

The characteristics of a standard solution are given in terms of its concentration, usually expressed as mass per specified volume, the unit generally being milligrams per liter (mg/L). An example would be a standard solution of a high-purity metal dissolved in a specified liquid. The concentration of the solution is given as

$$C = mP / V \text{ mg/L}$$

where m mg of metal is dissolved in V liters of liquid and P is the purity of the metal.

The following sources of uncertainty affect the uncertainty of concentration *C:*

- **Inaccuracy due to mass measurement:** To make a solution of specific concentration, the mass *m* mg of the metal to be dissolved in volume *V* is calculated. The amount of metal having a mass equal to *m* mg is weighed in an analytical balance very accurately. The uncertainty of the mass measurement *m* mg can be obtained from the accuracy specification of the balance.
- **Inaccuracy of volume measurement** ***V:*** The inaccuracy of the volume measurement is obtained from the accuracy specification as given by the manufacturer.
- **Repeatability of volume flask filling process:** There is a slight variability in the volumes of liquid while filling the flask repeatedly
- **Inaccuracy due to the purity of metal** ***P:*** The purity is generally expressed as the percentage *P.* The purity *P* also has an uncertainty associated with it, say ΔP. If the purity is expressed as $P \pm \sigma P\%$, this is expressed as $0.01P \pm 0.01\sigma P$ in terms of a fraction.

Suppose that we have to prepare 1 L of a standard solution of cadmium of purity 99.99 ±0.01% in dilute nitric acid with a concentration of 100 mg/L and that the measurement data are as follows:

- Mass of cadmium metal to be dissolved in 1 L of dilute nitric acid as weighed on an analytical balance = 100.05 mg
- Volume of liquid as measured by measuring flask = 1 L
- Purity of metal = 0.9999 ±0.0001

Then the concentration *C* of the standard solution is calculated as follows:

$$C = (100.05 \times 0.9999) / 1 = 100.04 \text{ mg/L}$$

Calculation of Standard Uncertainties

1. **Due to the measurement of mass:** The accuracy specification of the analytical balance states that its measuring accuracy is ±0.1% of the reading. Thus, the uncertainty of mass = 0.1 mg. If the frequency distribution of mass has a rectangular distribution, the standard uncertainty due to mass measurement $u(m) = 0.1 / \sqrt{3} = 0.06$ mg.
2. **Due to the measurement of volume:** This has two components.
   a. The calibration uncertainty of the 1-L measuring flask is 1 ml = 0.001L, having a triangular distribution. The standard uncertainty due to calibration $u_1(V) = 0.001 / \sqrt{6} = 0.0004$ L.

b. The second component is the standard uncertainty due to the repeatability of volume measurements. The volume of the measuring flask was measured accurately 10 times, and the standard deviation of the mean was calculated as type-A standard uncertainty, which was equal to 0.0002 L. The standard uncertainty due to repeatability $u_2(V) = 0.0002$ L.

3. **Due to the impurity of the cadmium metal:** The value of impurity uncertainty of 0.0001 can be assumed to have a uniform distribution. The standard uncertainty due to impurity $u(P) = 0.0001 / \sqrt{3} = 0.000058$.

Calculation of sensitivity coefficients: The mathematical relation is as follows:

$$C = mP / V$$

$$c(m) = \partial C / \partial m = P / V = 0.9999 / 1 = 0.9999$$

$$c(V) = \partial C / \partial V = mP / V^2 = 100.05 \times 0.9999 / 1 = 100.04$$

$$c(P) = \partial C / \partial P = m / V = 100.05$$

Uncertainty Budget: The various uncertainty components, their associated standard uncertainty and sensitivity coefficients are shown in Table 11.1.

Combined standard uncertainty $u_c(C) = \sqrt{\Sigma(c_i u_i)^2} = \sqrt{0.0056}$ = 0.075 mg/L

Expanded uncertainty with coverage factor ($k = 2$) = 0.15 mg/L

The value of the concentration of the standard solution can be stated as 100.04 ±0.15 mg/L at a 95% confidence level.

Table 11.1

Uncertainty Source	c_i	u_i	$c_i u_i$	$(c_i u_i)^2$
Mass measurement	0.9999	0.06	0.06	0.0036
Volume calibration	100.04	0.0004	0.04	0.0016
Volume repeatability	100.04	0.0002	0.02	0.0004
Purity of metal	100.05	0.00058	0.006	0.00036
			Total	**0.0056**

UNCERTAINTY CONSIDERATIONS IN COMPLIANCE TESTING FOR CONFORMITY ASSESSMENT

Product assurance aims at manufacturing and marketing products that meet the requirements of specifications that are acceptable to the customers. The products are also expected to meet regulatory requirements pertaining to safety for users and their surroundings, as well as electromagnetic interference/compatibility requirements according to which spurious electromagnetic signals emitted by a device have to be within specified limits. The product should also not degrade the environment during its manufacturing and use.

Product certification schemes are implemented by various product certification bodies in a number of countries. These third-party certification schemes are governed by rules and procedures of certification which are broadly based on international standards. The schemes provide third-party assurance through a third-party independent certification body that a product complies with the requirements of a specific standard. Products that comply are indicated with an identified mark. Compliance with certain requirements, including regulatory requirements, is also ensured through another type of scheme called the manufacturer's declaration of conformity. The European CE marking is one such scheme.

Testing laboratories play an important role in certifying the conformance of a product to certain technical specifications. Testing by these laboratories may be undertaken on behalf of the manufacturer, a user, a certification body, or a regulatory authority. A testing laboratory may be a third-party independent laboratory or it may form part of a bigger organization such as a manufacturer. Testing laboratories have to be technically competent and are expected to operate in accordance with the international standard ISO/IEC 17025:1999. These requirements of compliance with international standards for testing laboratories and certification bodies form part of the globally acceptable conformity assessment system. Products are tested in testing laboratories in accordance with standard test methods which are formulated by national and international standard-formulating bodies or professional scientific bodies such as the ASTM. The testing laboratory reports the test result to a client, which may be a user of the product, the manufacturer of the product, a certification body, or a regulatory authority. The test result is generally used to judge the compliance or otherwise of the product with the requirements of a standard specification. This is done by comparing the test result with the specification limits.

Requirements for a Test Result

A test result may also be used by an organization in making important decisions pertaining to its operations. It should therefore be fit for the desired purpose. ISO/IEC 17025:1999 also stipulates certain requirements as to how a result is to be reported:

- The result of each test carried out by the laboratory shall be reported accurately, clearly, unambiguously, and objectively and in accordance with any specific instructions in the test method
- Where relevant, a statement of compliance/noncompliance with requirements and/or specifications should be included
- Where applicable, a statement on the estimated uncertainty of measurement should be included

It is important to note that, unlike calibration results, a statement on the uncertainty of test results is to be reported where applicable. However, even if the uncertainty is not reported as a part of the test result, the laboratory still has to evaluate it and keep a record of it. In the case of a calibration laboratory, it is mandatory that either the uncertainty of a calibration result be reported or a statement of compliance with an identified metrological specification be reported.

The Effect of Measurement Uncertainty on Compliance Decisions

The compliance or otherwise of a test result is decided based on its comparison with specification limits. Specification limits may be one-way limits, such as >, ≥, <, or ≤, or they may be two-way (upper and lower) specification limits. If it is desired that the test report of a product also give a compliance/noncompliance statement with the stated requirements, the uncertainty of the test result should be taken into consideration. The uncertainty of a test result is expressed as an interval, and any value within that interval represents the result. Thus, compliance decision making becomes difficult if the limiting values of the specification fall within the uncertainty interval. If the test result and its uncertainty interval are much above or much below a one-way specification limit, there is no difficulty in making a compliance or noncompliance decision. Similarly, for two-way specification limits, if the test result and its uncertainty are either well within the specification limits or well beyond them, it is easy to make a compliance or noncompliance decision. In cases in which the specification limits fall within the uncertainty interval, it is not easy to state the compliance or noncompliance of the product with the specification.

A testing laboratory therefore expects its clients to provide the following information in respect to reporting test results while availing the services of the testing laboratory:

- Whether a statement of compliance/noncompliance with the specification requirements is to be stated along with the test results

- How the effect of the uncertainty of the test result on the compliance/noncompliance decision is to be taken into consideration

Often the test method or the specification requirements make no reference to taking measurement uncertainty into consideration. In such cases, the compliance/noncompliance decision can be made based on the test result without taking the uncertainty of the test result into consideration. This means that compliance will be decided if the test result is within specifications for two-way specification limits, and otherwise noncompliance will be decided. Similarly, a decision on compliance or otherwise can be made if the specification limits have been specified as > or <. However, in such a case, there is an inherent risk to the end user that the product may not meet its specification even though it has been tested using a standard and agreed-upon test method. This is particularly so when the test result lies close to the specification limits. In such a situation, it may be the case that the effect of the test result is not that critical and that the end user accepts the risk and agrees with the uncertainty of the test result.

The question of how the uncertainty of a test result should be taken into consideration in deciding the compliance or noncompliance of the test result with specification requirements has been a matter of concern to testing laboratories, certification bodies, and regulatory authorities. To overcome this difficulty, the guideline document ILAC G8:1996 was published by the International Laboratory Accreditation Cooperation (ILAC). The document was prepared basically to assist testing laboratories, but it can also be used by other agencies concerned with compliance decisions. The document is entitled *Guidelines on the assessment and reporting of compliance with specifications.*

ILAC GUIDELINES ON COMPLIANCE ASSESSMENT AND REPORTING

According to ILAC's guidelines, compliance is assessed based on how close the test result and its associated uncertainty are to the specification limits. If $\overline{x}$ is the test result and $2U$ is the expanded uncertainty interval such that the measurement result is expressed as $\overline{x} \pm U$, then ILAC G8:1996 consists of the following guidelines:

Case I: When the measurement result even when extended upward by half of the uncertainty interval falls below the upper specification limit (USL) or the measurement result even when extended downward by half of the uncertainty interval is above the lower specification limit (LSL), the product complies with the specification requirements. In this case, the interval around the measurement result represented by its expanded uncertainty falls within the specification limits

for compliance, $\bar{x} + U < \text{USL}$ or $\bar{x} - U > \text{LSL}$

Case II: When the measurement result even when extended downward by half of the uncertainty interval is above the upper specification limit (USL) or the measurement result even when extended upward by half of the uncertainty interval lies below the lower specification limit (LSL), the product does not comply with the specification requirements. In this case the interval around the measurement result represented by its expanded uncertainty falls outside the specification limits..

for noncompliance, $\bar{x} - U > \text{USL}$ or $\bar{x} + U < \text{LSL}$

where USL and LSL are the upper and lower specification limits.

Case III: When the measurement result is below the upper specification limit but by a margin less than half of the uncertainty interval or the measurement result is above the lower specification limit but by a margin less than half of the uncertainty interval, it is not possible to report compliance. However, the ILAC guidelines stipulate that if a confidence level of less than 95% is acceptable a compliance statement may be possible.

$\bar{x} < \text{USL} < \bar{x} + U$ or $\bar{x} - U < \text{LSL} < \bar{x}$, a compliance decision is possible

Case IV: When the measurement result is above the upper specification limit but by a margin less than half of the uncertainty interval or the measurement result is below the lower specification limit but by a margin less than half of the uncertainty interval, it is not possible to report noncompliance. However, according to the ILAC guidelines, if a confidence level of less than 95% is acceptable, a noncompliance statement may be possible.

$\bar{x} - U < \text{USL} < \bar{x}$ or $\bar{x} < \text{LSL} < \bar{x} + U$, a noncompliance decision is possible.

However, the ILAC guidelines recommend that a suitable statement to cover these situations, such as the following, be incorporated in the test report: "The test result is above (or below) the specification limit by a margin less than the measurement uncertainty; it is therefore not possible to state compliance/noncompliance based on a 95% level of confidence. However, where a confidence level of less than 95% is acceptable, a compliance/noncompliance statement may be possible."

Case V: When the measurement result is the same as the upper specification limit, that is, $\bar{x} = \text{USL}$, it is not possible to report either compliance or noncompliance. However, if a confidence level of less than 95% is acceptable, the ILAC guidance document advises the following:

- If the specification limit is stated as $\leq$, a compliance statement may be possible
- If the specification limit is stated as $<$, a noncompliance statement may be possible

Case VI: When the measurement result is the same as the lower specification limit, that is, $\bar{x}$ = LSL, it is not possible to report either compliance or noncompliance. However, if a confidence level of less than 95% is acceptable, the ILAC guidance document advises the following:

- If the specification limit is stated as $\geq$, a compliance statement may be possible
- If the specification limit is stated as $>$, a noncompliance statement may be possible

However, the ILAC guidelines recommend that a suitable statement to cover these situations, such as the following, be incorporated in the test report: "The test result is equal to the specification limits; it is therefore not possible to state either compliance or noncompliance at the stated level of confidence.

COMPLIANCE TESTING FOR ELECTROMAGNETIC INTERFERENCE

A large variety of electronic devices operate around us. These include household appliances such as microwave ovens, washing machines, mixers, and juicers as well as various entertainment and educational appliances such as televisions and personal computers. There are also complex devices such as communication equipment, navigational aids, and radar systems used by various regulatory and nonregulatory professional bodies. During their operation, these devices generate and propagate radio frequency signals covering a wide portion of the electromagnetic spectrum. The frequencies of various devices are allocated based on the technology of their operation and international consensus. Within a country, frequency allocation is done by regulatory bodies.

Due to the inherent nature of the phenomenon of electromagnetics, almost all devices generate and transmit electromagnetic signals either by radiation into free space around them or by conduction of these spurious signals into electric power lines. These radiated and conducted spurious signals interfere with the operation of other electronic devices in their proximity, and the interference is undesired. The signals behave like electromagnetic pollutants, and this phenomenon is called electromagnetic interference (EMI). To avoid this interference in the operation of other devices, electronic devices are designed so as to ensure that their radiated and conducted emissions are low enough not to interfere with the operation of other devices nearby. The safe limits of the radiated and conducted emissions are established by expert international professional bodies. The limits also depend upon the type of device in question. Detailed specification limits for the radiated and conducted spurious signals generated by various electronic devices have been worked out by the International Special Committee on Radio Interference (CISPR). These emissions should not exceed the specification limits worked out by CISPR, as a spurious signal below the specified level is not likely to interfere with the operation of other electronic devices nearby. These recommendations have been published as international standards jointly by CISPR and the International Electrotechnical Commission (IEC), which is an international standardization body in the area of electrotechnology. The recommendations have been adopted as national standards by various national standardization bodies and as community standards, as in the European Norms of the European Union.

EMI Testing and Associated Regulatory Aspects

Spurious radiated or conducted electromagnetic signals should be less than specified values. This is a mandatory requirement and is stipulated within countries through directives generally known as EMI directives. Regulatory authorities are designated by the state as having the responsibility of ensuring that these requirements are met. Electronic products are tested in EMI testing laboratories to ensure compliance to EMI directives.

Testing for EMI compliance is a very complex task, as the measurement process is susceptible to many influence factors due to the very nature of electromagnetic phenomena. Because of the influencing factors, the variability in repeat measurements increases. Thus, in making a compliance decision, the uncertainty of the measurement result due to various factors needs to be taken into consideration. The factors that contribute to the variability of measurement results are described in detail in section 1, "Application and Interpretation of Fundamental Definitions and Terms," of the document IEC 1000-1-1, *Electromagnetic compatibility technical report, part I—General,* published by the IEC. There are two important factors of variability which influence the compliance decision:

- Variability may arise due to test methods. Standardized test methods envisage a limited number of test situations, whereas the equipment has to function in a large number of actual situations.

- The spread in the characteristics from one product to another product will include a certain spread of EMI characteristics. Hence, there is a certain probability that a product randomly chosen from mass-produced items will comply with the specification requirements for EMI.

The products are tested according to the standard methods of measurement published by CISPR, IEC, or another professional body. These standards stipulate the limits of radio disturbance in addition to giving standard methods of measurement. The standards on EMI testing and specification requirements have been published for various product categories: for example, CISPR 22:1993 pertains to information technology equipment, and CISPR 13:1996 pertains to sound and television broadcast receivers and associated equipment.

Compliance Criteria: The 80%-80% Compliance Rule

Because of test method and product-to-product variability in the measured values of radiated and conducted emission levels, a compliance decision cannot be made based on a single measured value from one piece of equipment. Moreover, the measurement method specifies characteristics including the accuracy of the measuring equipment, the site conditions, and so on, but does not specify how the uncertainty of the measurement result is to be estimated. In addition, because of the regulatory nature of EMI compliance requirements, it is essential that products with spurious emissions within acceptable limits have a very low probability of being declared as noncomplying. At the same time, products with spurious emissions outside the acceptable limits should have a low probability of being declared as complying. This implies that the compliance criteria should be fair to the manufacturer of the product on the one hand and to the regulatory authorities, users of the product, and society at large on the other. To achieve this, most of the CISPR standards stipulate an 80%-80% compliance rule. The compliance criteria is as follows:

$$\bar{x} + ks \leq L$$

The spurious emission is measured on n randomly selected items of the product individually. $\bar{x}$ is the average of the n measurement results carried out on n items, and s is the standard deviation of these n measurements with $(n - 1)$ degrees of freedom. L is the specification limit for the radiated or conducted emission. The factor k is derived from a noncentral Student t distribution. Its value depends upon the number of items tested. For the 80%–80% compliance rule, the values of k for various values of n are as follows:

n	2	4	5	6	7	8	9	10	11	12
k	2.04	1.69	1.52	1.42	1.35	1.30	1.27	1.24	1.21	1.20

The 80%-80% compliance rule implies that if a decision is made based on this rule, 80% of the items will have an emission below the specified limits at an 80% confidence level.

The number of items to be tested in order to make a compliance decision depends upon the nature of the item in terms of its generation of spurious emissions. CISPR 22 stipulates the number of items to be tested as between 5 and 12, whereas CISPR 13 stipulates that more than 5 items must be tested. The product is reported as complying if the value of $\bar{x} + ks$ is less than or equal to the stipulated specification limit; otherwise it is reported as noncomplying.

REFERENCES

American Society for Testing and Materials (ASTM). 1990 (reapproved 1996). ASTM E177-90, *Use of the terms precision and bias in ASTM test methods.* West Conshohocken, PA: ASTM.

———. 1992. ASTM B193-92, *Standard test method for resistivity of electrical conductor materials.* West Conshohocken, PA: ASTM.

———. 1996. ASTM D4683-96, *Standard test method for measuring viscosity at high shear rate and high temperature by tapered bearing simulator.* West Conshohocken, PA: ASTM.

———. 1999. ASTM D26-99, *Standard test method for research octane number of petroleum products and lubricants.* West Conshohocken, PA: ASTM.

Eurachem/CITAC. 2001. *Guide to quantifying uncertainty in analytical measurement.* 2nd ed.: QUAM.

Guide to the Expression of Uncertainty in Measurement. 1995. Geneva Switzerland: ISO International Electrotechnical Commission (IEC). 1997. IEC 61189-1:1997, *Test methods for electrical materials, interconnection structures, and assemblies, part I—General test methods and methodologies:* IEC.

International Laboratory Accreditation Cooperation (ILAC). 1996. ILAC G-8-1996, *Guidelines on assessment and reporting of compliance with specification (based on measurements and tests in a laboratory):* ILAC.

International Organization for Standardization (ISO). 1994. ISO 5725:1994, *Accuracy (trueness and precision) of measurement methods and results, Part 1—General principles and definitions, Part 2—Basic method for the determination of the repeatability and reproducibility of a standard measurement method, Part 4—Basic method for the determination of the trueness of a standard measurement method,* and *Part 6—Use in practice of accuracy values.* Geneva, Switzerland: ISO.

International Organization for Standardization and International Electrotechnical Commission. 1990. ISO/IEC Guide 25:1990, *General requirements for the competence of calibration and testing laboratories.* Geneva, Switzerland: ISO.

———. 1999. ISO/IEC 17025:1999, *General requirements for the competence of testing and calibration laboratories.* Geneva, Switzerland: ISO.

International Special Committee on Radio Interference (CISPR). 1993. CISPR 22:1993, *Limit and methods of measurements of radio disturbance characteristics of information technology equipment:* CISPR.

———. 1996. CISPR 13-1996, *Limits and methods of measurement of radio frequency characteristics of sound and television broadcast receivers and associated equipment:* CISPR.

Kimothi, S. K., and U. K. Nandwani. 1999. Uncertainty considerations in compliance testing for electromagnetic interference. In *Proceedings of the Annual Reliability and Maintainability Symposium,* 99–103.

Technical Association of Pulp and Paper Industries (TAPPI). 2000. TAPPI T235 Cm-00:2000, *Test method for alkali solubility of pulp at 25°C:* TAPPI.

12

Measurement Uncertainty Considerations in Calibration Laboratories

CALIBRATION UNCERTAINTY AND RELATED ISSUES

Calibration is the process of verification of the accuracy of measurement standards and measuring equipment. Two metrological concepts with direct linkage to calibration are measurement traceability and measurement uncertainty. Calibration should ensure an unbroken chain of traceability of measurements to national standards. Information on uncertainty of calibration is used for deciding the fitness of measuring equipment or a measurement standard for a specific measurement application.

As regards uncertainty of calibration, a calibration laboratory is primarily concerned with the following:

- The evaluation and reporting of the uncertainty of calibration results for the measuring equipment and measurement standards being calibrated by the laboratory
- The use of the uncertainty information reported in the calibration certificates of the laboratory's own reference and measuring equipment

A calibration certificate must provide a statement about the bias in the measuring equipment or measurement standard, if any, and about the uncertainty associated with the assigned value. Knowledge about the bias helps us to apply corrections for known systematic effects. A calibration process starts with the definition of the measurand. The more information provided in the definition of the measurand, the less the uncertainty associated with its realization will be. The definition of the measurand should contain at least the following information:

- The parameter that is desired to be calibrated
- The calibration procedure, including corrections to be applied, if any
- The specified validity conditions

S. D. Phillips et al. and colleagues of the National Institute of Standards and Technology (NIST) have identified certain calibration-related issues, some of which are discussed below.

Specified Validity Conditions

The *International Vocabulary of Basic and General Terms in Metrology (VIM)* definition of *calibration* contains a qualifying phrase: "under specified conditions." The calibration certificate should state not only the calibration result but also the specified conditions under which the calibration result is valid. The validity conditions include all influence factors, such as temperature, humidity, and other conditions that influence the calibration process. These may even include the number of repeat measurements based on which the calibration result and its uncertainty have been arrived at, as the number of measurements influences the uncertainty due to the repeatability of observations. The number of degrees of freedom is generally specified in the uncertainty budget as well as in the report of the final result and its uncertainty. These specified validity conditions are derived from the definition of the measurand.

The uncertainty of a reference standard is assigned for specified validity conditions. If a reference standard is used as a reference for calibration, its uncertainty is included in the uncertainty budget of the calibration process. Sometimes it is not possible to realize the specified validity conditions as specified in the reference standard's calibration certificate while using it as a reference. This creates an additional source of uncertainty and should be additionally accounted for in the uncertainty budget. As we move down the traceability chain, the realization of the validity conditions specified in the definition of the measurand cannot be fully realized; therefore, the uncertainty increases as we move downward in the traceability chain.

The Relationship

The *VIM* definition of *calibration* is specific about "the relationship between values of quantities indicated by a piece of measuring equipment or a measuring system or value represented by a material measure or a reference material and the corresponding values realized by the standard." Two types of entities are calibrated in calibration laboratories. Some artifacts are representations of a single value of a parameter, such as a standard weight in a mass metrology laboratory, a gage block in a dimensional metrology laboratory, and a standard resistor in an electrical metrology laboratory. For a single-value standard, the calibration certificate should indicate its assigned value, its associated uncertainty, and its specified validity conditions. On the other hand, many instruments and artifacts either represent a large number of values of a specific parameter or measure the parameter as a continuous variable. Multifunction calibrators used in electrical calibration laboratories, digital multimeters, and micrometers are examples of this category. For such devices, it is impossible to calibrate all the possible values; during the process of calibration, engineering judgment must be used to assess the reasonable uncertainty figures for the parameter. There may be many uncertainty intervals in the entire range of the measurement capability of such a piece of equipment.

Extended Validity Conditions

It is common practice that a piece of measuring equipment is calibrated in specified validity conditions in accordance with the definition of the measurand but is actually used for measurement purposes in conditions that are different from the specified validity conditions. Thus, the relationship between a calibration result and its subsequent use with reference to its effects on measurement uncertainty needs to be considered. This situation is very frequently encountered in regard to measuring equipment used on the shop floor of industries. In industrial calibration, the influence factors are often quite different when a piece of equipment is used for measurement than they were when it was calibrated. Here one may not be interested in developing an uncertainty budget for every measurement performed but may be satisfied if the calibration function produces a calibration report that gives a reasonable uncertainty figure that takes into account the influence factors present at the time of actual measurement. In such a situation, the calibration laboratory should assess the uncertainty over this range of operating conditions. Two important factors that vary from industrial calibration to measurements made on the shop floor of industries are temperature and relative humidity.

The calibration certificate should state that the uncertainty has been evaluated for extended validity conditions. If the certificate states that the validity of the uncertainty statement applies only to the specified validity conditions of the calibration laboratory, it will be necessary to develop a separate uncertainty budget for subsequent measurements. One of the important sources of information for the additional uncertainty budget is the equipment manufacturer's performance specification for the equipment. The manufacturer may also specify the effect of influence factors on the measurement accuracy of the measuring equipment.

Reporting the Calibration Result

Some of the common practices prevailing in calibration laboratories regarding the reporting of calibration results are as follows:

- **Measurement result and uncertainty:** The measurement result is reported as the mean of a number of repeat observations. The uncertainty is evaluated as recommended in the *Guide to the Expression of Uncertainty in Measurement (GUM)* and is usually reported as an expanded uncertainty.
- **Deviation from the nominal value or reference value and uncertainty:** This format is applicable to artifacts that are assigned a nominal value or a reference value. For example, the result of a gage block calibration reports the deviation from the stated nominal length. Since the calibration process measures the deviation from the reference value represented by a reference gage block, the difference is measured through comparison. The quoted uncertainty is generally the uncertainty of deviation. There could be a situation in which the deviation is zero, but there will still be uncertainty associated with it.
- **Estimated systematic error and uncertainty:** This format is generally used in equipment calibration in which the estimated systematic error is to be employed to apply correction to the measurement result when the equipment is used for subsequent measurements.
- **Metrological requirements:** This format is typically used in industrial calibrations in which maximum permissible error is stipulated based on the measurement accuracy specification of the measuring equipment. Most of the equipment used on the shop floor of industries, such as dial gages, micrometers, and digital multimeters, fall in this category. Similarly, the calibration results of grade 0 and grade 00 gage blocks may fall in this category.

These calibration issues should be considered by calibration laboratories when they determine calibration results and assign uncertainties to them.

CALIBRATION RESULTS: UNCERTAINTY REQUIREMENTS

Calibration laboratories play an important role as part of the national measurement system that transfers SI units to end users in the traceability chain. The requirements for the technical competence of calibration laboratories are stipulated in ISO/IEC 17025:1999. This standard specifies certain requirements concerning the measurement uncertainty of calibration results, as follows:

- The result of calibration carried out by the laboratory shall be reported accurately, clearly, unambiguously, and objectively.
- The calibration certificate shall include the uncertainty of measurement and/or a statement of compliance with an identified metrological specification or clauses thereof where necessary for the interpretation of the calibration result.
- When a statement of compliance with a specification is made omitting the measurement results and associated uncertainties, the laboratory shall record those results and maintain them for possible future reference. (The standard also elaborates on the term *identified metrological specification,* stating that it must be clear from the calibration certificate which specification the instrument has been compared with, either by including the specification or by giving an unambiguous reference to the specification.)
- When statements of compliance are made, the uncertainty of measurement shall be taken into account.
- The calibration laboratory shall have and shall apply a procedure to estimate the uncertainty of measurement for all calibrations and types of calibrations.
- In estimating the uncertainty of measurement, all uncertainty components which are of importance in the given situation shall be taken into account using an appropriate method of analysis.

On the subject of the estimation of measurement uncertainty, the standard further refers to *GUM.*

The above requirements pertaining to measurement uncertainty can be categorized into two activities:

- Estimation of measurement uncertainty
- Reporting of measurement uncertainty

Estimating measurement uncertainty is a well-accepted practice of calibration laboratories. A number of laboratories still use traditional methods for uncertainty evaluation and reporting. The *GUM* document is now gaining popularity and is being used more extensively. The procedure for the evaluation and reporting of uncertainty according to *GUM* has been explained in chapter 8. As regards the reporting of measurement uncertainty, ISO/IEC 17025:1999 requires that either the uncertainty figure along with the statement of compliance with an identified metrological specification be reported in the calibration certificate or the statement of compliance with an identified metrological specification alone be reported.

An example of a compliance statement for the calibration of a nonautomatic weighing instrument is "The weighing instrument complies with the special accuracy class of International Organization of Legal Metrology (OIML) Recommendation OIML R76-1:1992." This example illustrates the unambiguous reference to an international standard. In the case that the calibration certificate is issued for compliance with an identified metrological specification which is not a standard specification, the specification can be included in the calibration certificate. This is especially relevant if the calibration certificate is issued for compliance with the accuracy specification of measuring equipment as claimed by its manufacturer.

Irrespective of whether a measurement uncertainty figure is included in the calibration certificate or only a statement of compliance with an identified metrological specification is made, it is mandatory for a calibration laboratory to evaluate the uncertainty of calibration results. The standard also requires that, even when a compliance statement is made that omits the calibration result and its uncertainty, the laboratory shall record these for possible future reference.

A calibration laboratory has to meet the above requirements to establish its competence in its area of operations. In addition to the evaluation and reporting of the uncertainty of calibration results, there are activities in a calibration laboratory in which uncertainty aspects need to be considered by the laboratory management. Some of these activities are as follows:

- Choosing reference standards for calibration
- Maintaining the laboratory's reference and working standards and assigning values and uncertainties to them
- Using past calibration data for uncertainty estimation

These aspects are explained later in this chapter.

THE UNCERTAINTY OF THE CALIBRATION PROCESS AND THE TEST UNCERTAINTY RATIO

Testing and measuring equipment and measurement standards are calibrated in calibration laboratories. During the process of calibration, their measuring accuracy is verified against reference standards. The value of a reference standard and the uncertainty associated with this value should be known. During this verification process, the standard value represented by the reference standard is compared with what is read on the measuring instrument and observations are made on its response. The decision as to whether the instrument is measuring within the desired accuracy or not depends upon the performance specifications provided by the instrument manufacturer. A decision on the instrument's compliance is made depending upon whether the response is within or outside of its specification tolerance. During the process of calibration, if the measuring instrument is found to be within its specification tolerance, it is accepted as valid for normal use; otherwise it should not be used for measurements until rectified. Sometimes a piece of equipment is found to be marginally within its specifications and another piece of equipment of the same type is found to be marginally out of specification. In such cases, the issue of compliance with specifications or otherwise is not easy, as it becomes a question of test confidence. The decision on compliance or noncompliance is generally influenced by the contribution of the uncertainty of the reference standard to the total calibration uncertainty. Thus, the uncertainty of the calibration process includes the uncertainty of the measuring equipment and also the uncertainty of the reference standard. The measuring equipment which is being calibrated is called the unit under calibration (UUC). The confidence placed in the decision of compliance with specifications or otherwise is influenced by the relative amount of uncertainty of the UUC and the uncertainty of the reference standard. This is an important consideration in calibration laboratories.

The ratio of the uncertainty of the UUC to the uncertainty of the reference standard is called the test uncertainty ratio (TUR). If a multifunction calibrator (MFC) is used to calibrate a digital multimeter (DMM) at 10 volts DC measurement, a reference voltage of 10.0 volts DC is set on the MFC and the same is measured on the DMM. The ratio of the measuring accuracy of the DMM at 10.0 volts DC as stated in its technical specification to the uncertainty of the calibrator at the 10.0-volt DC setting is the TUR of the calibration process. For example, if the technical specification of the calibrator stipulates its accuracy as ±5 ppm of its output setting, the uncertainty of the calibrator setting is 50 microvolts DC. Similarly, if the technical specification of the DMM stipulates its measuring accuracy as ±20 ppm, its uncertainty of measurement while measuring 10.0 volts DC will be 200 microvolts. As the TUR is the ratio of the uncertainty of the UUC to the uncertainty of the reference standard, in the present case the TUR will be 200:50, or 4:1. Alternatively, it can be stated that the reference standard is four times more accurate than the UUC. In the evaluation of the TUR, it is important that the uncertainty of the UUC and that of the reference standard are stated at the same confidence level. The value of the coverage factor which is used as a multiplier of standard uncertainty to evaluate overall uncertainty depends upon the confidence level. If the uncertainties of the UUC and the reference standard are stated at different confidence levels, a correction has to be made to evaluate the uncertainties at the same confidence level. If in the above example the uncertainty of the MFC were stated at the 95% confidence level and the uncertainty of the DMM at the 99% confidence level, the TUR would be evaluated as follows: The uncertainty of the MFC at a 10.0-volt setting at the 99% confidence level = (2.58/1.96) × 50 microvolts = 65.8 microvolts. The quantities 2.58 and 1.96 are coverage factors for normally distributed data at the 99% and 95% confidence levels, respectively. The TUR would thus be equal to 200:65.8, or 3:1.

The Effect of the TUR on the Uncertainty of the Calibration Process

Let U be the uncertainty of the UUC and S the uncertainty of the reference standard. Both the uncertainty figures have been stated at the same confidence level. If σ_u is the standard deviation for the uncertainty of the UUC and σ_s the standard deviation for the uncertainty of the reference standard, $U = k\sigma_u$ and $S = k\sigma_s$ where k is the coverage factor at the stated confidence level.

Let the TUR be equal to r:1.

Then = $U = r \cdot S$ and $\sigma_u = r \cdot \sigma_s$.

The overall calibration process uncertainty C is contributed to by the uncertainty U due to the UUC and the uncertainty S due to the reference standard. If σ_c is the standard deviation of the calibration process, $C = k\sigma_c$.

According to the law of propagation of uncertainty, the variance of the calibration process uncertainty is equal to the sum of the variances of the uncertainties due to the UUC and the reference standard:

$$\sigma_c^2 = \sigma_u^2 + \sigma_s^2 = r^2\sigma_s^2 + \sigma_s^2 = \sigma_s^2(1 + r^2)$$

$$\sigma_c = \sigma_s\sqrt{(1 + r^2)}$$

$$\sigma_u = \sigma_s.r$$

$$\frac{\sigma_c}{\sigma_u} = \frac{C}{U} = \sqrt{\frac{(1+r^2)}{r}}$$

$$C = U \times \frac{\sqrt{(1+r^2)}}{r}$$

Thus, the calibration process uncertainty is $\sqrt{(1 + r^2)}/r$ times that of the uncertainty of the UUC. Generally, the decision on compliance is made based

Table 12.1 Percentage increase in uncertainty for various TUR values.

TUR	r	$\sqrt{(1 + r^2)}/r$	% increase in uncertainty
10:1	10	1.005	0.5
4:1	4	1.031	3.1
3:1	3	1.054	5.4
2:1	2	1.118	11.8
1:1	1	1.414	41.4

on the calibration process uncertainty. Thus, for the compliance decision, the uncertainty of the UUC is enhanced by $\sqrt{(1 + r^2)}/r$ times. This amounts to specifying that the compliance decision is made based on a particular percentage increase in the tolerance specification of the UUC.

The apparent percentage increase in technical specification accuracy limits due to the calibration process depends upon the TUR. The percentage increase in accuracy specification for various values of the TUR is shown in Table 12.1.

THE EFFECT OF AN INCREASE IN UNCERTAINTY ON THE COMPLIANCE DECISION

During the process of calibration of measuring equipment, the UUC is accepted as complying if it measures the reference value set on the reference standard or reference equipment within the accuracy limits as claimed by its technical specification, that is, $\pm U$. Thus, the compliance decision is made based on $\pm U$, which is equal to $\pm k\sigma_u$. If it is presumed that there is no bias in the UUC, the mean of the UUC and the reference standard will have the same value. The probability distribution functions for the UUC and the reference standard for this situation are given in Figure 12.1.

The combined probability distribution function for the calibration process is also shown in the figure. The combined function is wider than that of the UUC due to the uncertainty of the reference standard. The width of the combined distribution function gives an indication of its standard deviation, which increases as the TUR decreases.

As the UUC has no bias, the mean of the UUC and that of the reference standard have the same value of μ. The UUC is considered to be in compliance if it reads the reference value μ within $\mu - U$ and $\mu + U$, or $\mu - k\sigma_u$ and $\mu + k\sigma_u$.

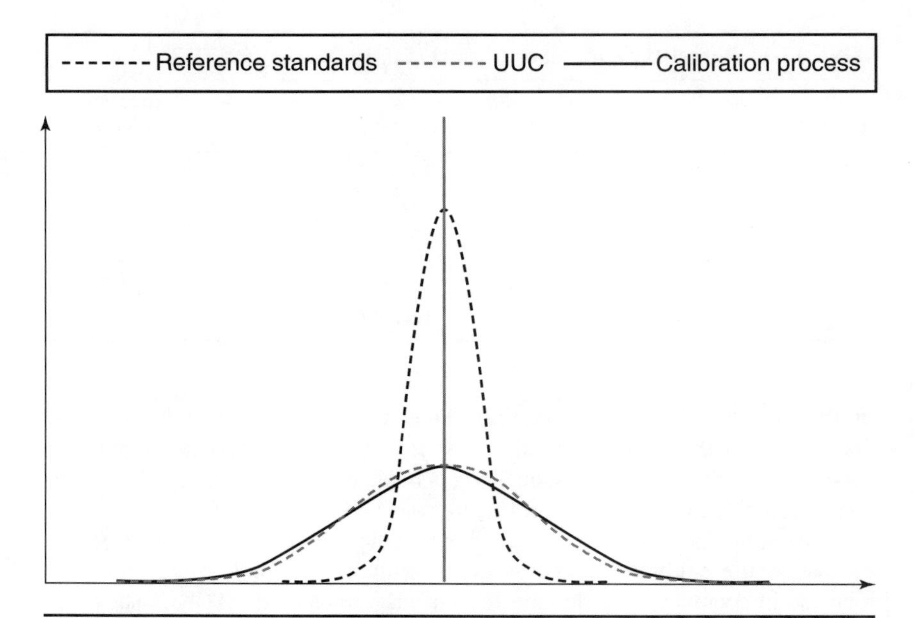

Figure 12.1 Probability distribution functions for the calibration process.

Case 1: Compliance Decision Based on $\pm 3\sigma_u$ Limits

The coverage factor 3 corresponds to a confidence level of 99.73%. This implies that even if the reference standard is absolutely error free, that is, its uncertainty S is zero, about 0.27% of the UUC will be categorized by the calibration process as noncomplying although it is complying due to the uncertainty U of the UUC. Thus, 27 units out of 10,000 units will be erroneously categorized as noncomplying.

If the uncertainty of the reference standard is taken into consideration, the standard deviation of the calibration process will be σ_c, which is equal to $\sigma_u \sqrt{(1 + r^2)}/r$.

$$\sigma_c/\sigma_u = \sqrt{(1 + r^2)}/r$$

or

$$\sigma_c > \sigma_u$$

The limits of the compliance decision which are fixed at $\pm 3\sigma_u$ will contain less than three multiples of the standard deviation σ_c in the interval $\pm 3\sigma_u$. Let the

$\pm 3\sigma_u$ interval be contained in k_1 times the standard deviation of the calibration process σ_c.

$$k_1\sigma_c = 3\sigma_u$$

$$k_1 = 3\sigma_u/\sigma_c = 3r/\sqrt{(1 + r^2)}$$

Thus, due to the uncertainty of the calibration process, it will categorize a certain percentage of equipment as noncomplying even though it does comply. The percentage of equipment erroneously categorized as noncomplying can be obtained from the standard normal table in Appendix A. This percentage is represented by the area under the normal curve beyond z values for which $z = \pm k_1$. As the TUR decreases, r decreases and the value of k_1 also decreases. Thus, the area representing the percentage of erroneous compliance decisions increases.

For example, corresponding to $z = 2.99$, the area under the normal curve according to the table in Appendix A is 0.4986, obtained from the entry in the row corresponding to $z = 2.9$ and the column corresponding to 0.09. The area under the curve corresponding to $z = \pm 2.99$ is equal to 2×0.4986, which is equal to 0.9972. The area under the curve beyond $z = \pm 2.99$ will be equal to $1.0000 - 0.9972$, which is equal to 0.0028, or 0.28%. The values of k_1 and the area beyond $z = \pm k_1$ are given in Table 12.2 for various values of the TUR.

Thus, as the uncertainty of the reference standard increases, the TUR decreases. With this, the probability of a false decision increases from 0.27% for a calibration process with an error-free reference standard to 0.28% for a TUR of 10:1, 0.36% for a TUR of 4:1, and so on. The probabilities of a false decision for other values of the TUR are given in Table 12.2.

Table 12.2 Effect of the TUR on the compliance decision at $\pm 3\sigma_u$.

TUR	r	$k_1 = 3r/\sqrt{(1 + r^2)}$	Area under normal curve beyond $\pm k_1\sigma_c$ (in percent)
∞	-	3.00	0.27
10:1	10	2.99	0.28
4:1	4	2.91	0.36
3:1	3	2.85	0.44
2:1	2	2.68	0.74
1:1	1	2.12	3.40

Case II: Compliance Decision Based on $\pm 1.96\ \sigma_u$ Limits

The coverage factor 1.96 corresponds to a confidence level of 95%. Thus, even if the reference standard is error free, 5% of the UUC will be categorized by the calibration process as noncomplying due to the uncertainty of the UUC. Based on an analysis similar to that given for Case I, the maximum probability of false decisions for various values of the TUR can be calculated. Let the compliance limits $\pm 1.96\sigma_u$ be contained in k_2 times the standard deviation of the calibration process σ_c. Thus,

$$k_2 = \frac{1.96r}{\sqrt{1+r^2}}$$

The value of the probabilities of a false decision for various values of the TUR for Case II are given in Table 12.3.

Thus, the probability of a false decision increases from 5% to 5.12% for a TUR of 10:1 and to 5.74% for a TUR of 4:1. Corresponding figures for other values of the TUR are given in Table 12.3.

Optimum Value of the TUR

As the uncertainty of the reference standard increases, the TUR decreases and the probability of a false compliance decision increases. Figure 12.2 shows a plot of the probability of a false decision for various values of the TUR for a compliance decision based on a 99.73% confidence level. It is observed that with a TUR of 10:1 there is a negligible increase in the probability of a false

Table 12.3 Effect of the TUR on the compliance decision at $\pm 1.96\sigma_u$.

TUR	r	$k_2 = \frac{1.96r}{\sqrt{(1+r^2)}}$	Area under normal curve beyond $\pm k_2\sigma_c$ (in percent)
∞	-	1.96	5.00
10:1	10	1.950	5.12
4:1	4	1.901	5.74
3:1	3	1.859	6.01
2:1	2	1.753	7.96
1:1	1	1.386	16.74

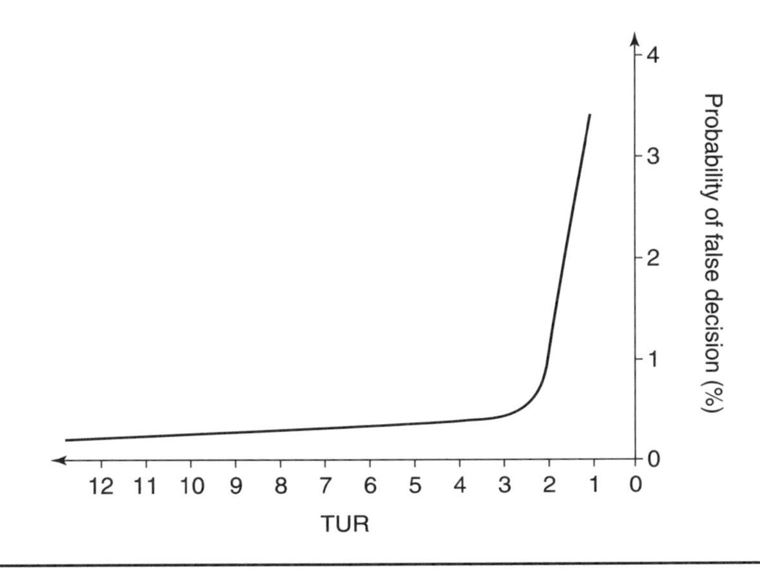

Figure 12.2 Plot of the probability of a false decision versus TUR.

decision, from an ideal situation of 27 per 10,000 to 28 per 10,000. Thus, a TUR of 10:1 is ideal for a calibration process.

However, technological advances in the design and manufacture of testing and measuring equipment have enhanced the available accuracy of measuring equipment considerably. Equipment of high accuracy is available at a moderate cost, particularly electronic equipment with digital readouts. Electronic equipment with high accuracy is extensively being used on the shop floor of medium- and small-scale industries. Maintaining a TUR of 10:1 for this equipment poses a challenge to calibration laboratories in choosing a reference standard. Until these accurate pieces of equipment became commonly available, a TUR of 10:1 was the accepted norm. However, today a TUR of 4:1 is an acceptable norm. It is also observed from Figure 12.2 that a TUR of 4:1 is optimum for most calibration applications, as the probability of a false decision marginally increases from 27 per 10,000 to 36 per 10,000, an increase of 0.09%. A TUR of 4:1 has been considered acceptable for calibration laboratories in accordance with the U.S. military standard MIL-STD-45662A:1980, *Calibration system requirements.* However, in certain measurement situations a TUR of 3:1 can also be considered acceptable. International standard ISO 10012-1: 1990, on quality assurance for measuring equipment, suggests that a minimum TUR of 3:1 should be aimed at for any measurement process.

The Effect of Bias of the Reference Standard on the Compliance Decision

In the preceding analysis, it was assumed that there was no bias in the UUC or the reference standard and that the means of the probability distribution functions of the UUC and the reference standard were the same. If there is a bias in the UUC, the probability of a false decision increases further.

Let us analyze a case with a TUR of 3:1. The uncertainty of the UUC U is equal to $3S$ where S is the uncertainty of the reference standard. If there is an unknown bias equal to σ_s in the UUC, the means of the probability distributions of the UUC and the reference standard will be separated by σ_s where σ_s is the standard deviation for the reference standard uncertainty. The probability distributions for the UUC and the reference standard for this situation are shown in Figure 12.3.

If there were no bias in the UUC, the compliance decision for the UUC would have been made at $\pm 3\sigma_u$ from the reference value. Because of bias equal to $\sigma_u /3$, the mean of the UUC is shifted by $\sigma_u /3$ with reference to the value of the reference standard. With reference to the shifted probability distribution

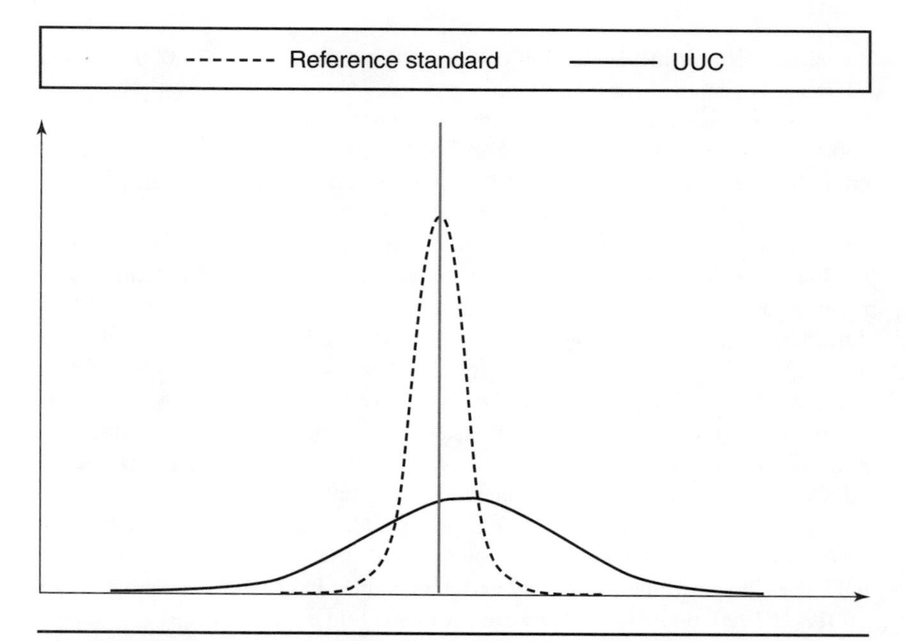

Figure 12.3 Probability distribution function with biased UUC.

function, the compliance point will now be at $z = 2.67$. Thus, the percentage of instruments lying between $z = 2.67$ and $z = 3$ which should have been declared complying will be declared noncomplying because of bias in the UUC. This value can be obtained from the standard normal table in Appendix A by subtracting the area corresponding to $z = 3$ from that corresponding to $z = 2.67$ that is, 0.4987 – 0.4962, which is equal to 0.0025, or 0.25%. Because of the shift in the probability density function of the UUC, there will also be about 0.09% false decision corresponding to the area between $z = -3.0$ and $z = -3.33$. Thus, about 0.34% of decisions will be wrong if the bias in the UUC is of the order of one standard deviation of the reference standard.

For a bias equal to 2σs, the probability of a false decision is equal to 0.85%, and for a bias equal to 3σs, the probability of a false decision increases to 2.14%. This is shown in Figure 12.4.

Similar calculations can be made for false decisions corresponding to various biases for TURs of 4:1 and higher. As the TUR improves, the probability of a false decision decreases considerably.

Thus, if there is even a small bias in the UUC, the chances of a false decision still increase compared with the values given in Table 12.2. The decision on the TUR is therefore crucial and should not be accepted as less than 4:1 in a calibration laboratory.

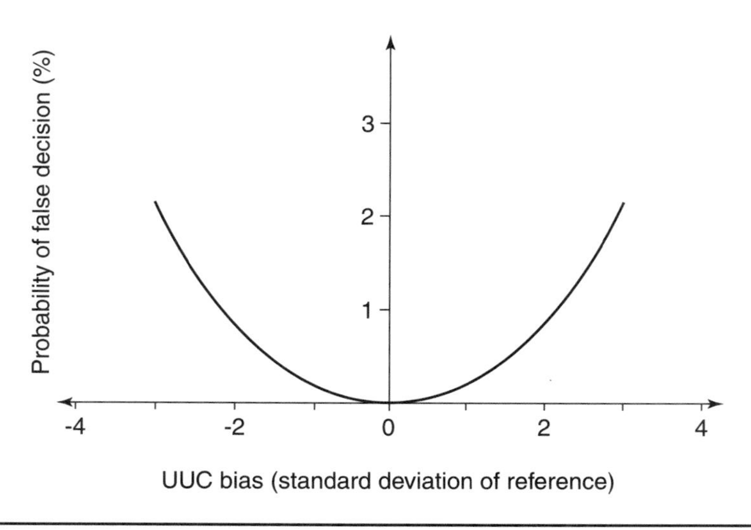

Figure 12.4 Effect of bias on the compliance decision.

ASSIGNING VALUES TO REFERENCE STANDARDS: THE USE OF TREND CHARTS

To establish the traceability of measurement results, calibration laboratories have their reference standards calibrated by higher-echelon calibration laboratories. A large number of calibration laboratories have their reference standards calibrated by national metrology laboratories, thus establishing direct traceability to national measurement standards. Thus, the values of the national standards are disseminated through the calibration process to the next level of calibration laboratories and through a chain of hierarchical calibration laboratories to the end user of measurement results. Each time a standard is sent for calibration, it is returned with a calibration report or certificate after it has been calibrated. The certificate indicates the assigned reference value of the artifact standard and its associated uncertainty. It is a common experience that the value assigned by the calibrating laboratory is not the same after each calibration of the standard. There is slight variation in the values assigned after different calibrations. Sometimes a reference standard is available to a calibration laboratory for many years. In such a case, a vast amount of information about its past assigned values and their uncertainties is maintained in the calibration laboratory. It is a common practice of many calibration laboratories to use the value of the reference standard and its associated uncertainty based on the most recent calibration. The reference values obtained in previous calibrations are not used for any purpose other than their archival value. A reference standard with a long known history of use is a precious asset for any calibration laboratory, as it is backed by a wealth of information about its past behavior. By ignoring a reference standard's past data, a laboratory not only deprives itself of knowledge about drifts in the standard but also at times will underestimate the uncertainty of the reference standard. Thus, it is not good practice to discard the valuable information about a standard which has been accumulated over years at a great cost of effort and resources.

Variability in Assigned Reference Values

Each time a calibrating laboratory calibrates a standard, the laboratory reports its value and the uncertainty associated with the value. The reported value and its uncertainty are derived as an outcome of a measurement process which is influenced by systematic and random effects. Due to this influence, the reported value may differ slightly from ones obtained earlier. With the passage of time, variation in the reported value may be due to a change in the value of the standard itself or due to various factors such as use, long storage periods in uncontrolled environmental conditions, instability, drift, and so on. It could also be due to changes in the measurement system of the calibrating laboratory with the passage of time due to varying systematic errors. The variability observed may be caused by any of these factors or by a combination of them working together.

Analyzing the History of a Standard

The information about a standard that has been accumulated by a calibration laboratory over time may be used to analyze the behavior of a standard in the long run. The information can be used to prepare a special type of chart called a trend chart. As the name suggests, it analyzes the trend of the changing values of the standard with the passage of time.

In a trend chart the abscissa represents calendar time and the ordinate represents the reference value of the standard. Each time a reference value is received from the calibration laboratory, it is plotted on the chart. Initially, when a standard is procured by a laboratory, it is assigned a value by the calibrating laboratory. After each calibration, a new value is assigned to the standard which is the mean of all the earlier reported values, including the most recently reported value. This new assigned value is used as the reference value of the standard until the standard is again sent for calibration and a new value is obtained. A new assigned value is then given to the standard which is once again the mean of all previously reported values including the most recent one. This practice is in sharp contrast to the practice followed in many calibration laboratories which use only the most recent reported value of calibration of the standard and ignore all the previously reported values. A line representing the value of the average of the previously reported values as the reference value is drawn on the chart. This line is called the trend line. After each calibration, a new trend line is drawn on the chart.

In addition to the trend line, two lines indicating the control limits for the behavior of the standard are drawn. The control limits are decided based on the variability of all the reported values in the past and on the average uncertainty of the standard reported by the calibrating laboratory. These limits are considered the bounds between which future values of the standard are expected to lie if the standard remains stable and the measurement process used for calibration remains unchanged.

The control limits indicate the variability in the value of the reference standard as observed during various calibrations. In order to draw the control limits, the standard deviation of the various reported values is calculated along with the mean value. Further, the standard deviation of the mean, or the standard error, is calculated. This standard error is multiplied by a Student t factor at a specific confidence level. This value is used to determine the control limits on the trend chart.

EXAMPLES OF THE DRAWING AND INTERPRETATION OF TREND CHARTS

The procedure for drawing trend charts is given in the following examples.

Example 1: Trend Chart for DC Reference Voltage Standard of Nominal Value 10.0 Volts

The reported values of the calibration results for this example are given in Table 12.4.

This standard was acquired in February 1993. Its value at that time is taken as the reference value. In April 1994, the second reported value is obtained and the standard is assigned a reference value which is the average of the two reported values. After the third calibration, the standard is assigned a reference value which is the average of the three reported values. This process is repeated every time the standard is calibrated.

To evaluate the uncertainty associated with the calibration values reported by the calibrating laboratory, the standard deviation is calculated along with the average. The upper and lower limits representing the between-calibration uncertainty are calculated using t distribution:

$$\text{Upper limit} = \overline{X} + t(s/\sqrt{n})$$

$$\text{Lower limit} = \overline{X} - t(s/\sqrt{n})$$

$\overline{X}$ is the assigned value on the nth calibration and is the average of n reported values. The variable s is the standard deviation of the above reported values, and t is the Student t factor for a specific confidence level.

In this example, the assigned values for various periods, the uncertainty associated with between-calibration variability, and the upper and lower

Table 12.4 Reported calibration results for DC reference voltage standard.

Date of calibration	Reported value (volts)	Measurement uncertainty (volts)
2/93	10.00000	0.000012
4/94	10.00000	0.00002
1/96	9.99989	0.00002
3/97	10.00001	0.00002
8/98	10.00000	0.00002
4/00	9.99999	0.00002

control limits due to this variability have been evaluated and are given in Table 12.5. The t factor has been taken for the 95% confidence level from the table in Appendix B.

It is observed that until the first four calibrations are completed the uncertainty due to between-calibration variability is zero, as the values reported by the calibrating laboratory are quite close among themselves. After the fifth and sixth calibrations, this uncertainty is greater than the uncertainty figure of 0.00002 volt reported by the calibrating laboratory. It is also observed that the values reported by the calibrating laboratory as given in Table 12.5 are within the upper and lower control limits. This implies that the standard has been stable and that the measurement process used for calibrating the standard has not changed over the years. In assigning the reference value to the standard, the uncertainty component due to between-calibration variability has also to be taken into consideration. Assuming that the uncertainty reported by the calibrating laboratory is also evaluated at the 95% confidence level, the overall uncertainty of the reference standard will be equal to the square root of the sum of squares of two uncertainty components:

$$U = \sqrt{(0.00002)^2 + (0.000047)^2} = 0.00005 \text{ as of April 2000}$$

$$U = \sqrt{(0.00002)^2 + (0.000062)^2} = 0.000065 \text{ as of August 1998}$$

The laboratory will thus assign the following reference values to the DC reference voltage standard:

9.99998 ±0.000065 volts, August 1998

9.99998 ±0.000050 volts, April 2000

Table 12.5 Assigned values and control limits for DC reference voltage standard.

Assigned value (volts)	From	To	Uncertainty (volts)	Upper control limit (volts)	Lower control limit (volts)
10.00000	2/93	11/94	-	-	-
10.00000	4/94	1/96	-	-	-
9.99996	1/96	3/97	-	-	-
9.99998	3/97	8/98	-	-	-
9.99998	8/98	4/00	0.000062	10.000042	9.999918
9.99998	4/00	-	0.000047	10.000027	9.999933

The trend chart for the reference standard is presented in Figure 12.5. It is observed that the uncertainty of the standard decreases as more calibration data is accumulated.

Example 2: Trend Chart for Standard Resistance of Nominal Value 1.0000Milliohms

The reported values of the calibration results for this example are shown in Table 12.6.

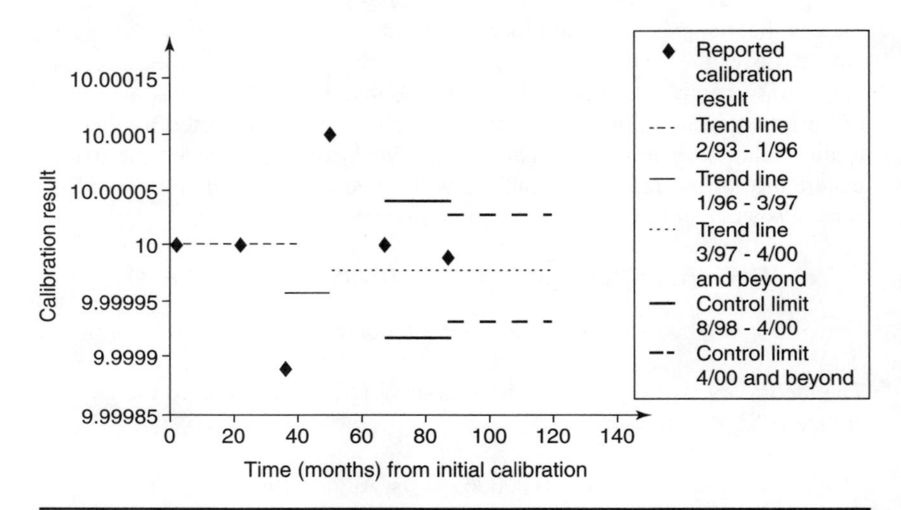

Figure 12.5 Trend chart for 10.0-volt DC reference standard.

Table 12.6 Reported calibration results for 1.0000-milliohm standard resistor.

Date of calibration	Reported value (milliohms)	Uncertainty (milliohms)
1/87	1.0000	0.0002
5/89	1.0000	0.0001
10/91	1.0000	0.0001
3/94	1.0000	0.0002
2/97	1.000036	0.000004
4/99	1.00008	0.00010

For the above standard, the assigned values for various periods, the uncertainty associated with between-calibration variability, and the upper and lower control limits due to this variability have been evaluated at the 95% confidence level and are given in Table 12.7.

The following can be observed from the scrutiny of the upper and lower control limits, the reported value of the reference standard, and the uncertainty associated with these values:

- The values reported by the calibrating laboratory during February 1997 and April 1999 are outside of the control limits
- The uncertainty reported by the laboratory for the February 1997 result is much lower than the values reported during earlier and later calibrations

Based on analysis of the trend chart, it is concluded that there is a discrepancy in the values reported during February 1997 and April 1999. These values, 1.000036 milliohms and 1.00008 milliohms, cannot be the outcome of the same measurement process as that used earlier for calibration and hence are considered to be outliers and not acceptable.

The trend chart for this example is shown in Figure 12.6.

The reason for this out-of-control condition is investigated, and it is observed that the first four calibration values were reported based on calibrations carried out at a national metrology institution, whereas the last two results were reported by another calibration laboratory. The trend chart thus clearly establishes that there is a bias in the calibration process of this laboratory and that the laboratory is not technically competent to undertake calibration of these types of reference standards.

Table 12.7 Assigned values and control limits for 1.0-milliohm standard resistor.

Assigned value (milliohms)	From	To	Uncertainty (milliohms)	Upper control limit (milliohms)	Lower control limit (milliohms)
1.0000	1/87	5/89	-	-	-
1.0000	5/89	10/91	-	-	-
1.0000	10/91	3/94	-	-	-
1.0000	3/94	2/97	-	-	-
1.000007	2/97	4/99	0.000020	1.000027	0.999987
1.000019	4/99	-	0.000035	1.000054	0.999985

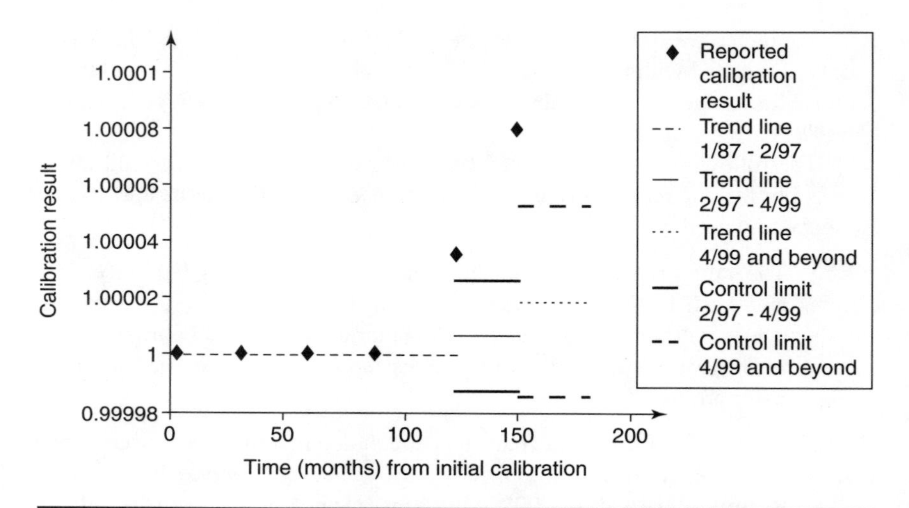

Figure 12.6 Trend chart for 1.00-milliohm reference standard.

Table 12.8 Reported calibration results for 1.00-volt reference standard.

Date of calibration	Reported value (volts)	Uncertainty (volts)
2/93	1.000000	0.000012
11/94	1.000000	0.000002
1/96	0.999993	0.000002
3/97	0.999983	0.000002
8/98	0.999991	0.000002
4/00	0.999992	0.000002

Example 3: Trend Chart for DC Voltage Reference Standard of Nominal Value 1.00 Volt

The reported values of the calibration results for this example are shown in Table 12.8.

For the above standard, the assigned values for various periods, the uncertainty associated with between-calibration variability, and the upper and lower control limits due to this variability have been evaluated at the 95% confidence level and are given in Table 12.9.

Table 12.9 Assigned values and control limits for 1.00-volt DC reference standard.

Assigned value (volts)	From	To	Uncertainty (volts)	Upper control limit (volts)	Lower control limit (volts)
1.000000	2/93	11/94	-	-	-
1.000000	11/94	9/96	-	-	-
0.999998	1/96	3/97	-	-	-
0.999994	3/97	8/98	0.000012	1.000006	0.999981
0.999993	8/98	4/00	0.000009	1.000002	0.999984
0.999992	4/00	-	0.000008	1.000000	0.999983

From the scrutiny of this information, the following observations can be made:

- All the values of the reference standard reported by the calibrating laboratory are within the upper and lower control limits. This implies that the standard has been stable and that the measurement process used for the calibration of the DC voltage reference standard periodically has not changed over the years. No additional bias has been introduced in the measurement process during the period from February 1993 to April 2000.
- The long-term uncertainty component due to between-calibration variability is almost four times the average uncertainty of the calibration, 0.00002 volt, reported by the calibrating laboratory. Using only this component of the uncertainty of the reference standard for the purpose of assigning values and uncertainties to other testing and measuring equipment will not give correct values. The calibration laboratory should therefore use both of the uncertainty components to assign total uncertainty to the reference standard.

The overall uncertainty of the DC reference voltage standard will therefore be as follows:

$$U = \sqrt{(0.000002)^2 + (0.000012)^2} = 0.000012 \text{ volt as of March 1997}$$

$$U = \sqrt{(0.000002)^2 + (0.000009)^2} = 0.000081 \text{ volt as of August 1998}$$

$$U = \sqrt{(0.000002)^2 + (0.000008)^2} = 0.000008 \text{ volt as of April 2000}$$

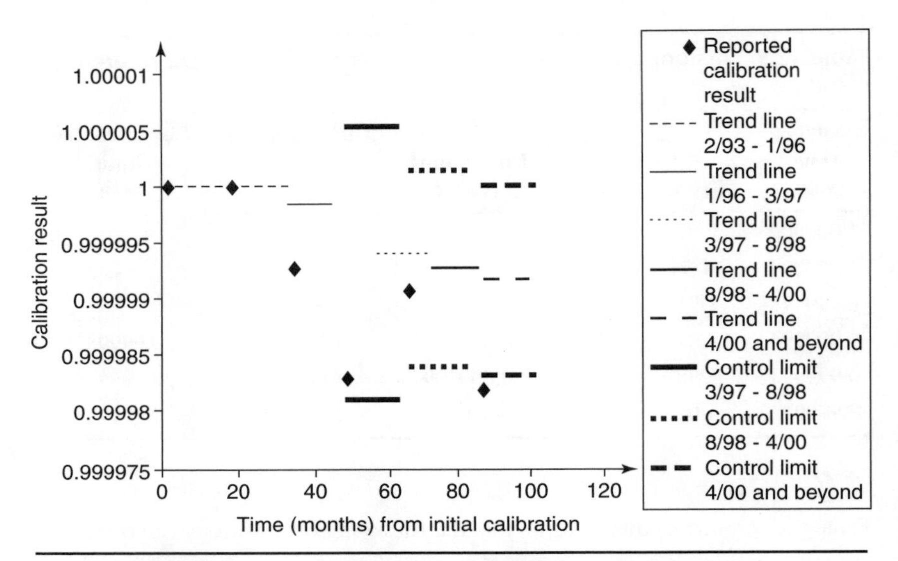

Figure 12.7 Trend chart for 1.0-volt DC reference standard.

It is important to note that the uncertainty component due to between-calibration variability is the dominant uncertainty component.

The following values will thus be assigned to the DC voltage reference standard of nominal value 1.000000 volt.

1.000000	±0.000012	Volts as of	February 1993
1.000000	±0.000002	volts as of	November 1994
9.99998	±0.000002	volts as of	January 1996
9.99994	±0.000012	volts as of	March 1997
9.99993	±0.000009	volts as of	August 1998
9.99992	±0.000008	volts as of	April 2000

The trend chart for the above reference standard is shown in Figure 12.7.

THE ADVANTAGES OF USING TREND CHARTS FOR REFERENCE STANDARDS

- Trend charts help us in assigning the best value to a reference standard. This value is indicated by the trend line on the trend chart. Calibration laboratories often make the mistake of assigning the latest value and its uncertainty as reported by the higher-echelon calibration laboratory to the reference standard. In doing this, they discard the past reported values as far as assigning a new value is concerned and do not make use of the valuable information that has been obtained at great cost and effort. The application of trend charts helps a calibration laboratory to assign the best value from all the available reported values.

- The use of trend charts also helps a calibration laboratory to make a realistic estimation of the uncertainty of a reference standard. The uncertainty reported by the higher-echelon calibration laboratory pertains to random variation and systematic effects, including the uncertainty of the higher-level reference standard, over the short time period during which the standard is under calibration. Long-term variability due to drift and constant use of the standard as well as variability in the measurement system used for the calibration process also need to be considered in assigning uncertainty to the reference standard.
- Drifts in the value of the standard can be predicted with the help of a trend chart that uses statistical techniques of regression analysis. For a reference standard, the predictability of the reference value is more important than its stability. Thus, trend charts based on regression analysis can be used to predict the values of reference standards.
- Trend charts are very effective tools in monitoring and controlling errors in the calibration of reference standards by higher-echelon calibration laboratories. There are many instances in which the values assigned by a higher-echelon laboratory have been questioned and new values assigned. The preceding example 2 concerning calibration results of 1.0000 milli-ohm standard resistor is an illustration of this.

SINGLE MEASUREMENT UNCERTAINTY

The evaluation of measurement uncertainty—particularly type-A evaluation, which entails taking repeat measurements—is an involved task. The equipment received for calibration by calibration laboratories falls into two categories:

- Equipment used as standards with which users calibrate their own measuring equipment, transfer parameters, and maintain traceability. For example, artifact standards such as standard resistors, standard cells, and DC voltage standards fall in this category. An appropriate evaluation of the uncertainty is necessary in these cases.
- Measuring equipment is generally used on the shop floor of industries for inspection of products. The purpose of calibration in this case is to verify compliance with the accuracy specification claimed by the manufacturer of the equipment. The objective is not to estimate the uncertainty of an actual measurement result. If compliance with its

> specification is verified, the purpose of the calibration is achieved. In such a situation, it may not be necessary to take repeat measurements solely for the purpose of type-A evaluation of measurement uncertainty. The uncertainty can be estimated based on the data available that the laboratory has accumulated during past calibrations of the same or similar equipment.

A calibration result is incomplete without a statement of its uncertainty. Laboratory accreditation bodies make it mandatory for accredited calibration laboratories to include the uncertainty of a calibration result in the calibration certificate. The demand for calibration services has also increased considerably. The main reason for the increase in this demand is that a large number of manufacturing and service organizations are implementing ISO 9000 quality management standards in their operations and becoming certified against these standards with a view to enhance their productivity and the quality of their products and services. At the same time, due to the global acceptability of conformity assessment practices and procedures, a large number of accredited calibration laboratories have come into operation.

In a cost-conscious calibration business environment, the management of calibration laboratories has to ensure that the quality of measurements provided conform to the requirements of the customers and the mandatory requirements of laboratory accreditation bodies. The estimation of uncertainty requires that a number of repeat observations be taken in order to calculate the random uncertainty associated with the imprecision of the measurement process. Taking repeat measurements of the same quantity costs the calibration laboratory in terms of operator hours and equipment hours and thus increases the cost of calibration. To reduce the cost of calibration, it is recommended to specify the calibration result as a single measurement observation; the uncertainty of calibration of the single measurement can then be specified based on the known standard deviation s of the calibration process obtained from past data. This technique could be used for reporting the uncertainty of routine types of testing and measuring equipment. This would reduce the cost of calibration considerably and make the operation of the calibration laboratory more cost competitive.

The 1995 NAMAS document NIS 3003, edition 8, entitled *The Expression of Uncertainty and Confidence in Measurement for Calibration,* recommends the practice of using the known standard deviation of the measurement process for the purpose of estimating random uncertainty. The document states,

> It may not always be practical to repeat the measurements many times during a calibration. In these cases a more reliable estimate of the standard deviation of a measurement system can be obtained from an earlier evaluation based on a large

> number of readings. The reliability of previous assessment will depend on the number of devices sampled and how well this sample represents all devices.

The document also stipulates that calibration can be performed based on a single observation in either of the following two situations:

- If it is known that random contributions in the measurements, including those for the device being calibrated, are negligible, a single observation may be used.
- Even if the measurement is known to have imperfect repeatability, reliance can be made on previous assessment of the repeatability of like devices.

If the objective of calibration is to verify compliance with the manufacturer's accuracy specification rather than to assign uncertainty to the standard, the uncertainty can be assigned based on measurement data accumulated during past calibrations. Calibration laboratories usually receive similar pieces of measuring equipment for calibration at periodic intervals from the same as well as from different clients. The data thus available to the calibration laboratory for a number of specific types of measuring equipment can be used to estimate the single measurement uncertainty. *GUM* also suggests a similar approach to the uncertainty evaluation based on single observations.

NAMAS document NIS 3003, edition 8, also states the following:

> Whenever possible at least two measurements should be made as part of the calibration procedure; however, it is acceptable for a single measurement to be made when it is known that the random contributions in the measurements including those for the device being calibrated are negligible. For some calibrations it may be desirable to make only one measurement on the device being calibrated, even though it is known to have imperfect repeatability, and to rely on a previous assessment of the repeatability of like devices. The reliability of previous assessment will depend on the number of devices sampled and how well this sample represents all devices. To avoid an underestimation of the random contribution it is preferred that the largest value of standard deviation is used rather than the average value. It is essential that data obtained from prior assessment should be regularly reviewed and updated if necessary. A previous estimate of standard deviation can only be used if there has been no subsequent change in the measurement system or procedure. If an apparently excessive spread in measurement values is found the cause should be investigated before proceeding further.

As single measurement uncertainty is based on an estimate of the standard deviation of past data, it is to be periodically ensured that the calibration process has not changed by a fresh estimation of the standard deviation based on repeat observations.

As an example, suppose that a parameter of reference value x_0 is measured in a piece of equipment. The uncertainty of x_0 is better than one-fourth of the allowable maximum error L according to the accuracy specification of the measuring equipment. The parameter x_0 is read by the equipment as x, which is a single observation measurement result.

Without consideration of the uncertainty of the single observation x, the compliance criteria for the equipment to meet its specification would be as follows:

$$x_0 + L > x > x_0 - L$$

Otherwise it does not comply with its specification.

The preceding compliance criteria cannot be applied in a situation in which the variability of x when a reference value of x_0 is being measured is not known. However, calibration laboratories have enough measurement data for a specific measurand when a number of pieces of identical equipment are calibrated. Let us suppose that the standard deviation of the measurement process for measuring reference parameter x_0 is σ, which has been calculated by the laboratory based on past data. Thus, we can say that $x = x_0 \pm k\sigma$, where k is a coverage factor whose value depends upon the confidence level and the number of observations based on which σ has been estimated. For a large number of observations, $k = 2$ at the 95% confidence level.

In this situation, the compliance criteria are modified to

$$x_0 + L > x_0 \pm k\sigma > x_0 - L$$

and thus $L > k\sigma$ becomes an additional compliance criterion.

If both the conditions are met, we can say that the equipment complies with its specification and also that if x is a single observation measurement result it can be assigned an uncertainty of $k\sigma$, as x cannot assume values greater than $x_0 + k\sigma$ or less than $x_0 - k\sigma$.

If the standard deviation is calculated as an estimate of σ based on a limited number of past observations, then we use a t factor instead of k. In such a situation, $x = x_0 \pm t{\cdot}s$

$$x_0 + L > x_0 \pm t{\cdot}s > x_0 - L$$

or $L > t{\cdot}s$ becomes an additional compliance criterion.

Since s is generally greater than σ because of a smaller number of observations and t is also greater than k, it is likely that the specification limit L will not be greater than the product $t{\cdot}s$. In such a situation, compliance cannot be ascertained and no uncertainty can be assigned to the single observation measurement result. It is therefore essential that the standard deviation be established based on a large number of past data.

Estimation of Pooled Variance

Harry H. Ku has commented upon the loss of information due to the inability of calibration laboratories to take into account the data accumulated from their past experience. For example, a reference gage block can be used to calibrate a number of gage blocks in a dimensional metrology calibration laboratory. The gage blocks that are calibrated in the laboratory have identical accuracy specifications. During the process of calibration, the dimensions of the gage blocks are compared against the dimensions of the reference gage block. The first gage block is compared against the reference gage block n_1 number of times resulting in n_1 repeat observations; let s_1^2 be the variance of these repeat observations. Similarly, the next gage block is compared n_2 times, with s_2^2 as the variance of these repeat observations, and so on.

Thus, s_1^2 variance is based on $(n_1 - 1)$ degrees of freedom, s_2^2 variance is based on $(n_2 - 1)$ degrees of freedom, and so on. The pooled variance of the calibration process can be estimated as follows:

$$S_p^2 = \frac{(n_1 - 1)s_1^2 + (n_2 - 1)S_2^2 + \ldots (n_x - 1)s_x^2}{(n_1 - 1) + (n_2 - 1) + \ldots (n_x - 1)}$$

$$S_p^2 = \Sigma(x_i - 1)s_i^2/\Sigma x_i - k$$

This pooled variance has $\Sigma n_i - k$ degrees of freedom. The square root of the pooled variance gives the standard deviation S_p of the calibration process. This value of the standard deviation can be used to estimate the uncertainty of the calibration process. Based on the knowledge of the degrees of freedom and the desired confidence level, the value of the Student t factor can be found from the t table. The uncertainty of a single measurement can be reported as $(t{\cdot}sp)$. However, it must be ensured that the single measurement result is within the accuracy specification of the measuring device. The reliability of the quoted data depends upon the number of devices sampled and how well they represent the population. It is also to be ensured that data obtained from prior assessment should be regularly reviewed and periodically updated. This method can generally be used only if there is no change in the calibration process, that is, if the same reference standard is being used for calibration under similar environmental conditions and with the same calibration method. The purpose of the calibration should also be basically to verify compliance with the accuracy specification of the measuring equipment rather than to assign uncertainties to laboratory standards.

Sometimes previous data are not available in the form of a number of variances of repeat observations but are available as a number of repeat measurements on the same type of device on a number of occasions and on a number of the devices. In this case, instead of the pooled variance, the variance of the available data and the standard deviation can be calculated. The information on the number of degrees of freedom is already available. Thus, *t* can be found out from a standard statistical table for the desired level of confidence and the available degrees of freedom.

The following example illustrates the procedure for evaluating single measurement uncertainty.

Example of the Evaluation of Single Measurement Uncertainty

A workbench is used in a calibration laboratory for the calibration of digital multimeters. A multifunction calibrator (MFC) is used as a reference standard. The uncertainty of the setting of various values on the MFC is on the order of parts per million (ppm) and is insignificant in comparison with the inaccuracy of the devices being calibrated. A reference value is set in the MFC, and the same value is read in the digital multimeters.

The following observations pertain to the calibration of a 3.5-digit digital multimeter of a specific model or type number. In accordance with the calibration procedure, a reference value of 100 millivolts is set in the MFC and repeat measurements on the UUC are taken. In the past, a number of 3.5-digit digital multimeters of the same make and model have been calibrated by the calibration laboratory. The data on repeat measurements on various units pertaining to the 100-millivolt reference voltage are available to the laboratory and have been collected on different occasions and on various units from different clients.

The data available to the calibration laboratory on five pieces of equipment during their calibration for 100-millivolt DC measurements are given below. The repeat observations, their standard deviations, and their degrees of freedom are presented in Table 12.10, and the calculations for their pooled variance and pooled standard deviation follow.

$$\text{Pooled variance} = \frac{(0.13)^2 \times 4 + (0.19)^2 \times 4 + (0.11)^2 \times 4 + (0.16)^2 \times 4 + (0.11)^2 \times 4}{4 \times 5}$$

$$= 0.02$$

Standard deviation = $(0.02)^{1/2}$ = 0.14 millivolt

For 95% confidence level and 20 degrees of freedom, *t* value as obtained from the table in appendix B = 2.09

Uncertainty = 2.09 × 0.14 = 0.29 millivolt

Table 12.10 Data available to the calibration laboratory on measurements on various UUCs.

Parameter	Equipment 1	Equipment 2	Equipment 3	Equipment 4	Equipment 5
Observation 1	99.9	100.1	99.9	100.0	99.8
Observation 2	99.8	100.0	99.8	100.1	99.7
Observation 3	100.0	99.7	99.9	100.2	100.0
Observation 4	100.1	99.9	100.0	99.9	99.9
Observation 5	99.8	100.2	100.1	99.8	99.8
Standard deviation	0.13	0.19	0.11	0.16	0.11
Degrees of freedom	4	4	4	4	4

Since these data are representative of measurements of 100 millivolts DC, a single measurement of a 100-millivolt reference value by the meter can be assigned an uncertainty of 0.29 millivolt.

The digital multimeter accuracy specification specifies its measurement accuracy as ±0.5% of the reading. The estimated uncertainty is less than the maximum permissible error of 0.5 volts. Thus, as long as the single measurement is within this limit, it can be categorized as complying with its specification, and the figure of uncertainty as estimated in this example can be assigned as the measurement uncertainty. This way the laboratory can save a great deal of effort and cost. However, the laboratory must analyze the past data at periodic intervals and ensure that there is no change in the measurement process.

REFERENCES

Capell, Frank. 1991. Calibration equipment: How good is your TUR. *Evaluation Engineering,* January.

Fluke Corporation. 1994. *Calibration philosophy in practice.* 2nd ed. Everett WA, Fluke Corporation.

Guide to the expression of uncertainty in measurement. 1995. Geneva, Switzerland: ISO and five other international organizations.

International Organization for Standardization (ISO). 1990. ISO 10012-1:1990, *Quality assurance requirements for measuring equipment, part I—Management of measuring equipment.* Geneva, Switzerland: ISO.

Istone. 1993. *IQN single measurement uncertainty.* Paper presented at colloquium, Uncertainty in Electrical Measurements. Institution of Electrical Engineers, London.

Ku, Harry H. 1967. Statistical concepts in metrology. In *Handbook of industrial metrology,* American Society of Tool and Manufacturing Engineers, 20–50. New York: Prentice Hall.

———. 1968. Expression of imprecision, systematic error and uncertainty associated with a reported value. *Measurement Data,* July–August, 73-72 to 77-76.

National Physical Laboratory of the United Kingdom (NPL). 1995. *The expression of uncertainty and confidence in measurement for calibration.* NIS 3003, ed. 8. Middlesex, England: NAMAS NPL.

Phillips, S. D., W. T. Estler, T. Doiron, K. R. Eberhardt, and M. S. Levenson. 2001. A careful consideration of calibration concepts. *Journal of Research of the National Institute of Standards and Technology* 106, no. 2:371–79.

Schumacher, Rolf B. F. 1981. Systematic measurement errors. *Journal of Quality Technology* 13, no. 1:10–24.

———. 1987. *Measurement uncertainty—Measurement assurance.* SanClemente, CA: Coast Quality Metrology Systems.

U.S. Department of Defense. 1980. MIL-STD-45662:1980, *Calibration system requirements.* Department of Defense.

Appendix A

Area under the standard normal probability distribution between the mean and positive values of *z*.

z	0.00	0.01	0.02	0.03	0.04	0.05	0.06	0.07	0.08	0.09
0.0	0.0000	0.0040	0.0080	0.0120	0.0160	0.0199	0.0239	0.0279	0.0319	0.0359
0.1	0.0398	0.0438	0.0478	0.0517	0.0557	0.0596	0.0636	0.0675	0.0714	0.0753
0.2	0.0793	0.0832	0.0871	0.0910	0.0948	0.0987	0.1026	0.1064	0.1103	0.1141
0.3	0.1179	0.1217	0.1255	0.1293	0.1331	0.1368	0.1406	0.1443	0.1480	0.1517
0.4	0.1554	0.1591	0.1628	0.1664	0.1700	0.1736	0.1772	0.1808	0.1844	0.1879
0.5	0.1915	0.1950	0.1985	0.2019	0.2054	0.2088	0.2123	0.2157	0.2190	0.2224
0.6	0.2257	0.2291	0.2324	0.2357	0.2389	0.2422	0.2454	0.2486	0.2517	0.2549
0.7	0.2580	0.2611	0.2642	0.2673	0.2704	0.2734	0.2764	0.2794	0.2823	0.2852
0.8	0.2881	0.2910	0.2939	0.2967	0.2995	0.3023	0.3051	0.3078	0.3106	0.3133
0.9	0.3159	0.3186	0.3212	0.3238	0.3264	0.3289	0.3315	0.3340	0.3365	0.3389
1.0	0.3413	0.3438	0.3461	0.3485	0.3508	0.3531	0.3554	0.3577	0.3599	0.3621
1.1	0.3643	0.3665	0.3686	0.3708	0.3729	0.3749	0.3770	0.3790	0.3810	0.3830
1.2	0.3849	0.3869	0.3888	0.3907	0.3925	0.3944	0.3962	0.3980	0.3997	0.4015
1.3	0.4032	0.4049	0.4066	0.4082	0.4099	0.4115	0.4131	0.4147	0.4162	0.4177
1.4	0.4192	0.4207	0.4222	0.4236	0.4251	0.4265	0.4279	0.4292	0.4306	0.4319
1.5	0.4332	0.4345	0.4357	0.4370	0.4382	0.4394	0.4406	0.4418	0.4429	0.4441
1.6	0.4452	0.4463	0.4474	0.4484	0.4495	0.4505	0.4515	0.4525	0.4535	0.4545
1.7	0.4454	0.4564	0.4573	0.4582	0.4591	0.4599	0.4608	0.4616	0.4625	0.4633
1.8	0.4641	0.4649	0.4656	0.4664	0.4671	0.4678	0.4686	0.4693	0.4699	0.4706
1.9	0.4713	0.4719	0.4726	0.4732	0.4738	0.4744	0.4750	0.4756	0.4761	0.4767
2.0	0.4772	0.4778	0.4783	0.4788	0.4793	0.4798	0.4803	0.4808	0.4812	0.4817
2.1	0.4821	0.4826	0.4830	0.4834	0.4838	0.4842	0.4846	0.4850	0.4854	0.4857
2.2	0.4861	0.4864	0.4868	0.4871	0.4875	0.4878	0.4881	0.4884	0.4887	0.4890
2.3	0.4893	0.4896	0.4898	0.4901	0.4904	0.4906	0.4909	0.4911	0.4913	0.4916
2.4	0.4918	0.4920	0.4922	0.4925	0.4927	0.4929	0.4931	0.4932	0.4934	0.4936
2.5	0.4938	0.4940	0.4941	0.4943	0.4945	0.4946	0.4948	0.4949	0.4951	0.4952
2.6	0.4953	0.4955	0.4956	0.4957	0.4959	0.4960	0.4961	0.4962	0.4963	0.4964
2.7	0.4965	0.4966	0.4967	0.4968	0.4969	0.4970	0.4971	0.4972	0.4973	0.4974
2.8	0.4974	0.4975	0.4976	0.4977	0.4977	0.4978	0.4979	0.4979	0.4980	0.4981
2.9	0.4981	0.4982	0.4982	0.4983	0.4984	0.4984	0.4985	0.4985	0.4986	0.4986
3.0	0.4987	0.4987	0.4987	0.4988	0.4988	0.4989	0.4989	0.4989	0.4990	0.4990

Example: To find the area under the curve between the mean and a point corresponding to 1.93 standard deviations to the right of the mean, look for the value in the row corresponding to 1.9 and the column corresponding to 0.03, which is 0.4732.

Appendix B

Values of *t* factors from Student *t* distribution for various degrees of freedom and confidence levels.

Degrees of freedom	Percent confidence level					
	68.27	**90**	**95**	**95.45**	**99**	**99.73**
1	1.84	6.31	12.71	13.97	63.66	235.80
2	1.32	2.92	4.30	4.53	9.92	19.21
3	1.20	2.35	3.18	3.31	5.84	9.22
4	1.14	2.13	2.78	2.87	4.60	6.62
5	1.11	2.02	2.57	2.65	4.03	5.51
6	1.09	1.94	2.45	2.52	3.71	4.90
7	1.08	1.89	2.36	2.43	3.50	4.53
8	1.07	1.86	2.31	2.37	3.36	4.28
9	1.06	1.83	2.26	2.32	3.25	4.09
10	1.05	1.81	2.23	2.28	3.17	3.96
11	1.05	1.80	2.20	2.25	3.11	3.85
12	1.04	1.78	2.18	2.23	3.05	3.76
13	1.04	1.77	2.16	2.21	3.01	3.69
14	1.04	1.76	2.14	2.20	2.98	3.64
15	1.03	1.75	2.13	2.18	2.95	3.59
16	1.03	1.75	2.12	2.17	2.92	3.54
17	1.03	1.74	2.11	2.16	2.90	3.51
18	1.03	1.73	2.10	2.15	2.88	3.48
19	1.03	1.73	2.09	2.14	2.86	3.45
20	1.03	1.72	2.09	2.13	2.85	3.42
25	1.02	1.71	2.06	2.11	2.79	3.33
30	1.02	1.70	2.04	2.09	2.75	3.27
35	1.01	1.70	2.03	2.07	2.72	3.23
40	1.01	1.68	2.02	2.06	2.70	3.20
45	1.01	1.68	2.01	2.06	2.69	3.18
50	1.01	1.68	2.01	2.05	2.68	3.16
100	1.005	1.660	1.984	2.025	2.626	3.077
∞	1.00	1.645	1.96	2.000	2.576	3.000

Note: The number of degrees of freedom is one less than the number of repeat observations.

Appendix C

Values of A_2, D_3, and D_4 factors for computing control chart limits.

Number of observations in the subgroup	A_2	Value of D_3	D_4
2	1.880	0	3.267
3	1.023	0	2.575
4	0.729	0	2.282
5	0.577	0	2.115
6	0.483	0	2.004
7	0.419	0.076	1.924
8	0.373	0.136	1.864
9	0.337	0.184	1.816
10	0.308	0.223	1.777

Source: Chrysler Corporation, Ford Motor Company, and General Motors Corporation, *Measurement System Analysis: A Reference Manual.* Automotive Industry Action Group, Southfield, Michigan.

Appendix D

Table 1 Repeatability calculations: Values of d_2 and K_1.

g (number of parts × number of appraisers)	**Number of trials $m = 2$**		**Number of trials $m = 3$**		**Number of trials $m = 4$**	
	d_2	**$K_1 = 5.15/d_2$**	**d_2**	**$K_1 = 5.15/d_2$**	**d_2**	**$K_1 = 5.15/d_2$**
10	1.16	4.44	1.72	2.99	2.08	2.48
11	1.16	4.44	1.71	3.01	2.08	2.48
12	1.15	4.48	1.71	3.01	2.07	2.49
13	1.15	4.48	1.71	3.01	2.07	2.49
14	1.15	4.48	1.71	3.01	2.07	2.49
15	1.15	4.48	1.71	3.01	2.07	2.49
>15	1.128	4.47	1.693	3.04	2.059	2.50

Notes:

- In R&R studies, generally two, three, or four trials are done; hence the table gives only these values.
- There are a minimum of two appraisers and at least five parts in any R&R study which covers the variation from part to part. The minimum value of g, which is the product of the number of parts and the number of appraisers, has therefore been taken as 10.
- The values given above suffice for most R&R studies; however, for different combinations of numbers of parts and appraisers and numbers of trials, the values of d_2 should be obtained from statistical tables.

Table 2 Reproducibility calculations: Values of d_2 and K_2.

Number of appraisers	**d_2**	**$K_2 = 5.15/d_2$**
2	1.41	3.65
3	1.91	2.70
4	2.24	2.30

Table 3 Part-to-part variation calculations: Values of d_2 and K_3.

Number of parts	d_2	$K_3 = 5.15/d_3$
2	1.41	3.65
3	1.91	2.70
4	2.24	2.30
5	2.48	2.08
6	2.67	1.93
7	2.83	1.82
8	2.96	1.74
9	3.08	1.67
10	3.18	1.62
11	3.27	1.57
12	3.35	1.54
13	3.42	1.51
14	3.49	1.48
15	3.55	1.45

Source: Chrysler Corporation, Ford Motor Company, and General Motors Corporation, *Measurement System Analysis: A Reference Manual.* Chrysler, Ford, and GMC, 1995.

Appendix E

Critical range factor *f*(*n*) for 95% confidence level for number of test results *n*.

n	*f*(*n*)	*n*	*f*(*n*)	*n*	*f*(*n*)	*n*	*f*(*n*)
3	3.3	9	4.4	19	5.0	45	5.6
				20	5.0	50	5.6
				21	5.0		
4	3.6	10	4.5	22	5.1	60	5.8
				23	5.1		
				24	5.1		
5	3.9	11	4.6	25	5.2	79	5.9
		12	4.6	26	5.2	80	5.9
				27	5.2		
6	4.0	13	4.7	28	5.3	90	6.0
		14	4.7	29	5.3		
				30	5.3		
				31	5.3		
				32	5.3		
7	4.2	15	4.8	33	5.4	100	6.1
		16	4.8	34	5.4		
				35	5.4		
				36	5.4		
				37	5.4		
8	4.3	17	4.9	38	5.5		
		18	4.9	39	5.5		
				40	5.5		

Appendix F

GLOSSARY OF TERMS USED

A number of terms pertaining to metrology have been used in this book. The definitions of the terms given in this glossary have been taken from the 1993 *International Vocabulary of Basic and General Terms in Metrology,* abbreviated as VIM, and the international standard ANSI/ISO/ASQ Q9000:2000, *Quality management systems—Fundamentals and vocabulary.*

Term	Definition
Measurement	A set of operations having the object of determining the value of a quantity Note: The operations may be performed automatically. *VIM, para. 2.1*
Metrology	The science of measurement Note: Metrology includes all aspects, both theoretical and practical, with reference to measurements, whatever their uncertainty and in whatever fields of science or technology they occur. *VIM, para. 2.2*
Method of measurement	A logical sequence of operations described generically, used in the performance of measurements *VIM, para. 2.4*
Measurement procedure	A set of operations, described specifically, used in the performance of particular measurements according to a given method *VIM, para. 2.5*
Measurand	A particular quantity subject to measurement Note: The specification of a measurand may require statements about quantities such as time, temperature, and pressure. *VIM, para. 2.6*

Term	Definition
Influence quantity	A quantity that is not the measurand but that affects the result of measurement *VIM, para. 2.7*
Result of measurement	The value attributed to a measurand obtained by measurement Notes: 1. When a result is given, it should be made clear whether it refers to: —the indication, —the uncorrected result, or —the corrected result and whether several values have been averaged. 2. A complete statement of the result of a measurement includes information about the uncertainty of the measurement. *VIM, para. 3.1*
Uncorrected result	The result of a measurement before correction for systematic error *VIM, para. 3.3*
Corrected result	The result of a measurement after correction for systematic error *VIM, para. 3.4*
True value (of a quantity)	A value consistent with the definition of a given particular quantity Notes: 1. This is a value that would be obtained by a perfect measurement. 2. True values are by nature indeterminate. 3. The indefinite article "a" rather than the definite article "the" is used in conjunction with "true value" because there may be many values consistent with the definition of a given particular quantity. *VIM, para. 1.19*
Conventional true value (of a quantity)	A value attributed to a particular quantity and accepted, sometimes by convention, as having an uncertainty appropriate for a given purpose *VIM, para. 1.20*

Term	Definition
Accuracy of measurement	The closeness of the agreement between the result of a measurement and a true value of the measurand *VIM, para. 3.5*
Repeatability (of results of measurements)	The closeness of the agreement between the results of successive measurements of the same measurand carried out under the same conditions of measurement Notes: 1. These conditions are called repeatability conditions. 2. Repeatability conditions include: —the same measurement procedure, —the same observer, and —the same measuring equipment, used under the same conditions, consisting of: —the same location and —repetition over a short period of time. 3. Repeatability may be expressed quantitatively in terms of the dispersion characteristics of the results. *VIM, para. 3*
Reproducibility (of results of measurements)	The closeness of the agreement between the results of measurements of the same measurand carried out under changed conditions of measurement Notes: 1. A valid statement of reproducibility requires specifications of the changes in conditions. 2. The changed conditions may include: —principle of measurement, —method of measurement, —observer, —measuring instrument, —reference standard, —location, —conditions of use, and —time. 3. Reproducibility may be expressed quantitatively in terms of the dispersion characteristics of the results. 4. The results are usually understood to be corrected results. *VIM, para. 3.7*

Term	Definition
Uncertainty (of measurement)	A parameter associated with the result of a measurement that characterizes the dispersion of the values that could reasonably be attributed to the measurand *VIM, para. 3.9*
Error (of measurement)	The result of a measurement minus a time value of the measurand Note: Since true value cannot be determined, in practice a conventional true value is used. *VIM, para. 3.10*
Random error	The result of a measurement minus the mean that would result from an infinite number of measurements of the same measurand carried out under repeatability conditions Notes: 1. Random error is equal to error minus systematic error. 2. Because only a finite number of measurements can be made, it is possible to determine only an estimate of random error. *VIM, para. 3.1*
Systematic error	The mean that would result from an infinite number of measurements of the same measurand carried out under repeatability conditions minus a true value of the measurand Notes: 1. Systematic error is equal to error minus random error. 2. Like true value, systematic error and its causes cannot be completely known. *VIM, para. 3.14*
Correction	A value added algebraically to the uncorrected result of a measurement to compensate for systematic error *VIM, para. 3.15*
Measurement system	A complete set of measuring instruments and other equipment assembled to carry out specified measurements Notes: 1. The system may include material measures and chemical reagents. 2. A measurement system that is permanently installed is called a measurement installation. *VIM, para. 4.5*

Term	Definition
Measurement standard	A material measure, measuring instrument, reference material, or measurement system intended to define, realize, conserve, or reproduce a unit or one or more values of a quantity to serve as a reference *VIM, para. 6.1*
Primary standard	A standard that is designated or widely acknowledged as having the highest metrological qualities and whose value is accepted without reference to other standards of the same quantity *VIM, para. 6.4*
Reference standard	A standard, generally having the highest metrological quality available at a given location or in a given organization, from which measurements made there are derived *VIM, para. 6.6*
Secondary standard	A standard whose value is assigned by comparison with a primary standard of the same quantity *VIM, para. 6.5*
Transfer standard	A standard used as an intermediary to compare standards Note: The term *transfer device* should be used when the intermediary is not a standard. *VIM, para. 6.8*
Traceability	The property of the result of a measurement or value of a standard whereby it can be related to stated references, usually national or international standards, through an unbroken chain of comparisons having stated uncertainties Notes: 1. This concept is often expressed by the adjective *traceable*. 2. The unbroken chain of comparison is called a traceability chain. *VIM, para. 6.10*
Calibration	A set of operations that establish under specified conditions the relationship between values of quantities indicated by a measuring instrument or measurement system or between values represented by a material measure or a reference material and the corresponding value realized by a standard *VIM, para. 6.11*

Term	Definition
Reference material	A material or substance one or more of whose property values are sufficiently homogeneous and well established to be used for the calibration of an apparatus, the assessment of a measurement method, or the assigning of values to materials *VIM, para. 6.13*
Measurement process	A set of operations to determine the value of a quantity *ANSI/ISO/ASQ Q9000-2000, para. 3.10.2*
Measurement control system	A set of interrelated or interacting elements necessary to achieve metrological confirmation and continual control of a measurement process *ANSI/ISO/ASQ Q9000-2000, para. 3.10.1*
Metrological confirmation	A set of operations required to ensure that measuring equipment conforms to the requirements for its intended use *ANSI/ISO/ASQ Q9000-2000, para. 3.10.3*
Measuring equipment	The measuring instruments, software, measurement standards, reference materials, auxiliary apparatus, or combination thereof necessary to realize a measurement process *ANSI/ISO/ASQ Q9000-2000, para. 3.10.4*
Metrological function	A function with organizational responsibility for defining and implementing a measurement control system *ANSI/ISO/ASQ Q9000-2000, para. 3.10.6*
Metrological characteristics	Distinguishing features which can influence the results of measurement Notes: 1. Measuring equipment usually has several metrological characteristics. 2. Metrological characteristics can be subject to calibration. *ANSI/ISO/ASQ Q9000-2000, para. 3.10.5*

Sources:

1. International Organization for Standardization (ISO), *International Vocabulary of Basic and General Terms in Metrology,* 2nd ed. (ISO, 1993).
2. ANSI/ISO/ASQ Q9000-2000, *Quality management systems—Fundamentals and vocabulary.* 2000. Milwaukee, WI: ASQ Quality Press.

Index

Note: Italicized page numbers indicate illustrations.

D

I

U

V

W